北京理工大学“双一流”建设精品出版工程

大学物理——近代物理

College Physics—Modern Physics

胡海云　缪劲松　冯艳全　吴晓丽 ◎ 编著

北京理工大学出版社
BEIJING INSTITUTE OF TECHNOLOGY PRESS

内 容 简 介

本书所介绍的近代物理学是大学物理的一部分。主要包括相对论和量子物理两部分内容。其中相对论部分着重介绍狭义相对论的基本原理、洛伦兹变换、相对论的时空效应以及动力学方面的内容，还简要介绍广义相对论的基本原理和时空观；量子物理部分着重介绍微观粒子的波粒二象性、薛定谔方程及其应用、固体中的电子、原子核物理等。各章后均有本章提要、思考题和习题。

本书适合作为高校理工科各专业的《大学物理——近代物理》英文教材或教学参考书，也可作为综合性大学和高等师范院校相关专业的教材或教学参考书。

图书在版编目(CIP)数据

大学物理 : 近代物理 = College Physics—Modern Physics : 英文 / 胡海云等编著. -- 北京 : 北京理工大学出版社, 2023.10

ISBN 978-7-5763-2932-2

Ⅰ. ①大… Ⅱ. ①胡… Ⅲ. ①物理学-高等学校-教材-英文 Ⅳ. ①O4

中国国家版本馆 CIP 数据核字(2023)第 188668 号

责任编辑：时京京　　**文案编辑**：时京京
责任校对：周瑞红　　**责任印制**：李志强

出版发行 / 北京理工大学出版社有限责任公司
社　　址 / 北京市丰台区四合庄路 6 号
邮　　编 / 100070
电　　话 / (010) 68944439（学术售后服务热线）
网　　址 / http://www.bitpress.com.cn

版 印 次 / 2023 年 10 月第 1 版第 1 次印刷
印　　刷 / 保定市中画美凯印刷有限公司
开　　本 / 787 mm × 1092 mm　1/16
印　　张 / 16
字　　数 / 376 千字
定　　价 / 52.00 元

Contents in brief

Modern physics introduced in this book is a part of college physics. It mainly includes two parts: relativity and quantum physics. The part of relativity focuses on the basic principles of special relativity, Lorentz transformation, space-time effect of relativity and dynamics, and also briefly introduces the basic principle and space-time view of general relativity. The part of quantum physics focuses on the wave-particle duality of microscopic particles, Schrödinger equation and its applications, electrons in solids, nuclear physics, etc. Each chapter is followed by a summary of the chapter, questions and exercises.

This book is suitable for the textbook in English or teaching reference book of "College Physics-Modern Physics" selected by all majors of science and engineering in colleges and universities, as well as the textbook or teaching reference book for related majors of comprehensive universities and normal colleges.

PREFACE 前 言

Quantum mechanics and relativity, collectively known as modern physics, are the greatest discoveries in physics in the 20th century. Without them, there would not be today's computer, laser, nuclear energy, nanotechnology, space exploration and other high and new technologies.

In the part of relativity, College Physics-Modern Physics focuses on the basic principles of special relativity and its important significance in physics. The discussion on Lorentz transformation, space-time view and space-time geometry, such as the relativity of simultaneity, time dilation and length contraction, as well as the introduction of dynamics, such as the relativistic mass and momentum, and the relativistic energy, will help readers deepen their understanding of special relativity. It also briefly introduces the basic principles and space-time view of general relativity.

In the part of quantum physics, we will follow the development history of human understanding of the nature of microscopic particles, through the law of blackbody radiation, photoelectric effect phenomenon, Compton effect, wave property of particle and so on, which are inscrutable to classical physics, introduce some basic concepts, laws and methods in quantum physics proposed by Planck, Einstein, de Broglie, Born, Heisenberg, etc. The wave-particle duality of microscopic particles, Schrödinger equation and its applications, and the foundation of Solid physics are emphatically introduced. Readers can experience and appreciate the harmony and beauty of the microcosmic world, and be inspired by innovative thinking and exploration spirit.

The authors of this textbook are all excellent teachers in College Physics teaching, with many years of experience in online and offline teaching. The courses taught by them have won the National Excellent Video Open Course (2016), the National Excellent Online Open Course (2018), the National Online First-Class Course (2020), the National Online-offline Blended First-Class Course (2020), and etc.

On the basis of the original provincial and ministerial high-quality-level textbooks, and with the support of the textbook project of the Beijing Institute of

Technology's 14th Five-Year Plan (2022) this textbook is written in English, and attention is paid to introducing novel and cutting-edge examples. The book is divided into five chapters, of which MIAO Jinsong is responsible for the writing of the first chapter, HU Haiyun is responsible for the writing of the second and the third chapters, FENG Yanquan is responsible for the writing of the fourth chapter, and WU Xiaoli is responsible for the writing of the fifth chapter. We thank the Academic Affairs Department of and Beijing Institute of Technology for their active support in the compilation and publication of this textbook.

The authors
February 2023

目　录
CONTENTS

Chapter 1　Mechanical Basis of Special Relativity ································ 001

1. 1　Basic Principles of Special Relativity ·· 002

1. 1. 1　Principle of Relativity of Mechanics, Absolute Space-time View and Galilean transformation ·· 003

1. 1. 2　Einstein's Principle of Relativity and the Principle of Constant Speed of Light ······ 005

1. 2　Effects of Space and Time in Relativity ·· 008

1. 2. 1　Measurement of Space and Time ·· 008

1. 2. 2　Time Dilation ··· 011

1. 2. 3　Length Contraction ··· 015

1. 2. 4　Relativity of Simultaneity ·· 017

1. 3　The Lorentz Transformation ·· 022

1. 3. 1　Derivation of the Lorentz Transformation ·· 022

1. 3. 2　Using the Lorentz Transformation to Verify the Spacetime Effect of Relativity ······ 028

1. 3. 3　Minkowski Space ·· 038

1. 4　The Relativistic Velocity Transformation ·· 043

1. 5　Basis of Relativistic Dynamics ·· 046

1. 5. 1　The Relativistic Momentum and Mass ··· 046

1. 5. 2　Mass-Energy Relation ·· 048

1. 5. 3　Energy-Momentum Relationships ··· 051

*1. 6　Introduction to General Relativity ·· 052

1. 6. 1　Basic Principles of General Relativity ·· 052

1. 6. 2　Several Experimental Verifications of General Relativity ··························· 055

Summary ·· 060

Questions ··· 061

Problems ······ 062

Chapter 2 Wave-particle Duality of Microscopic Particles ······ 065

2.1 Black Body Radiation and Planck's Energon Hypothesis ······ 066
2.1.1 Black Body Radiation ······ 066
2.1.2 Planck's Quantum Hypothesis ······ 072
2.2 Photoelectric Effect and Einstein's Photon Theory ······ 075
2.2.1 Photoelectric Effect ······ 075
2.2.2 Einstein's Photon Theory ······ 078
2.2.3 Wave-Particle Duality of Light ······ 084
2.3 Compton Effect ······ 086
2.3.1 Compton Effect ······ 086
2.3.2 Explanation of Compton Effect by Light Quantum Theory ······ 088
2.4 Hydrogen Atom Spectrum and Bohr's Theory of Hydrogen Atom ······ 093
2.4.1 Hydrogen Atom Spectrum ······ 093
2.4.2 Bohr's Theory of Hydrogen Atom ······ 095
2.5 The Wave Property of Particle and Born's Statistical Interpretation ······ 102
2.5.1 De Broglie Wave ······ 102
2.5.2 Experimental Verification of de Broglie Wave ······ 103
2.5.3 Born's Statistical Interpretation ······ 109
2.5.4 Wave Function of Free Particle ······ 117
2.6 Uncertainty Relation ······ 119
Summary ······ 127
Questions ······ 129
Problems ······ 130

Chapter 3 Schrödinger Equation and its Applications ······ 133

3.1 Schrödinger Equation ······ 133
3.1.1 Schrödinger Equation for a Free Particle ······ 134
3.1.2 Schrödinger Equation in General ······ 135
3.1.3 Steady-state Schrödinger Equation ······ 137
3.2 A Particle in a One-dimensional Square-well ······ 139
3.2.1 A Particle in a One-dimensional Infinite Square-well ······ 139
3.2.2 A Particle in a One-dimensional Finite Square-well ······ 144
3.2.3 One-dimensional Square Potential Barrier and Barrier Penetration ······ 145
3.3 The Simple Harmonic Oscillator ······ 149
3.4 Electrons in Atoms ······ 153
3.4.1 Hydrogen Atom ······ 153
3.4.2 The Stern-Gerlach Experiment and Electron Spin ······ 159

3.5 Four Quantum Numbers and Atomic Shell Structure ········ 164
3.5.1 Four Quantum Numbers ········ 164
3.5.2 The Pauli Exclusion Principle and the Principle of the Lowest Energy ········ 164
3.5.3 Shell Structure of Atoms ········ 165
*3.6 Laser ········ 169
3.6.1 The Generation of Laser ········ 170
3.6.2 Characteristics of Laser ········ 173
3.6.3 Application of Laser: Laser Cooling ········ 173
Summary ········ 174
Questions ········ 176
Problems ········ 177

Chapter 4 Electrons in Solids ········ 179

4.1 Free Electrons in Metals ········ 179
4.1.1 The Free-election-gas Model ········ 180
4.1.2 Fermi Energy of Free Electron Gas ········ 181
4.1.3 Density of States and Fermi-Dirac Distribution ········ 185
4.2 Band Theory of Solids ········ 189
4.2.1 Energy Bands of Solids ········ 189
4.2.2 Valence Band, Conduction Band and Forbidden Band ········ 193
4.2.3 Conductor, Insulator and Semiconductor ········ 194
4.3 Semiconductor Conduction ········ 197
4.3.1 Classification of Semiconductors ········ 198
4.3.2 PN Junction ········ 201
4.3.3 Semiconductor Devices ········ 202
Summary ········ 209
Questions ········ 210
Problems ········ 211

Chapter 5 Nuclear Physics ········ 213

5.1 Properties of the Nucleus ········ 213
5.1.1 Composition of the Nucleus ········ 213
5.1.2 Shape and Size of the Nucleus ········ 215
5.1.3 Spin and Magnetic Moment of the Nucleus ········ 216
5.2 Binding Energy and Nuclear Force ········ 219
5.2.1 Binding Energy ········ 219
5.2.2 Nuclear Force ········ 221
5.3 Radioactive Decay of the Nucleus ········ 223
5.3.1 Radioactivity ········ 223

5. 3. 2 Radioactive Decay Law ······ 224
5. 3. 3 Alpha Decay ······ 226
5. 3. 4 Beta Decay ······ 227
5. 3. 5 Gamma Decay ······ 229
5. 3. 6 Applications of Radioactivity ······ 230
5. 4 Nuclear Reactions ······ 231
5. 4. 1 Artificial Nuclear Reactions ······ 231
5. 4. 2 Nuclear Fission ······ 233
5. 4. 3 Nuclear Fusion ······ 236
Summary ······ 240
Questions ······ 242
Problems ······ 243

References ······ 245

Chapter 1

Mechanical Basis of Special Relativity

By the end of the 19th century, the theory of classical physics had been basically and completely established, including Newtonian mechanics, thermology, optics and electromagnetism. Using the theory of classical physics can not only explain the vast majority of natural phenomena actually encountered, but also directly trigger two industrial revolutions in the process of its establishment. Among them, steam engine and internal combustion engine were born in the first industrial revolution, and electrification and radio communication were realized in the second industrial revolution. All these have greatly promoted the development of productive forces and brought about great changes in human production and lifestyle.

At that time, it was thought that the understanding of the essence of physical phenomena had almost been completed, and the main task of physicists later seemed to be to do some piecemeal repair work. However, the results of the two experiments disturbed the physicists at that time. On April 27, 1900, the British physicist W. Thomson (named Sir Kelvin) gave a speech entitled "Nineteenth-Century clouds over the dynamic theory of heat and light" at the annual meeting of the Royal Society to welcome the new century, He said bluntly: "the theory of dynamics asserts that heat and light are both ways of movement. But now the beauty and clarity of this theory are eclipsed by two dark clouds... ". One of "two dark clouds" in the clear sky of classical physics is the zero result of the famous Michelson-Morley experiment looking for "Ether" ; The other is the "ultraviolet disaster" in the blackbody radiation theory. The former contradicts the space-time view of Newton's classical mechanics and Galilean transformation, while the latter cannot be explained by traditional thermodynamic or electromagnetic theories. It was these two dark clouds that gave birth to great changes in the history of physics. In the next three decades, the concepts and ideas of classical physics were broken through. The zero result of Michelson-Morey experiment gave birth to the establishment of Einstein's theory of relativity. The " ultraviolet disaster " in blackbody radiation directly led to the birth of quantum mechanics. Relativity and quantum mechanics are the two branches of modern physics. These two new disciplines are the greatest discoveries of physics in the 20th century. Without them, there would be no today's nuclear energy, space exploration, computers, lasers, nano and other high and new technologies.

If the establishment of quantum mechanics is the crystallization of the wisdom of many scientists, it can be said that the establishment of relativity is mainly due to personal achievements, this person is A. Einstein, the greatest physicist in the 20th century. In the late 19th century, after

Maxwell's electromagnetic theory was established, scientists realized that the speed of light predicted by Maxwell did not specify which reference frame it was relative to. In other words, light in vacuum propagates at the same speed c in any direction, independent of the reference frame. The null result of Michelson-Morley experiment also proves that the speed of light in vacuum is independent of the reference frame, but it is in contradiction with the Galilean velocity transformation of classical mechanics. To solve this contradiction, either Maxwell's electromagnetic theory should be modified, or the Galilean transformation should be abandoned. Einstein chose the latter based on the comprehensive analysis of the experimental results and the basic concepts of physics. He believed that all physical laws, not only mechanical laws, but also electromagnetic laws, are the same in all inertial reference frames. On this basis, he founded the theory of special relativity. Then Einstein further extended the applicability of physical laws to any reference frame including noninertial frame, and established the theory of general relativity.

The establishment of Einstein's theory of relativity means that people must give up the deeply-held concepts about time, space and motion, which have been formed in their minds long before, and establish a new and revolutionary concept of the space-time. Einstein's theory of relativity gave the physical laws satisfied by high-speed moving objects (close to the speed of light), revealed the internal relationship between mass and energy, and began the exploration of the essential relationship between universal gravitation and large-scale space. Now, relativity has become the theoretical basis for studying the interaction of matter, the origin of the universe, the evolution of galaxies and so on. Relativity includes two parts: special relativity and general relativity. This chapter is mainly about the mechanical basis of special relativity, and general relativity will be briefly introduced in the last section.

1.1 Basic Principles of Special Relativity

Time and space are the existing forms of matter. The movement of matter is closely connected with the nature of time and space. The description of material movement is relative. When discussing the specific material movement, we must first determine its reference frame. In all inertial reference frames, the laws of physics should have the same form, which is the principle of relativity. The principle of relativity is one of the most basic principles of physics, it means that the laws of physics established in one inertial reference frame are equally applicable to other inertial reference frames as long as they are transformed by appropriate coordinates (time and space transformation). Therefore, closely related to the principle of relativity is people's understanding of time and space, which is called time-space view. Historically, people's understanding of the principle of relativity has experienced the development from the principle of mechanical relativity to Einstein's principle of special relativity and general relativity. Accordingly, human's space-time view has also experienced the transformation from Newton's absolute space-time view to Einstein's relativistic space-time view.

1.1.1 Principle of Relativity of Mechanics, Absolute Space-time View and Galilean transformation

Newtonian mechanics (classical mechanics) based on Newton's three laws and the law of gravitation has the same form in all inertial reference frames, or Newton's laws hold in all inertial reference frames, which is the principle of relativity of mechanics. The Italian scientist Galileo was the first to describe the principle of relativity in history. In his book "Dialogue Concerning the Two Chief World Systems" published in 1632, he summarized the mechanical experiments carried out in a closed cabin moving at constant speed relative to the ground. He said: as long as the ship moves at constant speed and does not swing left or right, various mechanical phenomena observed, such as the jumping of people, the falling of water droplets, the swimming of fish and the flying of butterflies, happen the same as when the ship is stationary. People on ship can't judge whether the ship is moving or not from these phenomena. This vivid description contains an important truth: in any inertial reference frame, the same mechanical phenomenon occurs and evolves in the same way, that is, all inertial reference frames are equivalent. This is called Galileo's principle of relativity by Einstein. It is another expression of the principle of relativity of mechanics.

We know that Newtonian mechanics studies the mechanical motion of macro objects, and the spatial position of objects will continue to change with time in the process of motion. Therefore, the discussion of the motion law of macro objects is inseparable from the measurement of space and time. In daily life and production practice, people gradually realize that the description of the duration of time and the size of spatial range needs to be realized by clocks and rulers. The time experienced in a process is related to the angle of the clock pointer, and the displacement of an object in motion can be known only by comparing it with the ruler. In other words, the understanding of time and space should begin with the measurement of the time and space intervals experienced by the movement of objects. People's time-space view is formed in this process.

Before the establishment of special relativity, people's basic views on time and space formed in the long historical process were summarized as the space-time view of Newtonian mechanics: space is a three-dimensional Euclidean space that is uniform and isotropic everywhere, there is no connection between space and the movement of matter, and the distance between any two points in space is an absolute quantity independent of the observer's reference frame, that is, the length of space is absolute; Time passes evenly from the past to the present and future, in the whole universe, time is uniform and has nothing to do with the movement of matter. The time interval between two events does not change with the change of the reference frame, that is, the time interval is absolute; Time and space are absolute and independent of each other, which is the basis of object motion and of the first importance, while object motion is carried out within their framework and so secondary. This understanding of time and space in Newtonian mechanics is also known as Newton's absolute space-time view.

Based on the two basic assumptions of the principle of relativity of mechanics and Newton's absolute space-time view, Galilean transformation can be deduced. Galilean transformation describes the relationship between the temporal and spatial coordinates of the same event in two different inertial reference frames: there are two inertial reference frames S and S', which are represented by two rectangular coordinate systems $O-xyz$ and $O'-x'y'z'$ respectively. For simplicity, let x-axis of the frame S coincide with x'-axis of the frame S', y - and z - axis are parallel to y' - and z'-axis respectively. Let the frame S' moves at constant speed u with respect to the frame S along the x-axis, and the origin O' of the frame S' moves to coincide with the origin O of the frame S at $t=t'=0$. Suppose an event occurs at point P shown in Figure 1 - 1 at a certain time (for example, a particle just moves to point P), the observers in the frame S find that the event occurs at time t and space (x,y,z), while the observers in the frame S' find that the event occurs at time t' and space (x',y',z').

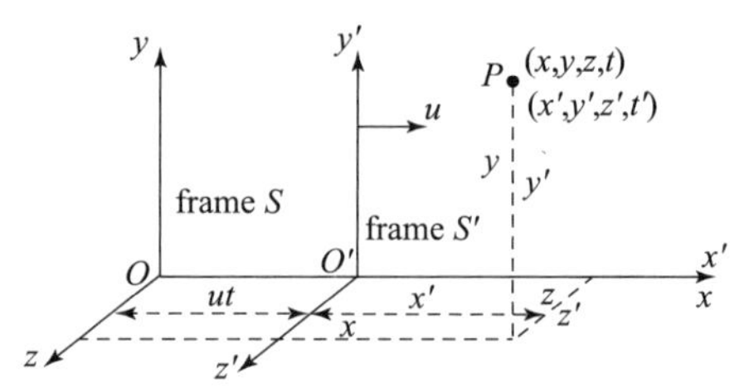

Figure 1 - 1 Galilean transformation: observed in frame S

According to the principle of relativity of mechanics and Newton's absolute view of space-time, that is, Newton's laws hold in both frame S and frame S' no matter viewed in which reference frame, the time experienced by the motion of the same object is the same in both frames, and the length of the same object is the same in both frames. It can be deduced that the following corresponding relationship can be satisfied between the two sets of measured values (x,y,z,t) and (x',y',z',t')

$$\left.\begin{aligned} x' &= x - ut \\ y' &= y \\ z' &= z \\ t' &= t \end{aligned}\right\} \tag{1-1}$$

The above equation (1 - 1) is called Galilean transformation. It can be seen that the measurement of time has nothing to do with space and relative motion. Time and space are independent and irrelevant to each other. Galilean transformation reflects Newton's absolute space-time view directly.

For the first three formula in equation (1 - 1), make the left side differential with t' and the right side differential with t, considering $t'=t$, Galilean velocity transformation can be obtained as

$$\left.\begin{aligned} \boldsymbol{v}'_x &= \boldsymbol{v}_x - u \\ \boldsymbol{v}'_y &= \boldsymbol{v}_y \\ \boldsymbol{v}'_z &= \boldsymbol{v}_z \end{aligned}\right\} \tag{1-2}$$

Galilean velocity transformation can also be written in vector form, i. e

$$\boldsymbol{v}' = \boldsymbol{v} - \boldsymbol{u} \tag{1-3}$$

The above formula has been derived in the first chapter of mechanics, where $\boldsymbol{v}$ and $\boldsymbol{v}'$ are the particle velocity measured in frame S and frame S' respectively. Then making both sides of the above formula differential with time, considering $\boldsymbol{u}$ is a constant vector, we get

$$\boldsymbol{a}' = \boldsymbol{a} \tag{1-4}$$

Where $\boldsymbol{a}$ and $\boldsymbol{a}'$ are the acceleration of the particle measured in frame S and frame S' respectively. The above formula shows that the acceleration of a particle measured in different inertial reference frames is the same.

In Newtonian mechanics, the mass and force of an object are considered to be independent of reference frame, that is, there is $m = m'$, $\boldsymbol{F} = \boldsymbol{F}'$, Therefore, if $\boldsymbol{F} = m\boldsymbol{a}$ is established in the inertial frame S, there is $\boldsymbol{F}' = m'\boldsymbol{a}'$ in the inertial frame S', which shows that Newton's law is invariant under Galilean transformation, and the mechanical laws hold in different inertial reference frames. It can be seen that Galilean transformation also reflects the principle of mechanical relativity.

From the above discussion, it can be seen that Newton's view of absolute space-time and the principle of mechanical relativity are connected through Galilean transformation. Because Newtonian mechanics was established when discussing the motion of almost all macro objects, and the concept of absolute space-time contained in Newtonian mechanics was also consistent with our understanding of time and space in real life, the concepts of absolute time and absolute space were also considered reasonable at that time. Of course, people did not think too much about them. It was not until the middle and late 19th century that James Clerk Maxwell established the electromagnetic theory, predicted the existence of electromagnetic waves, show that electromagnetic waves travel through vacuum at speed $c = 2.99 \times 10^8$ m/s. In fact, electromagnetic waves travel at the same speed in every inertial frame, regardless of the motion of the source or of the observer. This has led to the shaking of the concept of absolute time and absolute space. Based on the analysis of the experimental results, Einstein put forward Einstein's principle of relativity, and replaced the concepts of absolute space-time with the principle of constant speed of light. On this basis, Einstein established the theory of special relativity and realized the great change of the concept of space-time view.

1.1.2 Einstein's Principle of Relativity and the Principle of Constant Speed of Light

In the mid – 19th century, Maxwell put forward Maxwell's equations that describe electromagnetic fields. It consists of four differential equations describing the changes of electric fields and magnetic fields, which shows that the changing electric fields and magnetic fields depend on each other, excite each other in space and propagate far away. On this basis, Maxwell predicted the existence of electromagnetic waves, and calculated that the propagation speed of electromagnetic waves in vacuum is $c = 2.99 \times 10^8$ m/s, which is constant and the same as the speed of light in vacuum. Therefore, he further predicted that light is a kind of electromagnetic wave.

According to Maxwell's electromagnetic theory electromagnetic waves travel through vacuum at a constant speed c, which brought a crisis to the Newtonian mechanics at that time. First of all, Maxwell's equations do not have covariance under Galilean transformation, they have different forms in different inertial reference frames. Then, in what reference frame is Maxwell's electromagnetic theory established? Secondly, according to the Galilean velocity transformation light should travel at different speeds relative to different reference frames. Then, with respect to what kind of reference frame do electromagnetic waves travel at a constant speed c? The question is whether there is a special reference frame in which Maxwell's electromagnetic theory holds. Through comparing with mechanical waves such as sound wave and water wave, physicists at that time believed that electromagnetic waves or light was a vibration in an invisible, elusive medium called the ether, and the constant speed of electromagnetic waves c derived by Maxwell is the speed with respect to the ether. This means that Maxwell's electromagnetic theory is only established in this "ether" reference frame, which is regarded as Newton's absolute static reference frame.

physicists at that time believed that the universe was filled with the ether, in which celestial bodies were unimpeded. According to the formula of speed of mechanical waves $c=\sqrt{G/\rho}$, considering that the speed of electromagnetic waves c is very large, so the ether is a kind of matter with very small density ρ and very large elastic modulus G, which means that the ether is extremely thin but is thousands of times harder than steel. Although these properties of the ether seemed to be very strange, no one doubted "whether the ether really exists" under Newton's great aura. Countless scientists have also begun to look for what Newton called the absolute static reference frame, that is, looking for the ether.

Since the speed of light in vacuum with respect to the ether (absolute static reference frame) is a constant value c, according to Galilean velocity transformation (1 - 3), then in any inertial frame moving with respect to the ether, the speed of light should differ as the constant value c. The actual speed of light as measured on Earth should depend on the motion of the Earth through the ether. Therefore, to find the ether is to measure the actual speed of light different from the constant value c in different reference frames. and a large number of experiments have been designed, the most representative of which is the Michelson-Morley experiment.

The principle of Michelson-Morley experiment is like the Michelson interferometer mentioned in the optics. Its experimental device and working principle are shown in Figure 1 - 2. The whole interferometer is placed on the shockproof platform, which can rotate very smoothly relative to the ground. The light emitted by the monochromatic light source S is divided into two beams through the semitransparent mirror G_1. The two beams propagate in a direction perpendicular to each other, are reflected back by the two mirrors M_1 and M_2 respectively, and meet at the observation screen T. Since these two beams are coherent light, interference fringes will appear at the observation screen T. Since the earth's revolution speed with respect to the sun is about 30 km/s, no matter what the speed of the sun with respect to the ether is, the speed u of the earth with respect to the ether will exceed 30 km/s at some time in a year. In the experiment, first adjust the angle of the experimental platform so that one arm G_1M_2 is parallel to the direction of the relative velocity $\boldsymbol{u}$, and the other arm G_1M_1 is perpendicular to the direction of the relative velocity $\boldsymbol{u}$. Since the speed of light with respect to the ether reference frame is c no matter which direction it travels, according to the Galilean velocity addition as shown in Figure 1 - 2(c), it can be obtained that with respect to the earth the speed of light to and fro along the parallel arm G_1M_2 is $c-u$ and $c+u$ respectively, and the speed of light to and fro along the vertical arm G_1M_1 is $\sqrt{c^2-u^2}$.

Let the length of both arms be equal to l, so the round-trip time t_1 in arm G_1M_2 is

$$t_1=\frac{l}{c+u}+\frac{l}{c-u}=\frac{2l/c}{1-u^2/c^2}$$

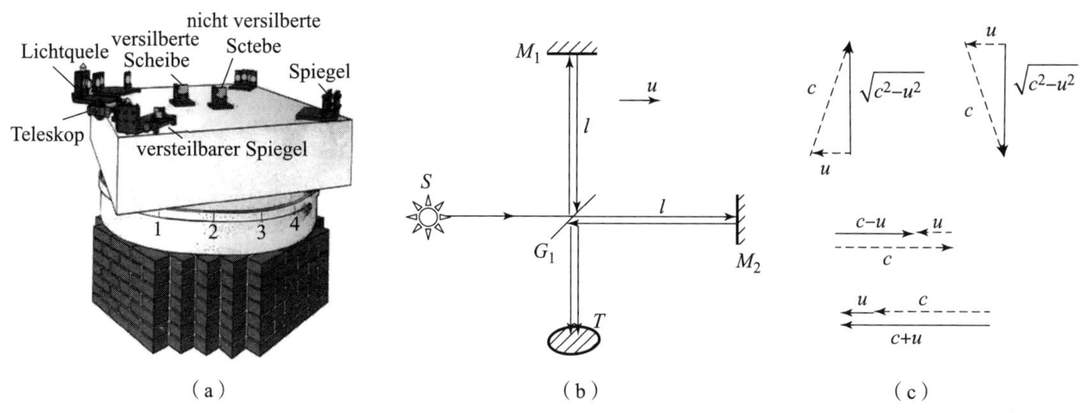

Figure 1-2 Michelson-Morey experiment

(a) Experimental device; (b) Schematic diagram of optical path; (c) Velocity transformation of different beams

The round-trip time t_2 in arm G_1M_1 is

$$t_2 = \frac{2l}{\sqrt{c^2 - u^2}} = \frac{2l/c}{\sqrt{1 - u^2/c^2}}$$

Thus, the optical path difference of the two beams when they meet at T is $c(t_1 - t_2)$.

After the interference pattern is observed under the above conditions, rotate the experimental platform by 90° so that the arm G_1M_1 becomes parallel to the direction of the relative velocity $\boldsymbol{u}$ and the arm G_1M_2 becomes perpendicular to the direction of the relative velocity $\boldsymbol{u}$. Accordingly, the optical path difference when the two beams meet at T becomes $c(t_2 - t_1)$. Because the optical path difference of the two beams changes before and after the rotation, the interference pattern will move. If the moving speed of the earth with respect to the ether is $u = 30\text{km/s}$ and the arm length is $l = 11\text{m}$, it can be calculated that interference pattern should have a movement of about 0.37 stripes.

However, no matter how the experimental conditions were changed (adjusting the angle, changing the light wavelength, measuring in different places and seasons), no change in the interference pattern was observed by Michelson and Morley, which is the "zero result" of the famous Michelson-Morley experiment in history. The Michelson-Morley experiment showed no observable change in the speed of light due to the motion of Earth with respect to the ether. So the speed of light in all inertial reference frames is the constant value c which has been producted by Maxwell. This led to the conclusion that there is no ether. Therefore, Galilean transformation is not correct when discussing the motion of light, and the principle of mechanical relativity and Newton's view of absolute space-time must be modified.

Einstein pointed out: "the reason why we can't find the ether is because the ether doesn't exist at all". On the basis of analyzing and summarizing the previous experimental results and ideas, Einstein published an epoch-making paper on electrodynamics of moving bodies in September 1905. In this paper, Einstein put forward the principle of relativity and the principle of constant speed of light as two basic postulates, and established the theory of special relativity, which now recognized as one of the cornerstones of modern physics.

1. Einstein's principle of relativity

The laws of physics are the same in all inertial reference frames, that is, for physical laws including electromagnetic laws, different inertial reference frames are equivalent, and there is no special inertial reference frame (such as ether reference frame). This is Einstein's principle of relativity.

It can be seen that Einstein's principle of relativity is a generalization of Galileo's principle of relativity, that is, the principle of relativity is applicable not only to mechanical phenomena, but also to all physical phenomena, including electromagnetic phenomena. This generalization contains profound physical connotation. It shows that in any inertial reference frame, it is impossible to determine whether the inertial reference frame is moving and its speed with respect to other inertial reference frame through any physical experiment. In this way, the concept of absolute motion or absolute stillness has been completely denied in the whole physics, which directly leads to the fundamental change in the understanding of the basic problem of physics——space-time view.

2. Principle of constant speed of light

In all inertial frames, the speed of light in vacuum is equal to c, regardless of how the light source and observer move. This is Einstein's principle of constant speed of light.

Einstein believed that "relativity" is the fundamental law of nature, which is also the essence of special relativity and the development of Galileo's principle of relativity. Einstein extended the principle of relativity to the field of electromagnetism, obtained the principle of constant speed of light, denied the ether reference frame, also denied Galilean transformation and Newton's absolute space-time view, and fundamentally shook the foundation of classical physics. Einstein established the theory of special relativity on the basis of the principle of relativity and the principle of constant speed of light, and formed a revolutionary space-time view of relativity. After Einstein established the theory of special relativity, he extended "relativity", the fundamental law of nature, to the noninertial reference frame, inherited the reasonable content of the theory of special relativity and established the theory of general relativity.

1.2 Effects of Space and Time in Relativity

Here, we will first introduce the important spacetime effects in relativity through thought experiments, such as time dilation, length contraction and the relativity of simultaneity.

1.2.1 Measurement of Space and Time

Before we begin to discuss these spacetime effects, we first explain some of the most basic concepts even in classical mechanics.

In the theory of relativity, the object we study is an event. The so-called event refers to an event that occurs at a certain position and at a certain time. To describe the event, we need to determine the time and spatial location of the event, that is, the spacetime coordinates of the event. Spacetime coordinates are composed of three independent parameters representing space position and one independent parameter representing time. For example, in inertial reference frame S, spacetime coordinates can be expressed as (x, y, z, t). Each event has definite spacetime coordinates in any inertial reference frame, and the spacetime coordinates of the same event in different inertial reference frames are different. The first thing that special relativity should discuss is the relationship between corresponding spacetime coordinates of one event in two different inertial reference frames. This corresponding relationship naturally reflects the spacetime view of special relativity.

According to Einstein's principle of relativity, the physical laws followed by the objective process of events are the same in different inertial reference frames, so it requires that there should be a transformation between the spacetime coordinates of one event measured in different inertial reference frames. According to the principle of constant speed of light, this transformation must not be the Galilean transformation reflecting Newton's absolute space-time view. To find this transformation, of course, it is inseparable from the accurate measurement of spacetime coordinates. In a certain inertial reference frame, how can we accurately measure the spacetime coordinates of an event?

The study of an event or process is essentially the measurement of spacetime coordinates. To accurately measure the spatial position coordinates of the event, it is not difficult to imagine that a set of coordinates system can be established to represent any inertial reference frame, the location of the event can be specified by three spatial coordinates (x, y, z). In order to accurately measure the time coordinate of the event, it is necessary to read the reading of the clock fixed at the location of the event, that is, no matter where the event occurs, there is a clock used to measure the time at the location of the event. The clock must also point to a certain scale when the event occurs. As long as the scale value is read, the time at which the event occurs is specified by t. In special relativity space and time is regarded as four-dimensional spacetime in which an event has four spacetime coordinates (x, y, z, t).

Note that measuring the time coordinates of events in relativity can no longer determine the time coordinates by "the observer reads the reading of the watch in his hand when he sees the event" as we are used to. This is because the motion of high-speed close to the speed of light is studed in special relativity. "Event occurrence" and "observer sees event occurrence" are actually two events with time difference. Although these two events can be connected at high speed through light propagation, the measurement error in time cannot be ignored when discussing the motion close to the speed of light. For example, in the 100m race in athletic meeting, when the referee sitting at the end sees a puff of smoke from the starting gun at the starting point, he starts the stopwatch to record the time coordinate of the athlete's "starting" event. When the athlete rushes to the line, the referee stops the stopwatch to measure the time coordinate of the athlete's "sprint" event. The difference between the two time coordinate is the time taken by the athlete to run 100m. This is the time measurement method we are familiar with in our daily life. Because the average speed of athletes running is only the order of 10m/s, while the speed of light is 3×10^8 m/s, the measuring error caused by starting the stopwatch later than the athlete's departure time can be ignored. However, if the "photon flyer" can move at the speed of 0.5c, measuring time in this way will obviously bring great errors. Then, to accurately measure the time taken by athletes to run 100m, we need to have two clocks aligned and synchronized with each other at starting and ending point respectively. Use the clock at the starting point to read out the time coordinate of the"starting"event, and use the clock at the ending point to read out the time coordinate of the"sprint"event. The difference between the two time coordinates is the accurate time interval of athletes to run 100m.

From the above discussion, it can be seen that the measurement of time coordinates is actually a problem of"simultaneity at the same place". "event occurrence" and"the clock at the place where event occurs points to a certain time" are two events that occur simultaneously at the same place. Therefore, the time coordinate of the event measured in this way are strictly accurate. In the theory of special relativity, the definition of simultaneity has been mentioned to a very important position and is the basis for discussing all problems. Einstein mentioned at the beginning of his paper "on electrodynamics of moving bodies": if we want to make a judgment on a problem related to time, the judgment is always associated with simultaneous events. For example, we say, "the train arrives here at 7 o'clock", which probably means, "the short needle of the clock on the platform points to 7 o'clock and the train arrives the platform at the same time".

Special relativity will involve two types of simultaneity: one is the"simultaneity at the same place", which is the basis of time measurement. The simultaneity at the same place is absolute, simultaneous events occurring at the same place are observed simultaneously in any inertial reference frame, which also means that the correctness of time coordinate measurement in one reference frame will be recognized by observers in other reference frames. The other is the"simultaneity at different places". The simultaneity at different places is relative, simultaneous events at different places in one reference frame do not occur simultaneously in other reference frames. The relativity of"Simultaneity at different places"is the most essential effect in the spacetime effect of special relativity.

Since the measurement of time coordinates requires the clock at the place where the event occurs, and the event may occur at any place in space, there is clock for measuring time at each coordinate position of a reference frame. These clocks are aligned and synchronized with each other, also known as "synchronous clocks". Each reference frame has its own series of "synchronous clocks". For example, if only events on the x-axis are discussed in the inertial reference frame S, a series of"synchronous clocks" need to be placed at different coordinates on the x-axis as shown in Figure 1 – 3. If the event occurs at x, you need to use the clock at x to measure the time coordinate of the event. Suppose that the clock at x point to time t, then the spacetime coordinate of the event is (x, t). Although it is impossible to place synchronous clocks at all coordinates of the reference frame in reality, it is obviously strict and accurate to imagine measuring time coordinates of events in the thought experiment.

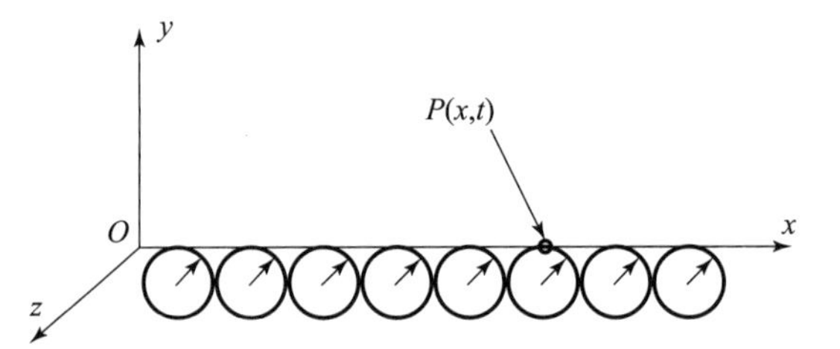

Figure 1 – 3 measuring time coordinates of an event with synchronous

With a correct understanding of spacetime coordinate measurement, we will discuss the important spacetime effects in special relativity on the basis of the principle of constant speed of light: time dilation, length contraction and the relativity of simultaneity.

1.2.2 Time Dilation

Einstein began to doubt Newton's concept of absolute time as early as his youth. He once imagined the famous "light chasing experiment": what would it look like if it could catch up with a beam of light? Can we see the still waves? Now we know that any particle with rest mass cannot move at the speed of light c, so such "light chasing experiment" can not be realized in practice. However, similar thought experiments can help us realize the relativity of time measurement.

We imagine that a spaceship moves at speed $0.5c$ with respect to the earth. At time $t=0$ in the earth frame, the spaceship emits a flash of light. At time t in the earth frame as shown in Figure 1-4, the spaceship flies forward by a distance of $0.5ct$, while the flash travel a distance ct (the speed of light with respect to the earth frame is c). At this time, the distance between the spaceship and the flash is $0.5\,ct$.

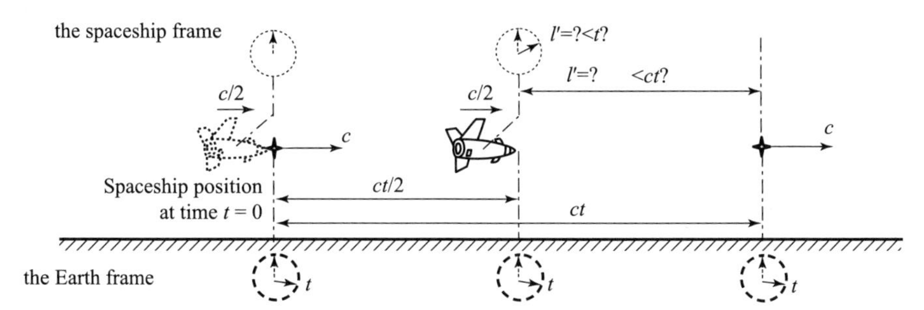

Figure 1-4 Thought experiment: the relativity of time measurement

The corresponding events of emitting flash and flash travelling can also be observed in the spaceship frame. Let the clock in the spaceship also points to $t'=0$ when the spaceship emits a flash of light. When the spaceship flies to the position shown in Figure 1-4, that is, time t in the earth frame, how much should the clock in the spaceship point to? According to Newton's view of absolute space-time, the clock in the spaceship should also point to time $t'=t$. However, it is obviously from Figure 1-4 that the distance l' travelled by the flash relative to the spaceship should be less than the travelled distance ct relative to the earth frame (at this time, we can't draw a conclusion whether l' measured in the spaceship frame is equal to $0.5ct$ measured in the earth frame, because the measurement of space distance may be different due to different frames. However, we can assume that the spaceship flies with respect to the earth at a faster speed (such as $0.8c$, $0.9c$), Then, at the same time t in the earth frame, the spaceship will fly farther forward, and the distance between the spaceship and the flash will be closer. If we change to the spaceship frame for observation, we should get the same change trend. In this way, $l' < ct$ can be qualitatively explained.). If the travelled distance of the flash relative to the spaceship $l' < ct$, according to the principle of constant speed of light, the speed of the light relative to the spaceship is still c, so $l' = ct'$, and the clock in the spaceship should point to a time t' less than t, namely $t' < t$. This means that Newton's concept of absolute time is no longer correct.

It can be seen from the above analysis that as long as the speed of light is recognized to be constant, the measurement of time is relative, that is, the time interval between the same two events measured in different inertial frames is different. In order to obtain the quantitative relationship describing the relativity of time measurement, let's discuss another thought experiment: let the inertial reference frame S' move at the speed u with respect to the inertial reference frame S along the positive direction of the x-axis. at time $t = t' = 0$ in the two frames, the coordinate origin O' of the frame S' coincides with the coordinate origin O of the frame S. At this time, as shown in Figure 1 – 5 (a), a flash is emitted at the origin O' of the frame S'. After the flash travels along the y' – axis for a distance d, it is reflected by a mirror M' and returns to the origin O'. We will discuss the time interval between the two events of "emitting light" and "receiving light" in the frame S and frame S' respectively.

First, observing in the frame S': it is not difficult to see from Figure 1 – 5 (a) that the flash travels a total distance of $2d$, and the speed of light remains unchanged as c. Therefore, the round-trip time $\Delta t'$ for the flash is

$$\Delta t' = \frac{2d}{c} \tag{1-5}$$

Since both "emitted light" and "received light" events occur at the origin O' of the frame S', the time interval $\Delta t'$ is measured by the clock located at the origin O'.

(a) Observe the time interval between "emitted light" and "received light" in the frame S'

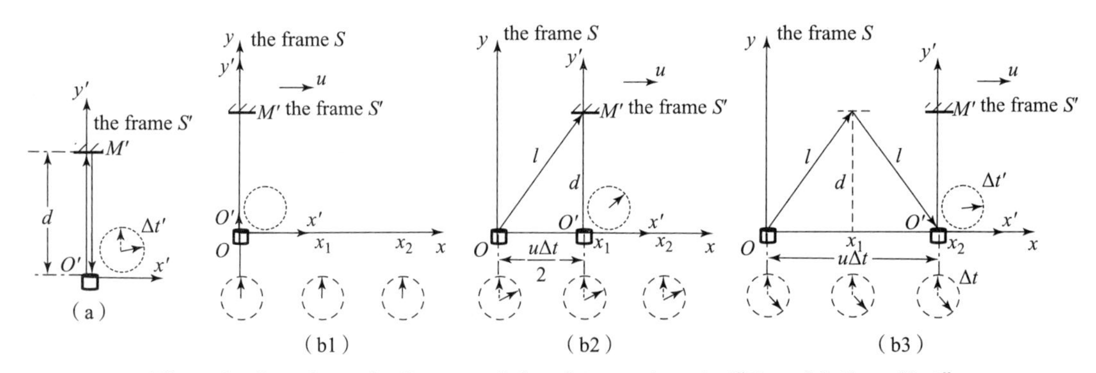

Figure 1 – 5 schematic diagram of thought experiment of "time delation effect"

(a) Observed the time interval between "emitted light" and "received light" in the frame S'; (b1) "Emitting light" event occurs; (b2) "reflecting light" event occurs; (b3) "Receiving light" event occurs

Then observe in the frame S: since the frame S' is moving with respect to the frame S, "emitting light", "reflecting light" and "receiving light" occur at different positions in the frame S as shown in Figure 1 – 5(b). These three positions are at the origin O, x_1 and x_2 respectively. It is not difficult to see that the travelling path of the flash is inclined up and down as shown in Figure 1 – 5(b3). It can be seen from the symmetry that the length of the path inclined up and down is the same, so let it be

l. According to the principle of constant speed of light, it can be obtained that the time interval Δt between "emitting light" and "receiving light" measured in the frame S is

$$\Delta t = \frac{2l}{c} \tag{1-6}$$

This time interval Δt is measured by two clocks located at the coordinate origin O and x_2 respectively, where $x_2 = u\Delta t$. In addition, the "reflecting light" occurs at the time of $\Delta t/2$, as shown in Figure 1-5(b2), and its corresponding coordinate $x_1 = u\Delta t/2$. By the Pythagorean theorem

$$l^2 = \left(\frac{u\Delta t}{2}\right)^2 + d^2 \tag{1-7}$$

Note that in the above formula, it is considered that the length $\overline{O'M'}$ in the frame S is still equal to d, because the length perpendicular to the relative motion direction is independent of the relative motion. This can be explained by the thought experiment of "train passing through the tunnel": we imagine that a train is driving towards a tunnel, and the height of the train and the tunnel is equal when they are at rest. If it is assumed that the length perpendicular to the relative motion direction will change due to the relative motion, whether it changes larger or smaller, contradictory results will be obtained in frames of the train and the tunnel. For example, assuming that the perpendicular length will be reduced due to the relative motion. The observer in the tunnel frame will think that the height of the train moving towards the tunnel will be lower than the height of the tunnel, so the train can pass through the tunnel. But the observer in the train frame will think that the tunnel moves towards the train, so the height of the tunnel becomes lower and the train cannot pass through the tunnel. However, whether the train can pass through the tunnel is a physical reality, which will not be different in different frames. Here, completely contradictory conclusions are obtained in two frames, which shows that the initial assumption is wrong. Therefore, the perpendicular length will not be reduced due to the relative motion. Similarly, the perpendicular length will not increase due to the relative motion. To sum up, the perpendicular length is independent of the relative motion, so it still exists $\overline{O'M'} = d$ in the frame S.

From equations (1-6) and (1-7), eliminating l

$$\Delta t = \frac{2d}{c} \cdot \frac{1}{\sqrt{1 - u^2/c^2}} \tag{1-8}$$

considering eq. (1-5), we get

$$\Delta t = \frac{\Delta t'}{\sqrt{1 - u^2/c^2}} \tag{1-9}$$

It can be seen that the time interval $\Delta t'$ between two events successively occurring at the same position in the frame S' is not equal to the corresponding time interval Δt measured in the frame S, and there is a factor difference $\sqrt{1 - u^2/c^2}$ between them. This shows that the measurement of time is relative, and Newton's concept of absolute time no longer holds.

Here, two events occur at the same place in the frame S', and the time interval $\Delta t'$ is measured by the same clock which is at rest. Thus, the time interval like $\Delta t'$ here which is measured in the rest frame of the clock is called the proper time interval, which also means the time interval between two events measured in an inertial frame in which the events occur at the same place. In the frame S two events occur at different places, the corresponding time interval Δt is measured by two clocks at different places. Since the relative speed u is always less than the speed of light c, there is $\sqrt{1-u^2/c^2} < 1$ and so $\Delta t' < \Delta t$, that is, the proper time interval $\Delta t'$ is the shortest and is always shorter than the time interval Δt measured in any other inertial frame. $\Delta t'$ is measured in the frame S', in which the clock is at rest, but the observers in the frame S find the rest clock in the frame S' is moving, it seems that the moving clock in the frame S' runs slowly. This effect is called "time dilation", also known as "clock slow effect", the time between ticks of the moving clock is dilated or expanded.

According to formula(1 – 9), the greater the relative speed u, the slower the S' clock runs, and the more significant the time dilation effect is. But when $u \ll c$, $\sqrt{1-u^2/c^2} \to 1$, there is $\Delta t \neq \Delta t'$, which means that the time interval measured in the two inertial frames is the same, that is, it returns to the concept of Newton's absolute time. Therefore, Newton's concept of absolute time is approximation of the concept of relativistic time when the relative speed is very small.

In addition, time dilation is a relative effect, that is, when a rest clock in the frame S moves relative to the frame S', the observers of the frame S' will also think that this clock in the frame S runs slowly. Therefore, when analyzing the time dilation effect, we should confirm which clock is at rest in its reference frame, then the time interval measured by it is the proper time interval.

Back to the thought experiment of "moving spaceship emitting light" discussed earlier, the rest clock in the spaceship moves with the spaceship relative to the earth frame, so the time t' measured by the clock in the spaceship is the proper time, while the corresponding time t is measured in the earth frame by two clocks. These two times satisfy the time dilation formula (1 – 9), that is $t = \dfrac{t'}{\sqrt{1-u^2/c^2}}$, which shows that $t' < t$, and explains "why the speed of light is still c when observed in the spaceship frame, but the distance l' travelled by the flash is less than the corresponding distance measured in the earth frame?". In addition, we can also calculate that the distance l' travelled by the flash measured in the spacecraft frame is $l' = ct' = ct\sqrt{1-u^2/c^2}$, which is also different from the $0.5ct$ measured in the earth frame. It can be seen that the measurement of distance or length is also relative.

Example 1 – 1 A spaceship flies at a constant speed $u = 9{,}000$m/s with respect to the ground. A rest clock in the spaceship has run for 5s. How much time has passed by in the ground frame?

Solution: the time interval $\Delta t' = 5$s measured by the rest clock in the spaceship is the proper time. Because the spaceship moves relative to the earth, it needs two clocks at different places on the ground to measure the corresponding time interval Δt, which is determined by the time dilation formula(1 – 9)

$$\Delta t = \frac{\Delta t'}{\sqrt{1-u^2/c^2}} = \frac{5\ \text{s}}{\sqrt{1-(9{,}000/3\times10^8)^2}} = 5.000{,}000{,}002\text{s}$$

it can be seen that for the reference frame with such a large speed relative to the ground, the time dilation effect has been difficult to measure in practice, which is why Einstein's view of spacetime of relativity is difficult to be understood and accepted by people.

1.2.3 Length Contraction

Time and space are closely related. Since the measurement of time is relative, does the measurement of space also have relativity? In other words, are the lengths of same object measured in different inertial frames the same? Here we will discuss the following example: When a train passes through a platform at a speed u, how to measure the length of the train in the train frame and in the platform frame? Are the lengths measured in two frames the same?

It is not difficult to imagine that since the train is stationary in the train reference frame, the observer in the train frame can measure the length l' of the train as long as he scales the positions of the head and the tail of the train respectively (positions can be scaled at different time). In the platform frame, the train is moving, and the positions of the train are constantly changing in the platform frame. Therefore, if the observer in the platform frame wants to measure the length l of the moving train accurately, he must scale the position of the head and the tail of the train at the same time. What is the relationship between the lengths measured in two frames?

In the platform frame, in addition to the method scaling both ends simultaneously, another method can also be used to measure the length of the moving train: as shown in Figure 1 - 6(a), when the front and tail of the train pass by a stationary observer on the platform, he records the time coordinates t_1 and t_2 of the two events "front passing" and "tail passing" respectively, so as to obtain the time interval $\Delta t = t_2 - t_1$ of train passing, the length l of the moving train can be calculated as $u\Delta t$, namely $l = u \cdot \Delta t$. The length of the moving train measured by these two methods should be the same.

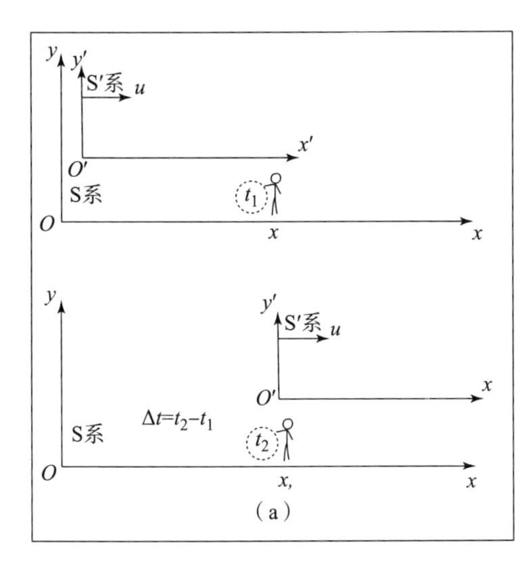

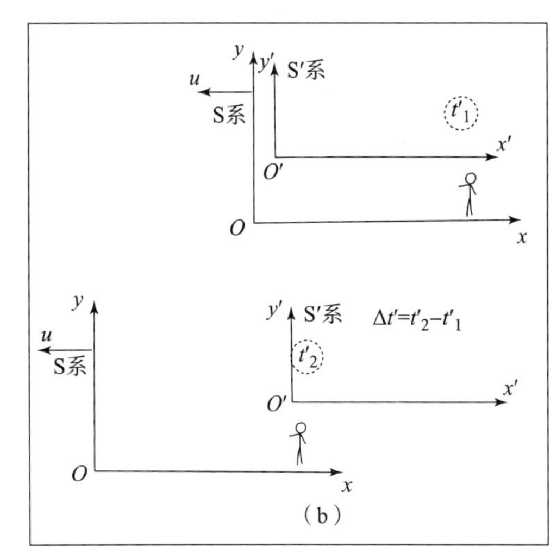

Figure 1 - 6 Thought experiment: measurement of the length of the train

(a) Observation in platform reference frame; (b) Observation in train reference frame

For the observers in the train, the train is stationary and the observer on the platform "runs" through the train at speed u as shown in Figure 1 - 6(b). The two events of "running over the front" and "running over the tail" in the train frame correspond to the two events of "front passing" and "tail passing" in the platform frame respectively. The time coordinates of the two events measured in the train frame are t_1' and t_2' respectively, and the time interval observer runing through is $\Delta t' = t_2' - t_1'$, so the travelling distance of the platform observer is $u \cdot \Delta t'$, which should be the train length l' measured in the train frame, i. e. $l' = u \cdot \Delta t'$.

It is not difficult to see that Δt and $\Delta t'$ here are the time intervals of the same two events measured in the platform and train frames respectively. In the platform frame, the time interval Δt between two events occurring at the same place is measured by one clock taken by the observer, so Δt can be regarded as the proper time interval. In the train frame, these two events occur at different positions (at the front and tail of the train), $\Delta t'$ is measured by two clocks at the front and tail of the train respectively. According to the time dilation formula (1 - 9), we have $\Delta t' = \dfrac{\Delta t}{\sqrt{1 - u^2/c^2}}$, considering the train length $l' = u \cdot \Delta t'$ and $l = u \cdot \Delta t$ measured in two frames, we get

$$l = l' \cdot \sqrt{1 - u^2/c^2} \tag{1-10}$$

Here l' is the length of the train in its rest frame (the train frame), so l' is called the proper length or rest length; and l is the length of the train measured in the platform frame, in which the train is moving. Since $\sqrt{1 - u^2/c^2} < 1$, there is $u > l'$, that is, the proper length of an object is the longest, the length measured in any other frame in which the object is moving is always shorter than the proper length. This means, an observer always finds that a moving object is contracted (shortened) along the direction of its motion, which is calles the length contraction effect.

According to the formula (1 - 10), the greater the relative velocity u is, the more significant the length contraction effect is. But when $u \ll c$, $\sqrt{1 - u^2/c^2} \to 1$, there is $l \neq l'$, which means that the length measured in the two inertial frames is the same, that is, it returns to the concept of Newton's absolute space. Therefore, the concept of Newton's absolute space is an approximation of the relativistic view of spacetime when the relative speed between two frames is very small.

Like time dilation, length contraction is also a relative effect. If observers in the frame S' measure the length of a rest rod in the frame S, it will also get the result that the length of the rod becomes shorter. Just as observers always find that a moving clock runs slowly compared with a stationary clock, observers always find that a moving object is contracted compared with its length in stationary state. Figure 1 - 7 shows the results obtained by the observers in one reference frame to measure the length of the metersticks moving with another reference frame. It can be seen that the length of the moving meterstick becomes shorter, no matter in which reference frame it is measured.

For two reference frames with relative motion, the coordinate axis along the direction of motion in one frame will also be regarded as shortened when observed in another frame due to the length contraction effect. Each coordinate reading on the axis of one frame is larger than its distance to the origin measured in other frames, while the distance is completely consistent with the coordinate

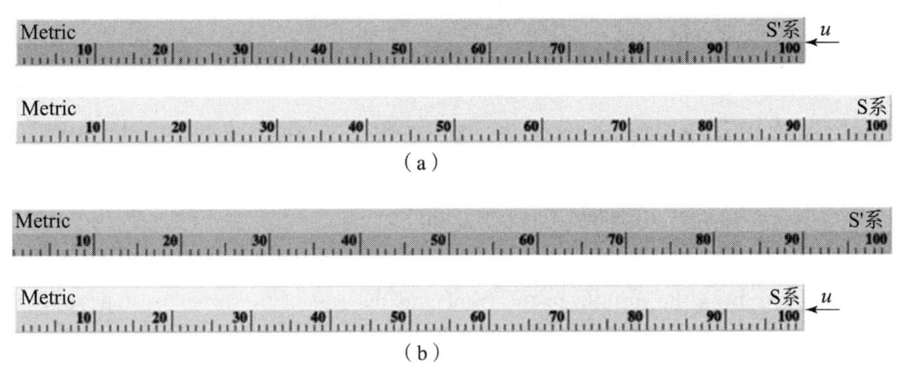

Figure 1 – 7 length contraction effect

(a) Observed in the frame S: the meterstick in the frame S' becomes shorter;

(b) Observed in the frame S': the meterstick in the frame S becomes shorter

reading in its own frame. This means that the moving coordinate axis is shortened. Therefore, when the same length or distance is measured in different frames, different results will be obtained, this is the relativity of spatial measurement.

In addition, length contraction is a longitudinal effect, that is, the length contraction can occur only in the direction of motion, lengths perpendicular to the direction of motion are not contracted (this has been explained by the thought experiment "train passing through the tunnel" earlier).

Example 1 – 2 The diameter of the earth is 1.27×10^4 km, its speed around the sun is about 30km/s. Both the earth frame and the solar frame are regarded as inertial frames. Calculate the contraction of the earth's diameter measured on the ecliptic plane (the plane of earth's orbit) in the solar frame.

Solution: the earth diameter d_0 measured in the earth frame is the proper length, according to the length contraction formula, the earth diameter d measured in the solar frame is

$$d = d_0 \sqrt{1 - u^2/c^2}$$

Thus, the contraction of earth's diameter is

$$d_0 - d = d_0 (1 - \sqrt{1 - u^2/c^2})$$

$$= 1.27 \times 10^4 \text{km} \times [1 - \sqrt{1 - (30/3 \times 10^5)^2}] = 6.39 \times 10^{-5} \text{km} = 6.39 \text{cm}$$

The results show that the length contraction effect caused by the movement of the earth at such a large speed relative to the sun is insignificant, which does not need to be considered even in space technology.

1.2.4 Relativity of Simultaneity

The relativity of simultaneity is the most essential spacetime effect in special relativity. time dilation and length contraction discussed above are the result of the relativity of simultaneity. The reason why we discuss the relativity of simultaneity after time dilation and length contraction is to express the relativity of simultaneity quantitatively, so as to deduce the Lorentz Transformation in next section.

Here we discuss it through the famous thought experiment of "Einstein train". Imagine that a train with the proper length L' moves at speed u relative to the ground reference frame. At a certain time, a flash is emitted from the middle of the train and travels to the front and tail of the train respectively. Here, we will discuss whether the two events of "flash reaching the front" and "flash reaching the tail" occur at the same time, as well as the sequence and time interval of two events in the train frame and in the ground frame respectively.

First, observing in the train frame: the train is stationary, the flash is emitted from the middle of the train, the distances from the middle to the front and the tail of the train are the same, the flashes arrive the front and the tail in the same time. Therefore, the two events of "flash reaching the front" and "flash reaching the tail" occur simultaneously in the train frame.

Then, observing in the ground frame: although the train moves forward at speed u, according to the principle of constant speed of light, the flashes travel to the front and the tail at the same speed as the speed of light in vacuum c as shown in Figure 1 – 8(a). However, since the tail moves in the direction opposite to the direction of the flash, the front of the train moves in the same direction as the flash travelling, when the backward flash reaches the tail, the forward flash has not caught up with the front, as shown in Figure 1 – 8(b). After that, the train continued to move forward, and after a period of time, the forward flash caught up with the front of the train as shown in Figure 1 – 8(c). Therefore, in the ground frame, the two events do not occur at the same time, "flash reaching the tail" occurs first and "light reaching the front" occurs later.

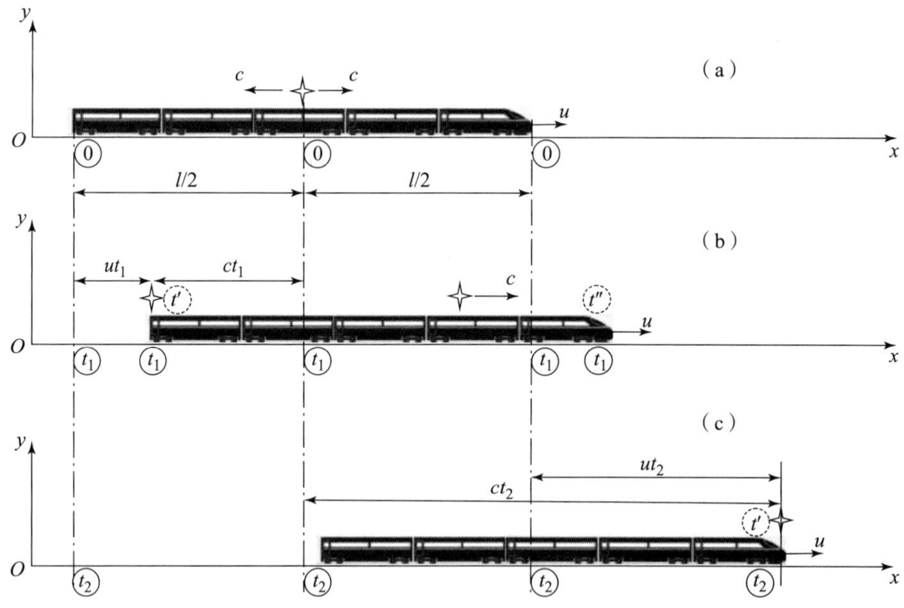

Figure 1 – 8　Einstein train: Relativity of simultaneity

(a) $t = 0$s in the ground frame; (b) time t_1 in the ground frame; (c) time t_2 in the ground frame

Through the above analysis, it can be seen that two events that occur simultaneously at different positions in one inertial frame (in the train frame, the flash reaches the front and the tail at the same time) do not occur simultaneously in another inertial reference frame with relative motion (in the ground frame, the flash reaches the tail first and then the front), this is the relativity of simultaneity.

Then, in the ground frame, what is the time difference between the two events of "flash reaching the tail" and "flash reaching the front"? For quantitative discussion, we set the three events of "flash emitted from the middle", "flash reaching the tail" and "flash reaching the front" correspond to time $0, t_1$ and t_2 in the ground frame respectively as shown in Figure 1 - 8. In addition, if the length of the train measured in the ground frame is l, the distance from the middle to the front and the tail of the train is $l/2$. Combined with Figure 1 - 8(a) and (b), it can be obtained

$$\frac{l}{2} = ut_1 + ct_1 \tag{1-11}$$

Combined with Figure 1 - 8(a) and (c), we can get

$$\frac{l}{2} = ct_2 - ut_2 \tag{1-12}$$

It can be obtained from the above two equations

$$t_2 - t_1 = \frac{l/2}{c-u} - \frac{l/2}{c+u} = \frac{lu}{c^2 - u^2} = \frac{lu/c^2}{1 - u^2/c^2} \tag{1-13}$$

Note that l is not equal to the proper length l' of the train. Since the train moves forward at speed u relative to the ground frame, according to the length contraction formula (1 - 10), $l = l' \cdot \sqrt{1 - u^2/c^2}$, it can be obtained

$$\Delta t = t_2 - t_1 = \frac{l'u/c^2}{\sqrt{1 - u^2/c^2}} \tag{1-14}$$

It can be seen from equation (1 - 14) that two events occurring simultaneously in the train frame do not occur simultaneously in the ground frame, and the time difference is determined by the relative speed u and the proper distance l' of the two events in the train frame.

We can judge whether different events in a reference frame occur at the same time by measuring their time coordinates using "synchronous clocks". In fact, the timing process of "synchronous clocks" at different positions, or the action of aligning clocks at different positions, is a series of simultaneous events in the reference frame. The "relativity of Simultaneity" means that synchronous clocks aligned with each other in one frame are no longer aligned with each other when observed in another frames. In the thought experiment of "Einstein train", suppose that the flash reaches the front and the tail at the same time t' in the train frame, which means both clocks at the front and the tail point to time t' when the flash reaches the front and the tail of the train. However, in the ground frame, the flash reaches the tail first and then the front, which means that the clock at the tail points to time t' first and the clock at the front points to time t' later. This means when observed in the ground frame, the two clocks in the train frame are not aligned and there is a time difference between them. When the flash reaches the tail, the clock at the tail points to t', but the clock at the front points to t'' as shown in Figure 1 - 8(b). $t' - t''$ is the time difference between the clocks at the front and the tail when observed in the ground frame.

By comparing Figure 1 - 8(b) and (c), it can be seen that when the clock at the front of the train moves from time t'' to time t', the corresponding clock in the ground frame moves from time t_1 to time t_2. Because the time interval $t' - t''$ is measured by one clock at the front of the train, it is proper time interval. According to the time dilation equation (1 – 9) the corresponding time interval $t_2 - t_1$ in the ground system is

$$t_2 - t_1 = \frac{t' - t''}{\sqrt{1 - u^2/c^2}} \tag{1-15}$$

Comparing equations (1 – 15) and (1 – 14), we get

$$\Delta t' = t' - t'' = \frac{l'u}{c^2} \tag{1-16}$$

The above equation (1 – 16) is the time difference between the clocks at the front and the tail of the train observed in the ground frame. Since $\frac{l'u}{c^2} > 0$, there is $t' > t''$, which means that the clock at the tail is ahead of the clock at the front. So the observers in the ground frame find that the clock behind the direction of motion is prior to the clock ahead in the train frame.

To sum up, due to the relativity of simultaneity, aligned clocks in one inertial frame are not aligned when observed in other inertial frames with relative motion. The time difference between clocks at two different positions is given by equation (1 – 16), which is proportional to the relative motion speed u and the proper distance between two clocks. In order to understand this spacetime image intuitively, we have set up two inertial frames S and S'. The frame S' moves in the positive direction of the x-axis relative to the frame S. Figure 1 – 9 shows the spacetime image observed in the frame S and S' respectively. When observed in the frame S: let's assume that it is the time $t = 0$ in the frame S, so all clocks are aligned and synchronized and point to time 0 as shown in Figure 1 – 9(a). but the observers of the frame S find that clocks at different positions in the frame S' are not aligned. If the clock at the origin O' point to time 0, the clocks at other positions do not point to time 0. The clocks at negative x' – axis (behind the direction of relative motion) are all ahead of the clock at the origin O', while the clocks at positive x' – axis (in front of the direction of relative motion) are all behind the clock at the origin O'. The time difference between the clock at the coordinate x' and the clock at the origin O' is $-\frac{ux'}{c^2}$. If the time difference is positive, it means ahead, and if it is negative, it means behind.

Although observers in the frame S believe that clocks in the frame S' are not aligned, observers in the frame S' do not. On the contrary, observers in the frame S' think that clocks in their own frame are aligned and synchronized, while clocks in the frame S are not aligned as shown in Figure 1 – 9 (b). When all clocks in the frame S' point to time 0, only the clock at origin O of the frame S points to time 0. The clocks at negative x – axis (in front of the relative motion) are behind the clock at origin O, and the clocks at positive x-axis (behind the relative motion) is ahead of the clock at origin O. The time difference between the clock at the coordinate x and the clock at origin O is $\frac{ux}{c^2}$.

The comparison of clocks observed in the frame S and S' shown in Figure 1 – 9(a) and (b) are

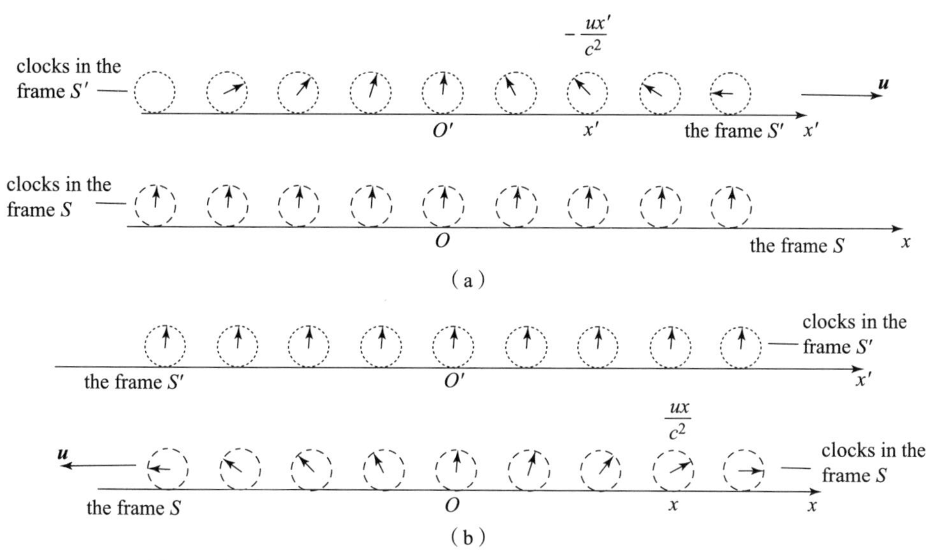

Figure 1 –9 space-time image observed in the frame S and S'

(a) observed in the frame S'; (b) observed in the frame S

consistent with each other, and there will be no contradictory results. when viewed in the frame S as shown in Figure 1 –9(a), it will be considered that clocks in the frame S' are not aligned, while all clocks in the frame S point to time 0. On the contrary, when viewed in the frame S', it is believed that the clocks at different positions in the frame S arrive at time 0 at different times, which means that the clocks in the frame S are not aligned. This is consistent with observation in the frame S' as shown in Figure 1 –9(b), all clocks of the frame S' point to time 0, while clocks in the frame S point to different times. "Clock misalignment observed by each other" is a measurement effect caused by the "relativity of simultaneity".

The reason why the relativistic spacetime effect is abstract is that it often interferes with the normal way of thinking that we are familiar with, so that we wrongly think whether there is something wrong with the ruler measuring the length or the clock measuring the time. In fact, the observers in each reference frame still measure length and time in the correct way they are familiar with, and the rulers and clocks used for measurement are still as accurate as before. All the cognition about length and time obtained by observers through measurement in their own reference frame are objective and have nothing to do with whether there are other reference frames or whether observers in other reference frames make the same measurement. However, if there exists other reference frame, in which the observers also make the same measurement, the observers in both frames will find that their measuring results are different from each other. The relationship between the different results does not conform to the view of absolute space-time and Galilean transformation. Observers in one reference frame will find a series of "strange phenomena" in other reference frame, such as "clocks at different positions are not aligned, each clock moves slowly, and the coordinate value is larger than its distance to the origin". It is these "strange phenomena" that cause the measuring results in other reference frame to be different from their own.

However, for people living on the earth, the relative motion speed between reference frames that can be encountered is much lower than the speed of light, and all kinds of "strange phenomena" in other reference frames cannot be directly observed. Therefore, people still think that Newton's view of absolute space-time and Galilean transformation are completely correct. However, Einstein revealed that when the speed of relative motion between reference frames is close to the speed of light, it is easy to show those "strange" spacetime effects, such as the relativity of simultaneity, time dilation and length contraction. On the basis of these spacetime effects, Einstein found the corresponding relationship between different measuring results obtained in different reference frames, which is the Lorentz Transformation reflecting spacetime view of relativity.

1.3 The Lorentz Transformation

In the last section, we have discussed the important spacetime effects in relativity: the relativity of simultaneity, time dilation and length contraction. On this basis, we will deduce the Lorentz Transformation reflecting Einstein's principle of relativity.

1.3.1 Derivation of the Lorentz Transformation

Similar to the discussion of Galilean transformation, we have two inertial reference frames S and S'. The x-axis of reference frame S coincides with the x' – axis of reference frame S', and the y – and z – axis are parallel to y' – and z' – axis respectively; the frame S' moves at a constant speed u with respect to the frame S along the x-axis; Initially, when $t = t' = 0$, the origin O' of the frame S' coincides with the origin O of the frame S. An event occurs at a certain point P in space at a certain time. Observers in the frame S and S' measure the spacetime coordinates of the event as (x, y, z, t) and (x', y', z', t') respectively. Let's discuss the relationship between the two set of coordinates.

As we have seen in the previous section, measurement of length perpendicular to the relative motion is independent of the reference frame. Since the relative motion is along the x-axis, we get $y' = y$, $z' = z$. Next, we will only discuss the x-coordinate transformation and time transformation.

First, we will discuss the x-coordinate transformation. Figure 1 – 10 shows the observed results in the frame S at the time t when the event occurred: the event occurred at coordinate x and time t. Since the origin O' coincided with the origin O at the time $t = 0$, the origin O' moved to the position $x = ut$ at time t. Since the event occurred at coordinate x' of the frame S', the coordinate x' of the frame S' coincided with the coordinate x of the frame S at time t. The observer of the frame S thinks that the distance between the coordinate x' and the origin O' of the frame S' is $x - ut$. Because the

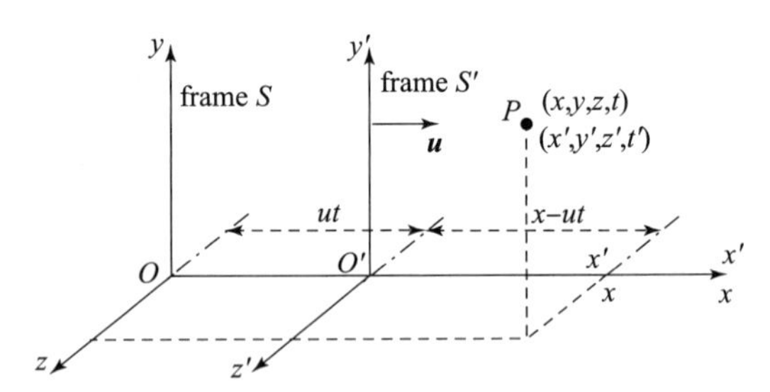

Figure 1 – 10 Deduction of Lorentz x coordinate transformation: observation in frame S

coordinate x' represents the length of $O'x'$ in the frame S', which can be regarded as the proper length, according to the length contraction formula(1 – 10), we have

$$x - ut = x' \cdot \sqrt{1 - u^2/c^2}$$

By deforming the above formula, we get

$$x' = \frac{x - ut}{\sqrt{1 - u^2/c^2}} \tag{1-17}$$

The above equation (1 – 17) is the x-coordinate transformation equation from the inertial reference frame S to S'. The equation(1 – 17) can be thought of as that the observer in the frame S measures the proper length $O'x'$ of the frame S' at time t, and the measured result is $x - ut$, so the equation(1 – 17) is actually a formula indicating the length contraction effect.

Similarly, the observer in the frame S' can also measure the proper length $O\,x$ of the frame S at time t', and the measured result is $x' + ut'$, so they also meet the relationship of length contraction, that is

$$x = \frac{x' + ut'}{\sqrt{1 - u^2/c^2}} \tag{1-18}$$

The above formula is the x-coordinate transformation formula from the inertial reference frame S' to S.

Next, we will discuss the time transformation between two inertial frames. Figure 1 – 11 (a) shows the time-space image observed at the time $t' = 0$ in the frame S': all clocks in the frame S' point to the time $t' = 0$, the clocks at the origin O', the coordinate x' and the position coinciding with coordinate x of the frame S, which all point to the time 0, are shown in the figure. The figure also shows the clocks at the origin O and the coordinate x of the frame S, which are not aligned. The clock at origin O points to time 0, and the clock at the coordinate x points to time $\frac{ux}{c^2}$ according to the equation(1 – 16). Figure 1 – 11(b) shows the time-space image observed at time t' in the frame S' when the event occurred: all clocks in the frame S' point to the time t'. The coordinate x' of the frame S' coincides with the coordinate x of the frame S, and the clock at the coordinate x of the frame S points to the time t.

It can be seen from Figure 1 – 11 that the observers in the frame S' believe that the event occurred at time t', and the time interval from the moment $t' = 0$ to event occurred is t'. In this period the clock at x of the frame S moves from $\frac{ux}{c^2}$ to t, so the corresponding time interval is $t - \frac{ux}{c^2}$ (in fact, all clocks of the frame S moves forward $t - \frac{ux}{c^2}$ in this period, such as the clock at origin O moves from 0 to $t - \frac{ux}{c^2}$). Since $t - \frac{ux}{c^2}$ is the time interval measured by one clock located at the coordinate x of the frame S, it is the proper time, according to the time dilation equation(1 – 9), we have

$$t' = \frac{t - \frac{u}{c^2}x}{\sqrt{1 - u^2/c^2}} \tag{1-19}$$

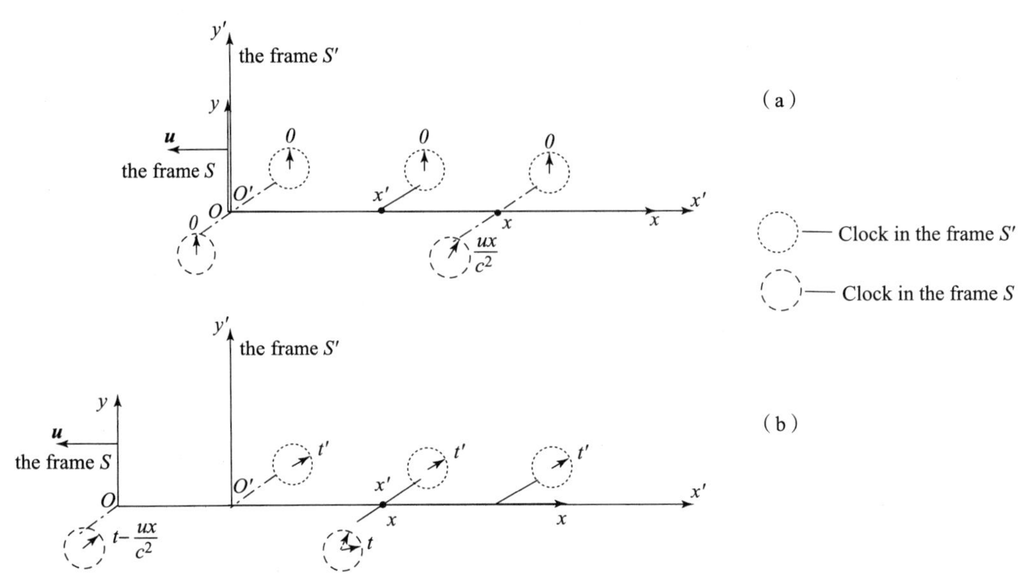

Figure 1 – 11 Deduction of Lorentz time t transformation: observed in the frame S'

(a) Observed at time $t' = 0$ in the frame S'; (b) Observed at time t' in the frame S'

The above equation (1 – 19) is the time transformation from the inertial frame S to S', which can be thought of as the process of correcting the reading of clocks of the frame S by the observers in the frame S': first, the observers in the frame S' consider that the clocks of the frame S are not aligned. At the time 0 in the frame S', the clock at coordinate x of the frame S pointed to the time $\frac{ux}{c^2}$. When the event occurred at the time t', the clock at coordinate x of the frame S passed actually time $t - \frac{u}{c^2}x$, which corresponds to the numerator in equation (1 – 19). After correcting the misalignment of the clocks of the frame S, the observers in the frame S' also find that all clock of the frame S go slowly, so they divide the time interval $t - \frac{u}{c^2}x$ by the factor $\sqrt{1 - u^2/c^2}$ to correct the clock-slow effect. After correcting the clock of the frame S twice in this way, the time t measured in the frame S is consistent with the time t' measured in the frame S'.

Similarly, when observed in the frame S, it will be considered that the clocks in the frame S' are not aligned [see Figure 1 – 9 (a)]. At the time 0 in the frame S, the clock at coordinate x' of the frame S' points to $-\frac{ux'}{c^2}$. When the event occurred at the time t, the clock at coordinate x' of the frame S' passed actually $t' + \frac{u}{c^2}x'$; In addition, the observers in the frame S also find that all clocks in the frame S' go slowly, so they divide $t' + \frac{u}{c^2}x'$ by the factor $\sqrt{1 - u^2/c^2}$. Thus the time t' measured in the frame S' is consistent with the time t measured in the frame S.

$$t = \frac{t' + \frac{u}{c^2}x'}{\sqrt{1 - u^2/c^2}} \tag{1-20}$$

The above equation (1 - 20) is the time transformation from inertial frame S' to S.

Combining the transformation equations (1 - 17), (1 - 19) and y-, z-coordinate transformation, we get the following Lorentz Transformation from the inertial frame S to S'

$$\left.\begin{aligned} x' &= \frac{x - ut}{\sqrt{1 - u^2/c^2}} \\ y' &= y \\ z' &= z \\ t' &= \frac{t - \frac{u}{c^2}x}{\sqrt{1 - u^2/c^2}} \end{aligned}\right\} \tag{1-21}$$

According to the above equations (1 - 21), if the spacetime coordinates (x, y, z, t) of an event in the frame S are known, the spacetime coordinates (x', y', z', t') of the event in the frame S' can be calculated.

Similarly, we can get the following Lorentz Transformation from the inertial frame S' to S.

$$\left.\begin{aligned} x &= \frac{x' + ut'}{\sqrt{1 - u^2/c^2}} \\ y &= y' \\ z &= z' \\ t &= \frac{t' + \frac{u}{c^2}x'}{\sqrt{1 - u^2/c^2}} \end{aligned}\right\} \tag{1-22}$$

According to the above equations (1 - 22), if the spacetime coordinates (x', y', z', t') of an event in the frame S' are known, the spacetime coordinates (x, y, z, t) of the event in the frame S can be calculated. In addition, considering that the frame S moves at the speed $-u$ with respect to the frame S', it is only necessary to replace u in equations (1 - 21) with $-u$, and exchange the quantities with and without prime to get equations (1 - 22).

The factor $1/\sqrt{1 - \frac{u^2}{c^2}}$ in the Lorentz Transformation is an important factor in relativity, so we assign it a symbol γ and call it the Lorentz factor

$$\gamma = \frac{1}{\sqrt{1 - u^2/c^2}} \tag{1-23}$$

Then, the Lorentz Transformation (1 - 21) and (1 - 22) above becomes

$$x' = \gamma(x - ut), y' = y, z' = z, t' = \gamma\left(t - \frac{u}{c^2}x\right) \tag{1-24}$$

$$x = \gamma(x' + ut'), y = y', z = z', t = \gamma\left(t' + \frac{u}{c^2}x'\right) \tag{1-25}$$

It can be seen from the Lorentz Transformation that the spacetime coordinates of an event in different inertial frames are related to each other, and the measurements of time and space in different inertial frames can no longer be separated from each other as in Galilean transformation. Therefore, in the theory of special relativity, we collectively refer to the customary three-dimensional space coordinates and one-dimensional time coordinate as four-dimensional spacetime coordinates.

It is easy to see from the Lorentz Transformation (1 – 24) that when $u \geqslant c$, the factor γ will become infinite or imaginary, which obviously has no physical significance. This shows that any inertial frame or actual object (excluding light) cannot move at a speed equal to or greater than the speed of light. The speed of light is the upper limit of the speed of any physical particle as well as energy and information. In addition, when $u \ll c$, the factor $\sqrt{1 - u^2/c^2} \to 1$, $\frac{u}{c^2} \to 0$, the Lorentz Transformation reduces to Galilean transformation shown in formula (1 – 1). This shows that Einstein's relativistic spacetime view is not a simple negation of Newton's absolute space-time view, but a development of the latter. Newton's absolute space-time view is approximation of the relativistic spacetime view when the relative speed u is very small. In the macro world, even the so-called high-speed moving objects, such as bullets and satellites, have velocities only about 10^3 m/s or 10^4 m/s, which are much smaller than the speed of light. Therefore, Newtonian mechanics is very accurate when the speed of motion is much lower than the speed of light. This is also the reason why it is difficult to establish and understand the theory of relativity.

Historically, the Lorentz Transformation was first obtained by the Dutch physicist H. A. Lorentz, but he was to explain the zero result of the Michelson-Morey experiment. On the premise of recognizing the absolute static inertial frame (the ether reference frame), he assumed that matter would be contracted when moving relative to the ether, thus obtaining the Lorentz Transformation. In Lorentz's theory, transformation is only regarded as a mathematical aid, and does not include the spacetime view of relativity. Einstein, on the basis of experimental facts, re deduced the Lorentz Transformation according to the principle of relativity and the principle of constant speed of light, and endowed the Lorentz Transformation with a new physical content.

Example 1 – 3: prove with the Lorentz Transformation that if the wavefront of a point light source is spherical in one inertial frame, it is also spherical in another inertial frame.

Prove: Assuming that the origin $O(O')$ of the inertial frame $S(S')$ coincides with each other at the time $t = t' = 0$, then the point light source at the origin emit an optical signal. The shape of the wavefront at any time t in the inertial frame S is spherical meeting the equation

$$x^2 + y^2 + z^2 = c^2 t^2$$

Substitute the Lorentz-transform (1 – 22) into the above equation,

$$\left(\frac{x' + ut'}{\sqrt{1 - u^2/c^2}}\right)^2 + y'^2 + z'^2 = c^2 \left(\frac{t' + \frac{u}{c^2}x'}{\sqrt{1 - u^2/c^2}}\right)^2$$

Then we get

$$x'^2 + y'^2 + z'^2 = c^2 t'^2$$

The above equation is also a spherical equation, which shows that the shape of the wavefront at any time t' in the inertial frame S' is also spherical.

The equation of the wavefronts in two inertial frames have identical mathematical forms, which means that the wave equation would remain invariant and the same value of c would appear for all inertial observers. This is also a manifestation of Einstein's principle of relativity. Similarly, Maxwell equations would also remain invariant in different inertial frames under the Lorentz Transformation, which is called the invariance of the Lorentz Transformation.

Example 1 – 4: The spaceship flies at speed $0.8c$ relative to the ground, and the observer on the spaceship measures that the length of the spaceship is 30m. A flash travels from the tail to the head of the spaceship. When observed on the spaceship and on the ground, what is the distance and time required for the flash to travel?

Solution: Let the ground reference frame be the frame S and the spacecraft reference frame be the frame S'. They can all be regarded as inertial frames, and the frame S' moves at a constant speed $u = 0.8c$ relative to the frame S along the positive direction of the x – axis.

This question involves two events:

Event 1: The flash is emitted from the tail of the spaceship, its spacetime coordinates in the frame S and S' are (x_1, t_1), (x'_1, t'_1) respectively;

Event 2: The flash reaches the head of the spaceship, and its spacetime coordinates in the frame S and S' are (x_2, t_2), (x'_2, t'_2) respectively;

According to the known conditions, the length of the spaceship is measured in the frame S', so

$$x'_2 - x'_1 = 30\text{m}$$

In the frame S', the spaceship is stationary, so the distance that the flash travels is 30m, and the time required to travel this distance is

$$t'_2 - t'_1 = \frac{x'_2 - x'_1}{c} = \frac{30\text{m}}{3 \times 10^8 \text{m/s}} - 10^{-7}\text{s}$$

By substituting the above results into the Lorentz Transformation (1 – 22), it can be obtained that the space interval and time interval of two events in the frame S are

$$x_2 - x_1 = \frac{(x'_2 - x'_1) + u(t'_2 - t'_1)}{\sqrt{1 - u^2/c^2}} = \frac{30\text{m} + 0.8c \times 30\text{m}/c}{\sqrt{1 - 0.8^2}} = 90\text{m}$$

$$t_2 - t_1 = \frac{(t'_2 - t'_1) + \dfrac{u}{c^2}(x'_2 - x'_1)}{\sqrt{1 - u^2/c^2}} = \frac{10^{-7}\text{s} + 0.8 \times 10^{-7}\text{s}}{\sqrt{1 - 0.8^2}} = 3 \times 10^{-7}\text{s}$$

so the distance travelled by flash in the frame S is 90m, and the time required is 3×10^{-7}s. In the frame S, the speed of light is still $(x_2 - x_1)/(t_2 - t_1) = 3 \times 10^8$ m/s, so the speed of light remains constant in any inertial frame.

According to Galilean transformation formula (1 – 1), the space interval and time interval of two events in the frame S are

$$x_2 - x_1 = (x_2' - x_1') + u(t_2' - t_1') = 30\text{m} + 0.8 \times 30\text{m} = 54\text{m}$$

$$t_2 - t_1 = t_2' - t_1' = 10^{-7}\text{s}$$

From this, it should be concluded that the speed of light in the frame S is greater than that in the frame S', which is inconsistent with the experimental results of constant speed of light. From this example we can see that the theory of special relativity is essential to study the motion with speed close to the speed of light.

1. 3. 2 Using the Lorentz Transformation to Verify the Spacetime Effect of Relativity

In the theory of special relativity, the Lorentz Transformation is the most basic expression. The spacetime properties and kinematic conclusions of special relativity, such as the relativity of simultaneity, length contraction, time delation and speed transformation, can be directly derived from the Lorentz Transformation. In the following discussion, two inertial frames will be involved: the frame S and S', where the frame S' moves at a constant speed u relative to the frame S along the positive direction of the x-axis.

1. Relativity of Simultaneity

Two events occur simultaneously at different positions in the frame S'. The spacetime coordinates of these two events in the frame S' are (x_1', t') and (x_2', t') respectively. Do these two events occur simultaneously in the frame S?

Let the spacetime coordinates of these two events in the frame S be (x_1, t_1) and (x_2, t_2) respectively. Here we will discuss the time relationship of these two events in the frame S. According to the time transformation formula in Lorentz Transformation (1 – 22)

$$t_1 = \frac{t' + \frac{u}{c^2}x_1'}{\sqrt{1 - u^2/c^2}}, \quad t_2 = \frac{t' + \frac{u}{c^2}x_2'}{\sqrt{1 - u^2/c^2}}$$

From the above two equations, we get

$$t_2 - t_1 = \frac{(t' - t') + \frac{u}{c^2}(x_2' - x_1')}{\sqrt{1 - u^2/c^2}} = \frac{\frac{u}{c^2}(x_2' - x_1')}{\sqrt{1 - u^2/c^2}} \tag{1 – 26}$$

In equation (1 – 26) above, $t_2 \neq t_1$ can be obtained due to $x_2' \neq x_1'$, which means that the two events do not occur at the same time when observed in the frame S, and the time difference is proportional to the distance of the two events in the frame S'. This result is consistent with the result (1 – 14) obtained in the thought experiment "Einstein Train" in the previous section.

On the basis of the Lorentz Transformation, the relativity of simultaneity is verified by simple derivation. The relativity of simultaneity is the most essential spacetime effect in relativity, which lead to establishment of Einstein's theory of special relativity. To help us further understand the nature of time, let's compare the following two descriptions about the space and the time:

(1) Two events occurring at the same place and at different times in one reference frame will occur at different places when observed in another reference frame.

(2) Two events occurred at the same time and at different places in one reference frame will occur at different times when observed in another reference frame.

The only difference between the two descriptions is that the positions of "place" and "time" are exchanged, which reflects that "time" and "space" are in the equivalent status in special relativity. The time interval is transformed into the space interval in the first description, which is very easy to be understood and accepted, while the space interval is transformed into the time interval in the second description, which is very difficult to be understood and recognized. This is because the speed of macroscopic objects we are familiar with is far less than the speed of light. Einstein, with his profound insight, realized that the problem of time is essentially a problem of simultaneity. On the basis of the principle of constant speed of light, he deduced the relativity of simultaneity, making time and space equivalent and inseparable. Space and time can be regarded as two different components of the same uniform "spacetime continuum".

In order to describe the equivalence of time and space in special relativity, the German mathematician H. Minkowski added time as the fourth dimensional coordinate to the three-dimensional space coordinate and established the Minkowski space. He emphasized that only the unity of space and time exists in reality. In Minkowski space, various spacetime effects of relativity will be displayed intuitively and vividly. We will introduce Minkowski space in section 1. 3. 3.

2. Time Dilation

Event 1 and event 2 occur successively at the same place in the inertial frame S', and their spacetime coordinates are (x', t_1') and (x', t_2') respectively. $t_2' > t_1'$, the time interval of two events in the frame S' is $\Delta t' = t_2' - t_1'$. Find the time interval Δt of two events in the inertial frame S.

Let the spacetime coordinates of these two events in the frame S be (x_1, t_1) and (x_2, t_2) respectively. Similarly, according to the time transformation formula in Lorentz-transform (1 – 22), we have

$$t_2 - t_1 = \frac{(t_2' - t_1') + \frac{u}{c^2}(x' - x')}{\sqrt{1 - u^2/c^2}} = \frac{t_2' - t_1'}{\sqrt{1 - u^2/c^2}}$$

Therefore, the time interval between these two events in the frame S can be expressed as

$$\Delta t = \frac{\Delta t'}{\sqrt{1 - u^2/c^2}} \tag{1-27}$$

Here, two events occur successively at the same position in the frame S', and the corresponding time interval $\Delta t'$ is measured by a clock at x', so $\Delta t'$ is the proper time. From the x coordinate transformation formula in Lorentz inverse transformation, we have

$$x_2 - x_1 = \frac{(x' - x') + u(t_2' - t_1')}{\sqrt{1 - u^2/c^2}} = \frac{u(t_2' - t_1')}{\sqrt{1 - u^2/c^2}}$$

It is obvious that $x_2 \neq x_1$, which means that two events occur at different positions in the frame S, and the time interval Δt is measured by two clocks at x_2 and x_1 respectively.

For all u, the denominator in Formula (1 – 27) is always less than 1, so there is always $\Delta t > \Delta t'$, that is, the proper time is the shortest. This is the time delation effect discussed in Section 1. 2. 2.

The time interval measured by this clock in the frame S' is smaller than the time interval measured in the frame S. In fact, this clock is not really slow, all clocks in two frames are the same standard clock, and they all tick at the same rate. But when one clock in the frame S' compares constantly with the clocks of the frame S it meets, the result is that its reading is always smaller than the reading of these clocks. However, the observers of the frame S' will think: The clocks at different positions in the frame S are not aligned, and the readings of clocks encountered later are always ahead, their readings are always larger when meeting with the clock of the frame S'. Therefore, the observers of the frame S get the conclusion that the clock of the frame S' is ticking slowly. It can be seen that time delation is actually caused by the relativity of simultaneity.

Example 1 – 5: It has been observed in modern physical experiments that the average passing distance of muon flying freely at speed $u = 0.91c$ during its lifetime is 17.14m, what is its lifetime?

Solution: in the reference frame of ground, the average time that the muon flies is

$$\Delta t = \frac{d}{u} = \frac{17.14\text{m}}{0.91 \times 3 \times 10^{8}\text{m/s}} = 6.28 \times 10^{-8}\text{s}$$

But Δt here is not the lifetime required. The lifetime $\Delta t'$ required refers to the surviving time of muon in the reference frame in which the muon is at rest, so the lifetime $\Delta t'$ is the proper time. According to the time delatiom formula (1 – 27), we get

$$\Delta t' = \Delta t\sqrt{1 - u^2/c^2} = 6.28 \times 10^{-8}\text{s} \times \sqrt{1 - 0.91^2} = 2.60 \times 10^{-8}\text{s}$$

This result is in good agreement with the experimental results.

Example 1 – 6: A star is far away from the earth at speed 0.6c, it emits flashes regularly with a certain period. The period of the flash is measured by a clock on the earth to be 5s. Find the flash period of the flash on the star.

Solution 1: In the earth frame, the star leaves the earth at speed 0.6c and emits two flashes successively as shown in Figure 1 – 12. Let the star emit the previous flash at time t_1' and the next flash at time t_2' in its own reference frame. The time interval $\Delta t' = t_2' - t_1'$ between the two flashes is the required flash period on the star. Since t_1' and t'_2 are measured by the same clock located on the star, the flash period $\Delta t'$ is the proper time. According to the time delation formula (1 – 27), it can be obtained that the time difference Δt of the two flash events in the earth frame is

$$\Delta t = \frac{\Delta t'}{\sqrt{1 - u^2/c^2}} = \frac{\Delta t'}{\sqrt{1 - (0.6)^2}}$$

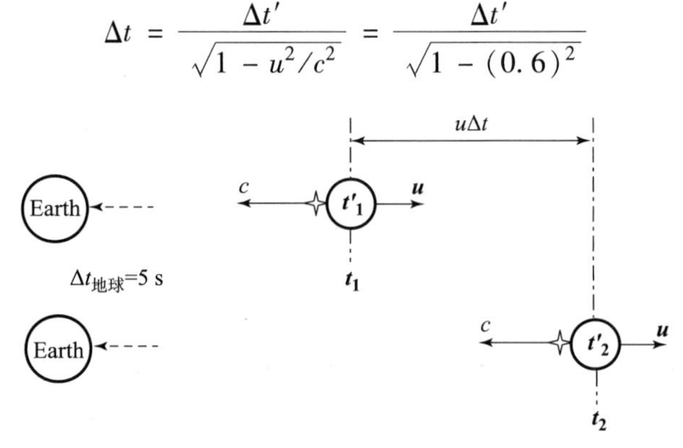

Figure 1 – 12 Example 1 – 6

Because the star moves forward a distance of $u\Delta t$ relative to the earth between the two flashes, the distance travelled by the last flash to the earth is $u\Delta t$ longer than that travelled by the previous flash, and the time for flash to travel this distance is $u\Delta t/c$. So the period $\Delta t_E = 5\text{s}$ measured by the clock on the earth can also be expressed as

$$\Delta t_E = \Delta t + \frac{u\Delta t}{c} = 1.6 \cdot \frac{\Delta t'}{\sqrt{1-(0.6)^2}}$$

we get

$$\Delta t' = 2.5\text{s}$$

That is, the flash period on the star is 2.5s.

Solution 2: In the solution 1, we solved the flash period on the star from the perspective of the observer in the earth frame, and we can also calculate it from the perspective of the star frame.

In the star frame the earth leaves the star at speed 0.6c, the star emits two flashes travelling to the earth successively in period $\Delta t'$. The time interval $\Delta t_E = 5\text{s}$ between the arrival of two flashes measured by the clock on the earth is the proper time, and the corresponding time interval measured in the star frame can be obtained by the time delation formula as follow

$$\Delta t'_E = \frac{\Delta t_E}{\sqrt{1-u^2/c^2}} = \frac{5\text{s}}{\sqrt{1-(0.6)^2}} = 6.25\text{s}$$

Therefore, the last flash will travel $u\Delta t'_E$ more distance to reach the earth than the previous flash, and the time taken for the last flash to travel more distance is $u\Delta t'_E/c$, so there is

$$\Delta t'_E = \Delta t' + \frac{u\Delta t'_E}{c}$$

There is

$$\Delta t' = 2.5\text{s}$$

From this question, we can see that when discussing the spacetime effect of relativity, we must analyze the events clearly. Here, although there is a relationship between the period Δt_E measured by the clock on the earth and the period $\Delta t'$ on the star, they do not meet the time delation formula. This is because the events that the flashes arrive at the earth successively and the events that the star emit flashes successively are four different events. Only for the two events which occur successively at same place in a reference frame, if they are observed in another reference frame, the corresponding time intervals measured in these two frames meet the time delation formula.

Twin effect: After Einstein put forward the theory of special relativity, because its spacetime view is different from the widely accepted Newton's absolute spacetime view, there have been many debates about Einstein's spacetime view of relativity, one of which is the so-called "twin paradox".

The "twin paradox" was first proposed by French scientist P. Langevin to question the time delation effect in special relativity. The "twin paradox" is described as follows: There is a pair of twin brothers A and B, A stays on the earth, and B travels in space at speed close to the speed of light by spaceship. If observed in the Earth frame, the spaceship leaves at a high speed and returns at a high speed. Due to the time delation effect, the clock on the spaceship slows down. After B returns to the

Earth, A finds that B is younger. However, the observers in the spaceship frame will observe that the earth leaves and returns at a high speed, the clock on the earth moves slowly due to the time delation effect. When the earth returns to the spaceship, B finds that A is younger. Here, oberved in different frames, the opposite conclusion will be obtained, but there can only be one physical reality, so this is also called "paradox". The above discussion uses the time delation effect of special relativity. Does this mean that there is a problem with the time delation effect?

The discussion of this paradox has gone beyond the category of special relativity. Special relativity is only applicable to inertial frames, and the relative speed between different frames is constant. The earth frame here can be considered as an inertial frame, in which the theory of special relativity is true; However, the spaceship leaves first and then returns to the earth, during which it must undergo the process of deceleration, turning around and re acceleration. Therefore, the spaceship frame can no longer be regarded as an inertial frame in this process, and the theory of special relativity is no longer valid in the spacecraft frame. In this sense, the two frames can no longer be considered symmetrical, and the contradiction caused by symmetry will no longer exist.

So, if the spaceship returns to the Earth, who is younger? As it involves noninertial frames, this problem needs to be explained by general relativity. Since the spaceship frame is noninertial frame in the process of turning around, and noninertial frame is equivalent to the existence of a gravitational field, which can bend space and delay time, so B is younger.

With the development of science and technology, the results of the discussion on the "twin paradox" were also verified by experiments: in 1966, in the experiment that muon moves along a circle with a diameter of 14 meters and returns to the starting point, the results show that the life of moving muon is longer than that of stationary muon. In 1971, the US Naval Observatory installed four cesium atomic clocks on the aircraft which flew eastward and westward around the Earth's equator respectively. After the aircraft returned to its original location, it was found that the clock flying eastward (in the same direction as the earth's rotation) was 59 ns slower than the stationary clock on the ground, and the clock flying westward (opposite to the rotation direction of the earth) is 273ns faster than the stationary clock on the ground, which shows that the clock with higher speed relative to the inertial frame (the earth frame revolving around the sun) moves slower, and the experimental results are consistent with special relativity. As verified by experiments, the twin paradox is also known as the twin effect.

In the modern global positioning system (for example the BeiDou Satellite Positioning), in order to locate accurately, we must consider the time delation effect. The basic working principle of satellite positioning is that the user terminal (such as mobile phone) will simultaneously receive the spacetime coordinates (x_i, y_i, z_i, t_i) of four satellites sent by corresponding satellites, so that four propagation equations $\sqrt{(x-x_i)^2+(y-y_i)^2+(z-z_i)^2}=c(t-t_i)$ can be obtained. The spacetime coordinates (x, y, z, t) of the user terminal can be obtained by solving the propagation equations, so as to achieve the purpose of positioning and timing. In order to position more accurately, the relativistic effect must be considered to correct the reading of the clock on the satellite to the corresponding time of the earth frame. As the satellite moves at a high speed (about 4000m/s), considering the time delation effect of special relativity, the atomic clock on the satellite is about 7μs slower than the clock on the earth every day; Because the satellite flies in the high altitude (about 20000 km) where the gravity of the earth is weak, considering the clock slow effect of general relativity, the atomic clock on the satellite is about 45μs faster than the clock on the earth every day. Thus, the clock on satellite is about 38 μs faster than the clock on the earth every day. Considering this time difference, the reading of the clock on the satellite can be corrected to the time coordinate t_i in the earth frame, so that the precise position of the user terminal on the earth surface can be calculated. Because the signals are transmitted between the satellites and the earth with the speed of light c, and $38\mu s \times c = 11.4km$, if the relativistic effect was not considered, the position of the user terminal would accumulate an error of ±11.4km per day, which makes accurate positioning impossible.

3. Length Contraction

A stationary rod in the frame S' is placed along the x' – axis, and its length is l' which is the proper length. The rod moves at speed u with respect to the frame S. What is the length l of the rod measured in the frame S?

To measure the length of the moving rod in the frame S, it is necessary to scale the left and right ends of the rod at the same time. Let the observers in the frame S scale the left and right ends of the rod at time t, and obtain their coordinates as x_1 and x_2 respectively, then the length of the moving rod measured in the frame S is $l = x_2 - x_1$.

The spacetime coordinates of the two events scaling the motional rod in the frame S are (x_1, t) and (x_2, t) respectively. Two events are also observed in the frame S', and their position coordinates are assumed to be x_1' and x_2' respectively. Because these two events occur at the left and right ends of the rod, there is $x_2' - x_1' = l'$. According to the x coordinate transformation formula of the Lorentz Transformation, there is

$$x_1' = \frac{x_1 - ut}{\sqrt{1 - u^2/c^2}},\ x_2' = \frac{x_2 - ut}{\sqrt{1 - u^2/c^2}}$$

subtract the above two equations to get

$$x_2' - x_1' = \frac{(x_2 - x_1) - u(t - t)}{\sqrt{1 - u^2/c^2}} = \frac{x_2 - x_1}{\sqrt{1 - u^2/c^2}}$$

therefore

$$l = l' \sqrt{1 - u^2/c^2} \qquad (1-28)$$

Obviously, $l < l'$. This shows that the proper length l' of the rod is larger than the length l measured by observers for whom the rod is moving. The proper length of an object is the longest compared with the length measured in other frame in which the object is moving. This is the length contraction effect obtained in Section 1.2.3. Just as observers always finds that a moving clock goes slow compared with stationarys clock, observers always finds that a moving object is contracted along the direction of its motion.

Just like time delation, length contraction is also an inevitable result of the relativity of simultaneity. In the frame S, the length of the moving rod is measured by simultaneously scaling the left and right ends of the rod. However, it was observed in the frame S' that two scaling events did not occur at the same time: the event at the right end (i. e., the front end of the moving rod, corresponding to the coordinate x_2) occurred first, and the event at the left end (i. e., the rear end of the moving rod, corresponding to the coordinate x_1) occurred later. When the left end was scaled to get x_1, the coordinate x_2 had moved a distance with the frame S, this leads to the length $l = x_2 - x_1$ measured in the frame S being less than the proper length of the rod.

Example 1 – 7: When the train passes a platform at speed $0.6c$, two manipulators on the platform 10m apart draw two marks on the train at the same time, calculate the distance between the two marks measured by observers on the train.

Solution: Here we use three different methods to solve.

Method 1: Observed in the platform reference frame

Two manipulators on the platform draw marks on the train at the same time, which are two simultaneous events in the platform frame. If there were two marks on the train itself, the observers on the platform should measure the distance between the marks, and they should also be required to measure positions of two marks at the same time. Through the comparison of the above two different narrations, it can be seen that $\Delta x = 10\text{m}$ (the distance between two manipulators on the platform) can also be understood here as the measuring result of the distance of two marks on the moving train (assuming that there were marks first, which is not contradictory to drawing marks on the train at the same time), so the required distance $\Delta x'$ between the two marks on the train is the proper length in the platform frame. From the length contraction formula (1 – 28), there is

$$\Delta x' = \frac{\Delta x}{\sqrt{1 - u^2/c^2}} = \frac{10\text{m}}{\sqrt{1 - 0.6^2}} = 12.5\text{m}$$

Method 2: Observed in train reference frame

The distance between the two manipulators on the platform $\Delta x = 10\text{m}$ is the proper length in the platform frame, according to the length contraction formula (1 – 28), the corresponding distance $\Delta x''$ measured in the train frame

$$\Delta x'' = \Delta x \cdot \sqrt{1 - u^2/c^2} = 10\text{m} \times \sqrt{1 - 0.6^2} = 8\text{m}$$

In the train frame it is observed that the two drawing events do not occur at the same instant. The manipulator near the front of the train scratches first, and the manipulator near the tail of the train scratches later. According to the formula (1 – 26), the time difference between the two drawing events is

$$\Delta t' = \frac{u\Delta x/c^2}{\sqrt{1 - u^2/c^2}}$$

After the manipulator near the front of the train scratched, the manipulator near the tail of the train moved forward $u\Delta t'$ before scratching. Therefore, the distance between the two marks measured in the train frame $\Delta x'$ is

$$\Delta x' = \Delta x'' + u\Delta t' = 8\text{m} + \frac{10\text{m} \times 0.6^2}{\sqrt{1 - 0.6^2}} = 12.5\text{m}$$

Method 3: using the Lorentz Transformation

The spacetime coordinates of the two drawing events are (x_1, t_1) and (x_2, t_2) in the platform frame, (x_1', t_1') and (x_2', t_2') in the train frame. It is known that $\Delta x = x_2 - x_1 = 10\text{m}$, $t_1 = t_2$. From the x coordinate transformation formula of the Lorentz Transformation, it can be obtained that the distance $\Delta x' = x_2' - x_1'$ between the two marks measured in the train frame is

$$\Delta x' = x_2' - x_1' = \frac{x_2 - x_1 - u(t_2 - t_1)}{\sqrt{1 - u^2/c^2}} = \frac{10\text{m}}{\sqrt{1 - 0.6^2}} = 12.5\text{m}$$

Here, we obtained the same result by three different methods. From this example we can see: when we discuss a distance between two events in different frames, and the two events occur in one frame simultaneously, even if the two events have no relationship with length measurement, the length contraction formula (1 - 28) can also be used to solve the problem, and the distance in the frame in which the two events don't occur at the same instant can be regarded as the proper length.

Example 1 - 8: Planets A and B are close to each other at speed $0.8c$: (1) When the planet A emits a flash, the observers in the frame of planet A (hereinafter referred to as frame A) measure the distance between the two planets to be 3×10^8 m. What is the distance between the two planets measured in the frame of planet B (hereinafter referred to as frame B)? (2) When the planet B emits a flash, the observers in the frame A measure the distance between two planets to be 3×10^8 m. What is the distance between the two planets measured in the frame B?

Solution: What is known in this problem is the distance between two planets measured at a moment when an event occurs in the frame A, and what is calculated is the corresponding distance between two planets in the frame B when the event occurs. To measure the distance between the two planets, the observers in the frame A are required to scale the position of the planet B at some moment, which can be also referred to an event. The event can occur anywhere at a certain moment in the frame A, for example, a flash is emitted at the planet A or at the planet B. No matter where the event occurs, the measured result will be the same for the observers in the frame A, because the moment at which the event occurs and the event of scaling the position of the planet B must occur simultaneously. However, when observed in the frame B, the simultaneous events at different position of the frame A, such as emitting a flash at the Planet A or B, do not occur at the same instant. Since there is relative motion between planets A and B, it is necessary to make clear whether it is at the moment when Planet A emits a flash or at the moment when Planet B emits a flash, when we calculate the corresponding distance between planets A and B in the frame B.

(1) It is known that the distance between the two planets measured in the frame A is 3×10^8 m when Planet A emits a flash, which means that the position of the planet B is scaled (called the event 2) at the moment when Planet A emits a flash (called the event 1). The two events occur simultaneously in the frame A, with a distance of $l = 3 \times 10^8$ m, as shown in Figure 1 - 13.

However, in the frame B, Planet B is stationary, the planet A moves towards the planet B at speed $0.8c$, and the event 1 which was observed at time t_1' occurred before the event 2 which is observed at time t_2'. In the frame B, the position of the planet A is scaled at time t_1' when the event 1

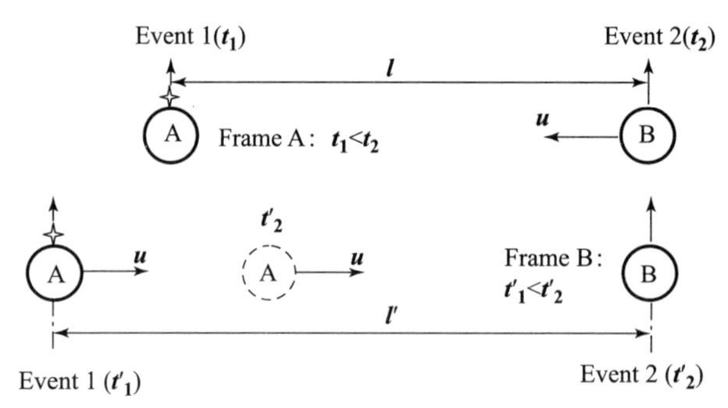

Figure 1 – 13 Example 1 – 8(1)

occurred. After a period of time, the event 2 occurred at time $t_2'(t'_2 > t_1')$ at Planet B. During this period the planet A had moved forward for a distance. Since the planet B is stationary in the frame B, the distance between the planet A and B when the event 1 occurred is actually the spatial distance between the event 1 and the event 2.

Since the event 1 and the event 2 occur at the same time in the frame A, according to the discussion about Example 1 – 7, the distance l' between the event 1 and the event 2 measured in the frame B corresponds to the proper length. According to the length contraction formula (1 – 28), there is

$$l' = \frac{l}{\sqrt{1 - u^2/c^2}} = \frac{3 \times 10^8}{\sqrt{1 - 0.8^2}} = 5 \times 10^8\,(\mathrm{m})$$

Therefore, the observers in the frame B measure that the distance between the two planets is 5×10^8 m when the planet A emits a flash.

(2) It is known that the distance between two planets measured in the frame A is 3×10^8 m when the plant B emits a flash, which can be considered that the positions of the planet A and B are scaled at the same time in the frame A when the Planet B emits a flash. $l = 3 \times 10^8$ m is also the spatial distance between the event of scaling the position of the planet A (called the event 1) and the event of emitting the flash at the planet B (called the event 2).

If observed in the frame B, the planet B is stationary, the planet A moves towards the planet B at speed 0.8c, and the event 1 occurred before the event 2. After the event 1 occurred, the planet A continued to move forward. When the planet B emits a flash, that is, when Event 2 occurs, the planet A has moved forward for a certain distance. At this time, if the observer in the frame B scales the position of the planet A (called the event 3), as shown in Figure 1 – 14, the distance l'' between the two planets can be measured, it can be seen that l'' is the spatial distance between two simultaneous events (the event 2 and the event 3) in the frame B.

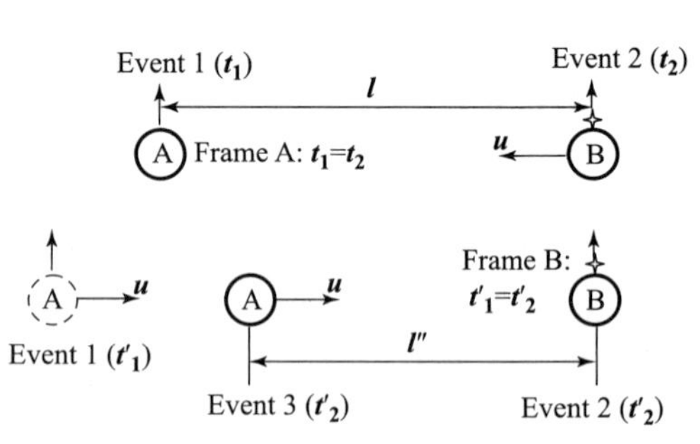

Figure 1 – 14 Example 1 – 8(2)

Here, although the event 1 and the event 3 occur at different positions in the frame B, both events occur at the planet A, which is stationary in the frame A, so the event 1 and the event 3 occur at the same position (the planet A) in the frame A. This shows that the spatial distance between the event 2 and the event 3 is equal to that between the event 2 and the event 1 in the frame A, that is, $l = 3 \times 10^8$ m.

Since the event 2 and the event 3 occur at the same time in the frame B, according to the discussion about Example 1 - 7, the spatial distance l between the event 2 and the event 3 in the frame A is the proper length. According to the length contraction formula (1 - 28), there is

$$l'' = l \cdot \sqrt{1 - u^2/c^2} = 3 \times 10^8 \times \sqrt{1 - 0.8^2} = 1.8 \times 10^8 (\mathrm{m})$$

Therefore, the observers in the frame B measure that the distance between the two planets is 1.8×10^8 m when the planet B emits a flash.

Here, the length contraction formula is used in both cases. In the first case (the planet A emits a flash as a time reference), the required distance between two planets measured in the frame B can be regarded as the proper length, which is $\dfrac{l}{\sqrt{1 - u^2/c^2}}$. In the second case (the planet B emits a flash as a time reference), the known distance l between two planets measured in the frame A can be regarded as the proper length, and the required distance in the frame B is $l \cdot \sqrt{1 - u^2/c^2}$.

To sum up, when discussing the distance between two objects with relative motion (close or far) in the frames of these objects, we must take a specific event as the time reference. If an event occurring on one object is taken as the time reference, the distance between two objects measured in the frame of another object is the proper length, and the corresponding distances in two frames meet the length contraction formula (1 - 28).

4. Law of Causality

According to the relativity of simultaneity, it can be known that two events that occur simultaneously in one inertial frame can not occur simultaneously when observed in another inertial frame, and the time sequence of these two events can also be changed when observed in different reference frames. Does this mean that the sequence of two causal events will also be different when observed in different reference frames? That is, will the time sequence of cause and result be reversed?

Here we consider that two events with causality occur successively in the inertial frame S, the event 1 is the cause and the event 2 is the result, their spacetime coordinates are (x_1, t_1) and (x_2, t_2) respectively. Because the cause occurs first and the result occurs later, there must be $t_1 < t_2$. In another inertial frame S' with relative motion, let the spacetime coordinates of these two events be (x_1', t_1') and (x_2', t_2') respectively. Let's discuss the time sequence of these two events in the frame S'. According to the time transformation formula of the Lorentz Transformation (1 - 21)

$$t_2' - t_1' = \frac{(t_2 - t_1) - \dfrac{u}{c^2}(x_2 - x_1)}{\sqrt{1 - u^2/c^2}} = \frac{t_2 - t_1}{\sqrt{1 - u^2/c^2}}\left[1 - \frac{u}{c^2}\frac{x_2 - x_1}{t_2 - t_1}\right] \tag{1-29}$$

set

$$\nu_s = \frac{x_2 - x_1}{t_2 - t_1} \tag{1-30}$$

which can be understood as the transmission speed of the contact signal between the cause and the result. For example, for the corresponding events in the process of projectile motion, the contact signal is the thrown object; for the corresponding events in the process of light propagation, the contact signal is the propagating light wave. Here, in the frame S, the contact signal takes the time $t_2 - t_1$ to transmit the distance $x_2 - x_1$. In practice, no matter whether the contact signal is an object or a light wave, its transmission speed must not exceed the speed of light c, namely $\nu_s \leqslant c$. Bring the equation (1-30) into equation (1-29), we get

$$t_2' - t_1' = \frac{t_2 - t_1}{\sqrt{1 - u^2/c^2}}\left[1 - \frac{u}{c^2}\nu_s\right] \tag{1-31}$$

because $\nu_s \leqslant c$ and $u < c$, there is $1 - \frac{u}{c^2}\nu_s > 0$, so $t_2' - t_1'$ always get the same sign with $t_2 - t_1$, which means that the time sequence of the two events will not be reversed in the frame S', and Event 1 also occurs before Event 2.

To sum up, the time sequence of two events with causality observed in an inertial frame will not be reversed in any other inertial frame. This conclusion is very natural in classical physics, and is also true in special relativity. Therefore, we say that special relativity is subject to the law of causality, and the idea of trying to use the theory of special relativity to go back to the past or make the dead back to life cannot be realized.

1.3.3 Minkowski Space

Shortly after Einstein established special relativity, Minkowski, a mathematics teacher in his college days, established a four-dimensional space to describe Einstein's spacetime of special relativity. This space was later called the Minkowski space. The Minkowski space also provided the basis for Einstein to establish the theoretical framework of general relativity. In order to help us deepen our understanding of the spacetime concept of special relativity and the Lorentz Transformation, we will briefly introduce the Minkowski space here.

The Minkowski space is a four-dimensional flat space, which is used to describe the spacetime coordinates of events in different inertial frames. The four-dimensional space includes three spatial dimensions and one time dimension. In the Minkowski space, each inertial frame will be represented by a coordinate system composed of four orthogonal coordinate axes x, y, z and ict. As long as the coordinate system representing one inertial frame is determined, the coordinate systems representing other inertial frames will be also determined (obtained by rotation or translation of the coordinate axes). The transformation between different coordinate systems follows the Lorentz Transformation. Any event can be represented by a point in the Minkowski space, whose coordinate (x, y, z, ict) correspond to the spacetime coordinates of the event. Of course, the spacetime coordinates of the event in different inertial frames also meet the Lorentz Transformation.

Different from the familiar three-dimensional Euclidean space, the Minkowski space contains not only three spatial coordinate axes x, y, z, but also an independent time axis, which reflects the equivalence of spatial coordinates and time coordinate. The time axis is expressed as ict, which can be understood as transforming the time coordinate into the same dimension as the space coordinate. So why is the time axis expressed as an imaginary number like ict?

In the examples 1 – 3 in section 1.3.1, we have proved using the Lorentz Transformation that the shapes of the optical wavefronts observed in the frame S and S' are both spherical under certain conditions, and the spherical equation can be expressed as the following general form

$$x^2 + y^2 + z^2 = c^2 t^2$$

Similarly, the Lorentz Transformation can also be used to prove that the shape of the wavefront of the light emitted at any time by a point light source at any position (the spacetime coordinates of the luminous event in the frame S and S' are (x_0, y_0, z_0, t_0) and (x'_0, y'_0, z'_0, t'_0) respectively) is still spherical in two frames, that is, the wavefront shape in two frames meets the following equations

$$\left.\begin{aligned} \Delta x^2 + \Delta y^2 + \Delta z^2 &= (c\Delta t)^2 \\ \Delta x'^2 + \Delta y'^2 + \Delta z'^2 &= (c\Delta t')^2 \end{aligned}\right\}$$

In which $\Delta x = x - x_0$, $\Delta y = y - y_0$, $\Delta z = z - z_0$, $\Delta t = t - t_0$; $\Delta x' = x' - x'_0$, $\Delta y' = y' - y'_0$, $\Delta z' = z' - z'_0$, $\Delta t' = t' - t'_0$. Here, (x, y, z, t) and (x', y', z', t') can be understood as the spacetime coordinates of the event of light travelling to any point on the wavefront in the frame S and S' respectively. Move the items on the right side of the above equations to the left side to get

$$\left.\begin{aligned} \Delta x^2 + \Delta y^2 + \Delta z^2 - (c\Delta t)^2 &= 0 \\ \Delta x'^2 + \Delta y'^2 + \Delta z'^2 - (c\Delta t')^2 &= 0 \end{aligned}\right\} \tag{1-32}$$

Since the Minkowski space is a four-dimensional flat space, the distance between any two points in the space can be expressed as

$$S^2 = \Delta x^2 + \Delta y^2 + \Delta z^2 + (ic\Delta t)^2 \tag{1-33}$$

Because $i^2 = -1$, S^2 may be negative, its root may get an imaginary number which has no physical meaning. Therefore, the distance between any two points in the Minkowski space is expressed in the form of S^2, which is called the four-dimensional spacetime interval of events.

By comparing formulas (1 – 33) and (1 – 32) we can see why the time axis in the Minkowski space should be expressed as an imaginary number like ict. In Minkowski space, the luminous event is represented by a point whose coordinates in the frame S and S' are (x_0, y_0, z_0, ict_0) and $(x'_0, y'_0, z'_0, ict'_0)$ respectively; The event of the light travelling to a certain position in space (a point on the front of the wave) is represented by another point whose coordinates in the frame S and S' are (x, y, z, ict) and (x', y', z', ict') respectively. Because the time coordinate of the event is expressed as ict, the left sides of the two formulas in (1 – 32) represent the four-dimensional spacetime interval between the luminous event and the event of light travelling to a certain position in the frame S and S' respectively, which are both equal to 0. This means that the four-dimensional spacetime interval between two events linked by propagation of the light is independent of the reference frame in Minkowski space and is a constant value of 0.

The Lorentz Transformation can also be used to prove that not only for two events linked by propagation of the light, but also for any two events, the four-dimensional spacetime interval between them defined by the formula(1 – 33) remains unchanged in different inertial frames, that is, the four-dimensional interval S^2 between the same two events is an invariant in different inertial frames. This also means that if the space interval $\Delta x^2 + \Delta y^2 + \Delta z^2$ between two events changes due to the change of the reference frame, the time interval Δt will inevitably change accordingly. It can be seen that the space and time intervals between events which we are familiar with are unified into the four-dimensional space and time interval of Minkowski space. Space and time are unified as a whole, the measurement of space and time are interrelated and indivisible, the concept of the absolute time and the absolute space cannot be held.

In order to deepen the understanding of the relationship between the Minkowski space and the relativistic spacetime view, here we use "four-dimensional spacetime interval is invariant" to discuss the time delation effect described in section 1. 2. 2. Two events which occur successively at the same place in the frame S' have the spacetime interval of $-(c\Delta t')^2$; These two events occur at different places in the frame S, and the corresponding spacetime interval was $(x_2 - x_1)^2 - (c\Delta t)^2$. because of the invariance of spacetime interval, there is $-(c\Delta t')^2 = (x_2 - x_1)^2 - (c\Delta t)^2$. It can be seen that there is $\Delta t' < \Delta t$ because of $(x_2 - x_1)^2 > 0$. In addition, it is easy to get the expression of the time delation $\Delta t = \dfrac{\Delta t'}{\sqrt{1 - u^2/c^2}}$ due to $x_2 - x_1 = u\Delta t$.

As we have seen before, the spacetime interval between events associated with propagation of the light is 0, which is an inevitable result of the principle of constant speed of light. In addition, the spacetime interval between events can also be greater than 0 or less than 0, which represent different spacetime attributes between events. $S^2 = 0$ is called the light-like interval, and the corresponding events can be connected through the light; $S^2 < 0$ is called the time-like interval, and the corresponding events (such as two events occurring successively at the same place of a reference frame) can be connected by a signal with a transmission speed lower than the speed of light; $S^2 > 0$ is called the space-like interval, and the corresponding events (such as two events occurring at the same time and at different places of a reference frame) must not be connected by a signal with a transmission speed less than or equal to the speed of light. Since events with the time-like or light-like interval can be connected by signals with transmission speeds less than or equal to the speed of light, there can be causality between these events, and the time sequence of their occurrence cannot be changed when observed in different reference frames. For events with the space-like interval, because they cannot be connected by a signal with a transmission speed less than or equal to the speed of light, there is no causality between these events. Whether these events occur at the same time or the order of occurrence can be different when observed in different reference frames.

The Minkowski space can also be used to show more intuitively the relativistic spacetime view and the spacetime relationship between events in the form of spacetime diagram. Minkowski space has four independent dimensions, which are difficult to be intuitively expressed in the three-dimensional world. However, if we only discuss one-dimensional relative motion (in the x direction), Minkowski space can be represented on a two-dimensional plane composed of the time axis ct (omit i) and the space axis x, which is called Minkowski diagram. In the spacetime diagram, the inertial frame S (stationary) can be represented by the mutually perpendicular ct – axis and x-axis as shown in Figure 1 – 15. Each point on the spacetime diagram can represents an event, the coordinates (ct, x) of the point are the spacetime coordinates of the event in the frame S. Each point on the spacetime diagram is called the world point. For example, the origin O in the spacetime diagram represents the event taking place at $t = 0$ and $x = 0$ in the frame S. Here we call it the Origin event; The different points on the ct-axis correspond to the events occurring at $x = 0$ at different times in the frame S; The different points on the x-axis correspond to the events occurring at different positions at $t = 0$ in the frame S.

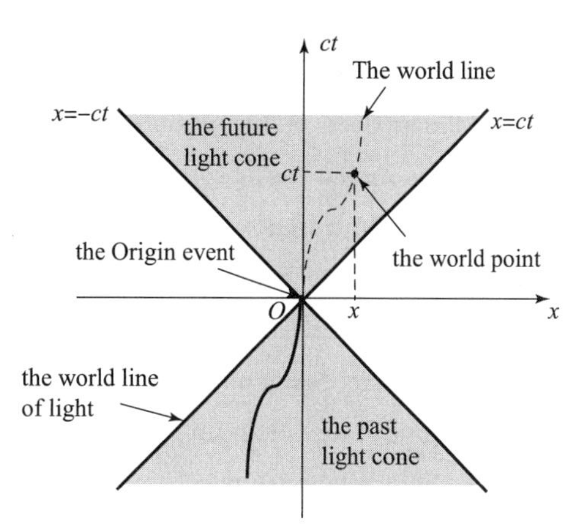

Figure 1 – 15 Two-dimensional Minkowski spacetime diagram

The x-axis in the spacetime diagram (corresponding to the equation $ct = 0$) divides the spacetime diagram into areas of $t < 0$ and $t > 0$, where the world points in the area of $t < 0$ represent the events that have occurred in the past, and the world points in the area of $t > 0$ represent the events that will occur in the future. A series of world points representing a series of events in a process can be connected into a line, which is called the world line. The process associated with the origin event can be expressed as a world line through the origin O (the corresponding event in the area of $t > 0$ has not yet occurred, so it is drawn as a dotted line).

If the units of the ct – axis and the x-axis are the same, the two lines that bisect the angle between the ct – axis and the x-axis can be expressed by the equation $x = \pm ct$. It is easy to see that these two lines correspond to the world line of the process of light travelling through the origin O (at the time $t = 0$, the light travels to the position of $x = 0$). The positive and negative signs in the equation correspond to light travelling in the positive and negative direction of the x-axis respectively. The world lines intersecting at the origin O form two cones, which are called light cones associated with the origin event. The light cone located in the area $t < 0$ is called the past light cone, and the light cone located in the area $t > 0$ is called the future light cone. When different origin events are considered, the world lines associated with them will also be different. Since the speed of an object cannot exceed the speed of light, the world line associated with any origin event must be within the corresponding light cone.

The light cone divides the spacetime diagram into three regions: on the light cone, inside the light cone and outside the light cone. If the origin event is taken as the reference, the spacetime interval between the world point on the light cone and the origin O is the light-like interval ($S^2 = 0$), the spacetime interval between the world point in the light cone and the origin O is the time-like interval($S^2 < 0$), and the spacetime interval between the world point outside the light cone and the origin O is the space-like interval($S^2 > 0$). Since there can be causality between two events with the light-like interval or the time-like interval, the time sequence of their occurrence will not change when observed in different reference frames. Therefore, no matter in any other reference frames, the events on the surface of the past light cone or in the past light cone must occur before the origin event, which is called the absolute past of the origin event; Events on the surface of the future light cone or in the future light cone must occur after the origin event, which is called the absolute future of the origin event. For events with the space-like interval from each other, since there is no causality between them under any circumstances, whether they occur at the same time and the sequence of occurrence can vary when observed in different reference frames. In addition, when any event occurs, you can imagine it as the origin event and make a corresponding light cone to facilitate the discussion of the spacetime relationship between this event and other events.

In the spacetime diagram shown in Figure 1 – 15, only the coordinate system (mutually perpendicular ct – axis and x-axis) representing the observer's inertial frame S is drawn. How can we determine the coordinate system representing the inertial frame S' (moving at a constant speed relative to the frame S)? Suppose that the coordinate system representing the frame S' in the spacetime diagram is composed of the time axis ct' and the space axis x'. According to the Lorentz Transformation(1 – 21), set $x' = 0$, the equation of the ct' – axis in the spacetime diagram of the frame S is obtained as $ct = \frac{c}{u}x$; Set $t' = 0$, the equation of the x' – axis in the spacetime diagram of the frame S can be obtained as $ct = \frac{u}{c}x$. As shown in Figure 1 – 16, the ct' – axis and x' – axis representing the frame S' is determined in the spacetime diagram of the frame S (for the observers in the frame S', the coordinate system is composed of mutually perpendicular ct' – axis and x' – axis, and the equation of ct – axis and x-axis representing the frame S can be determined according to the Lorentz Transformation(1 – 22) in similar way).

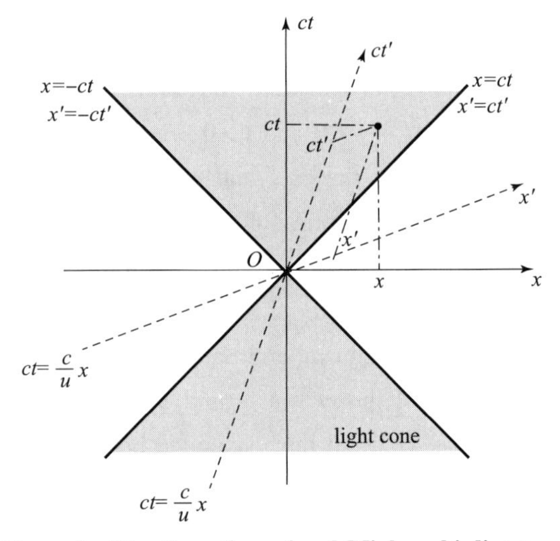

Figure 1 – 16 Two-dimensional Minkowski diagram: the coordinate systems of the frame S and the frame S' m

It can be seen that the origins of two coordinate systems coincide with each other, indicating that the origin event of the frame S is also the origin event of the frame S', which is consistent with the given conditions(when $t = t' = 0$, the origin O of the frame S coincides with the origin O' of the frame S'). In addition, the slopes of the x' – axis and the ct' – axis are reciprocal to each other, which indicates that they rotate the same angle in opposite direction with respect to the x – axis and the ct – axis respectively. The size of the angle is related to the relative velocity u. It can be seen that the x' – axis and ct' – axis are symmetrical with respect to the world line of light travelling through the origin. Since the speed of light is constant, the equation of the world lines of light propagation in the frame S' is $x' = \pm ct'$, it means that the world lines of light propagation should also bisect the included angle between the x' – axis and the ct' – axis of the frame S'. It can be seen that the shape of the light cone does not change in the two reference frames, which is the inevitable result of the principle of constant speed of light, and is also an important feature of the Minkowski diagram.

After determining the coordinate axes representing the frame S and the frame S', we can easily obtain the spacetime coordinates of an event by drawing straight lines parallel to the coordinate axes of the frame S and the frame S' as shown in Figure 1 – 16. It can be seen that it is very convenient to read the spacetime coordinates of the same event in different reference frames by using the Minkowski diagram, so it is also very intuitive to discuss the different spacetime effects of special relativity.

1.4 The Relativistic Velocity Transformation

Obviously, in the special relativity, Galilean velocity transformation (1 – 2) is no longer valid, so what kind of velocity transformation should replaced it?

Consider the case where a particle moves relative to the inertial frame S and the inertial frame S'. Let the spacetime coordinates of the particle in the frame S and frame S' be (x, y, z, t) and (x', y', z', t') respectively. According to the definition of velocity, the velocity component of the particle relative to the frame S and frame S' is

$$\nu_x = \frac{dx}{dt}, \qquad \nu_y = \frac{dy}{dt}, \qquad \nu_z = \frac{dz}{dt}$$

$$\nu'_x = \frac{dx'}{dt'}, \qquad \nu'_y = \frac{dy'}{dt'}, \qquad \nu'_z = \frac{dz'}{dt'}$$

It can be obtained from the Lorentz Transformation (1 – 21)

$$\frac{dx'}{dt'} = \frac{\frac{dx'}{dt}}{\frac{dt'}{dt}} = \frac{\gamma\left(\frac{dx}{dt} - u\right)}{\gamma\left(1 - \frac{u}{c^2}\frac{dx}{dt}\right)}$$

$$\frac{dy'}{dt'} = \frac{\frac{dy'}{dt}}{\frac{dt'}{dt}} = \frac{\frac{dy}{dt}}{\gamma\left(1 - \frac{u}{c^2}\frac{dx}{dt}\right)}$$

$$\frac{dz'}{dt'} = \frac{\frac{dz'}{dt}}{\frac{dt'}{dt}} = \frac{\frac{dz}{dt}}{\gamma\left(1 - \frac{u}{c^2}\frac{dx}{dt}\right)}$$

Then we get

$$\left.\begin{aligned} \nu'_x &= \frac{\nu_x - u}{1 - \dfrac{u\nu_x}{c^2}} \\ \nu'_y &= \frac{\nu_y \sqrt{1 - u^2/c^2}}{1 - \dfrac{u\nu_x}{c^2}} \\ \nu'_z &= \frac{\nu_z \sqrt{1 - u^2/c^2}}{1 - \dfrac{u\nu_x}{c^2}} \end{aligned}\right\} \tag{1-34}$$

This is the relativistic velocity transformation formula. It should be noted when $u \ll c, \nu_x \ll c$, the above formula(1 − 34) can be simplified as $\nu'_x = \nu_x - u$, $\nu'_y = \nu_y$, $\nu'_z = \nu_z$, which is Galilean velocity transformation formula. Galilean velocity transformation is an approximation of the relativistic velocity transformation when the relative speed is much lower than the speed of light. If the velocity of photon in the frame S $\nu_x = c$, it can be obtained from the relativistic velocity transformation formula(1 −34).

$$\nu'_x = \frac{c - u}{1 - uc/c^2} = c$$

That is, the speed of photons in the frame S' is still c. It can be seen that the relativistic velocity transformation conforms to the principle of constant speed of light.

The inverse transformation of relativistic velocity transformation can be obtained by replacing u with $-u$ and exchanging quantities with and without prime in relativistic velocity transformation formula(1 −34).

$$\left.\begin{aligned} \nu_x &= \frac{\nu'_x + u}{1 + \dfrac{u\nu'_x}{c^2}} \\ \nu_y &= \frac{\nu'_y \sqrt{1 - u^2/c^2}}{1 + \dfrac{u\nu'_x}{c^2}} \\ \nu_z &= \frac{\nu'_z \sqrt{1 - u^2/c^2}}{1 + \dfrac{u\nu'_x}{c^2}} \end{aligned}\right\} \tag{1-35}$$

Example 1 −9: Two particles moving in opposite direction are emitted from the high-energy accelerator, they both have the speed of 0.9c observed in the laboratory reference frame. Find the relative velocity between two particles.

Solution: As shown in Figure 1 − 17, suppose that the laboratory frame is the frame S, the reference frame of particle A is the frame S'. It is known that the frame S' moves at speed $u = -0.9c$, the particle B moves at speed $\nu_x = 0.9c$ relative to the frame S. From the relativistic velocity transfor-mation (1 −

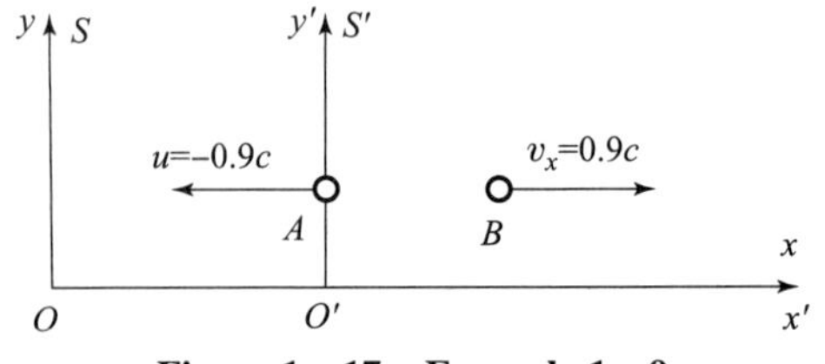

Figure 1 −17 Example 1 −9

34), the velocity of particle B relative to particle A (that is, the frame S') is

$$\nu'_x = \frac{\nu_x - u}{1 - \frac{u\nu_x}{c^2}} = \frac{0.9c - (-0.9c)}{1 - (-0.9) \times 0.9} = \frac{1.8}{1.81}c = 0.994c$$

It can be seen that the relative velocity between any two particles is always less than c in special relativity, which indicates that the speed of any physical particles relative to any inertial frame cannot exceed the speed of light in vacuum. If Galilean velocity transformation is used, the wrong relative velocity $\nu'_x = \nu_x - u = 1.8c$ will be obtained, which means that Galilean velocity transformation is not suitable for discussing the motion of high speed close to the speed of light.

Example 1 – 10: In the solar reference frame, a beam of starlight shines vertically on the ground at speed c, while the earth moves perpendicular to the light at speed 30 km/s. Find the magnitude and direction of starlight velocity observed in the ground reference. frame.

Solution: As shown in Figure 1 – 18, the solar reference frame is taken as the frame S, and the ground reference frame is taken as the frame S'. The speed of the frame S' relative to the frame S along the x-axis is $u = 30\text{km/s}$. In the frame S, the light is vertically directed to the ground, so we have $\nu_x = 0, \nu_y = -c, \nu_z = 0$, according to the relativistic velocity transformation (1 – 34), we get

$$\nu'_x = -u, \qquad \nu'_y = -c\sqrt{1 - u^2/c^2}, \qquad \nu'_z = 0$$

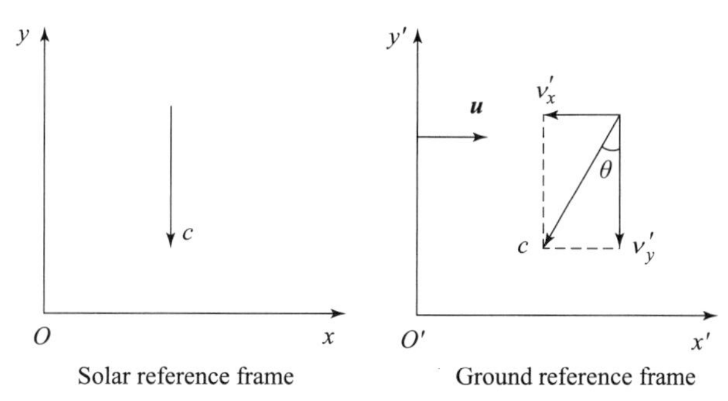

Figure 1 – 18 Example 1 – 10

Therefore, the magnitude of starlight velocity relative to the ground frame is

$$\nu' = \sqrt{\nu'^2_x + \nu'^2_y + \nu'^2_z} = c$$

It can be seen that the speed of light observed in the ground frame is still c, which indicates that the speed of light in vacuum is independent of the motion of the reference frame; If the angle between the starlight direction and the vertical direction is θ, then

$$\tan\theta = \frac{|\nu'_x|}{|\nu'_y|} = \frac{u}{c\sqrt{1 - u^2/c^2}} \approx \frac{u}{c} = 10^{-4}, \qquad \theta \approx 20.6''$$

In the observation of famous "optical aberration" in the 18th century, the velocity of light was calculated by accurately measuring the aberration angle θ and using the earth's revolution speed u and the equation $c = u/\tan\theta$, and the obtained velocity of light was very close to the accurate value.

1.5 Basis of Relativistic Dynamics

The previous sections discussed the kinematics content of special relativity, from which we have learned that the spacetime properties in high-speed motion are different from in Newtonian mechanics, so the laws of motion of high-speed particles must also be changed accordingly, which include redefinition of the physical quantities such as momentum, mass, energy, and establishment of new laws. These new definitions and new laws should go beyond those corresponding to Newtonian mechanics.

1.5.1 The Relativistic Momentum and Mass

We have proved in section 1.1.1 that the follow theorem of momentum (or Newton's law) remains unchanged under the Galilean transformation.

$$\boldsymbol{F}=\frac{\mathrm{d}\boldsymbol{p}}{\mathrm{d}t}=\frac{\mathrm{d}(m\boldsymbol{v})}{\mathrm{d}t}=m\frac{\mathrm{d}\boldsymbol{v}}{\mathrm{d}t} \tag{1-36}$$

However, the momentum theorem cannot remain unchanged under the Lorentz Transformation. Einstein believed that the momentum theorem is a basic and universal law in nature, and also can be applied to the field of high-speed motion, so it should remain unchanged under the Lorentz Transformation. To incorporate the momentum theorem into special relativity, we must modify these laws and related concepts.

The revision of the momentum theorem can be begun with the discussion of the principle of momentum conservation, which is a universal law of nature and should be true in all inertial reference frames. On the premise of acknowledging the Lorentz Transformation and the relativistic velocity transformation, through the analysis of the momentum conservation in simple collision process, if the momentum conservation is valid in different inertial reference frames, and the momentum $\boldsymbol{p}$ of particle remains in the form of mv, it is easy to get that the mass of particle must be redefined as follow

$$m=\frac{m_0}{\sqrt{1-\dfrac{v^2}{c^2}}} \tag{1-37}$$

Where m_0, which is called the rest mass, is mass of the particle measured in its rest reference frame; m, which is called the relativistic mass, is the mass of the particle moving at velocity v. The above formula(1-37) is called mass-velocity relationship.

It can be seen from the mass-velocity relationship that the relativistic mass of a particle is related to its speed, that is, when observed in different inertial frames, the particle will have different relativistic mass. Figure 1-19 shows the relationship between the relativistic mass m of a particle and its velocity v. It can be seen that in the case of $m_0 \neq 0$, when v increases, m increases. As v approaches the speed of light, the relativistic mass m increases without bound. the relativistic mass m can get as large as you want without the speed ever reaching the speed of light. In other words, it is impossible to accelerate a particle to the speed of light. The fact has been verified in the experiments of particle accelerators in high-energy physics research. Particles such as electrons and protons are accelerated to the speeds closer to—but never exceed—the speed of light. For photon, it is generally believed that its rest mass $m_0 = 0$, its velocity can reach c, and then its relativistic mass m is a finite value. It can also be seen from Figure 1 – 19 that when $v \ll c$, $m \approx m_0$, that is, the mass of a particle is independent of its speed and equal to its rest mass. Therefore, m_0 is actually the mass independent of motion in Newtonian mechanics.

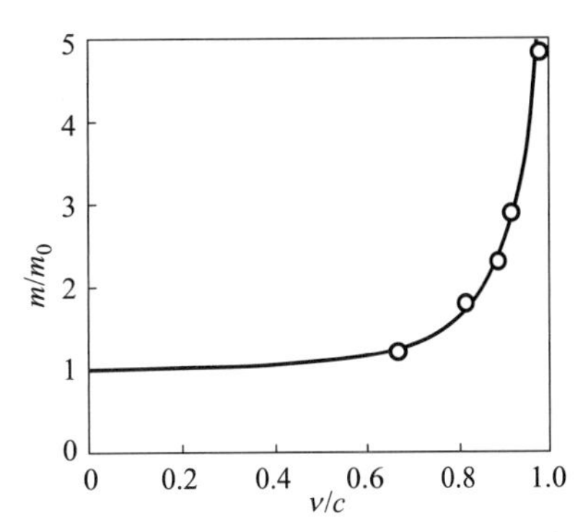

Figure 1 – 19　The relativistic mass of a particle changes with moving velocity

As early as 1901, the German physicist W. Kaufmann had found that the relativistic masses of particles increased with their speed in the experiments of measuring the charge-mass ratio e/m of β-ray (electron ray). Soon after, the mass-velocity relationship was verified by the accurate measurement of the charge-mass ratio of β-rays.

After modifying the constant mass in classical mechanics to the relativistic mass, the momentum of particles is correspondingly modified to

$$\boldsymbol{p} = m\boldsymbol{v} = \frac{m_0 \boldsymbol{v}}{\sqrt{1 - u^2/c^2}} \tag{1-38}$$

It is called the relativistic momentum.

In the special relativity, the force on particles is still defined by the time change rate of momentum. After defining the relativistic mass and the relativistic momentum, the momentum theorem in special relativity can be written as

$$\boldsymbol{F} = \frac{\mathrm{d}\boldsymbol{p}}{\mathrm{d}t} = m\frac{\mathrm{d}\boldsymbol{v}}{\mathrm{d}t} + \frac{\mathrm{d}m}{\mathrm{d}t}\boldsymbol{v} = m\boldsymbol{a} + \frac{\mathrm{d}m}{\mathrm{d}t}\boldsymbol{v} \tag{1-39}$$

This is the basic equation of the relativistic mechanics. Unlike Newton's second law $\boldsymbol{F} = m\boldsymbol{a}$ in classical mechanics, the effect of force in relativity is not only to produce acceleration, but also to increase the relativistic mass of particle. Therefore, there is an essential difference between the relativistic mechanics and Newtonian mechanics on the problem that particles are accelerated: In Newtonian mechanics, the particle's velocity increases continuously with the work done by external forces, which can reach infinity; However, in the theory of relativity, as the external force does work, not only the speed but also the mass will increase, so the inertia of the particle will become larger and larger, the acceleration will get smaller and smaller as the particle's speed approaches the speed of light c, and the speed of particles can never exceed the speed of light. It can be seen that when dealing with motion of high speed near the speed of light, Newton's second law $\boldsymbol{F} = m\boldsymbol{a}$ can no longer be valid, but the basic equation (1 – 39) of the relativistic mechanics must be used for discussion.

In addition, when $v \ll c$, $m \approx m_0$, the basic equation of the relativistic mechanics (1 – 39) reduces to $\boldsymbol{F} = m_0\boldsymbol{a}$, which is the form of Newton's second law in classical mechanics. It can be seen that classical mechanics is a special case of the relativistic mechanics when the speed is much lower than the speed of light. The relativistic mechanics are more general and are true for any speed.

1.5.2 Mass-Energy Relation

Starting from the basic equation of the relativistic mechanics (1 – 39), we can get the most important and meaningful relationship in special relativity——Mass-Energy relation.

1. Relativistic Kinetic Energy

First, let's discuss how to express the kinetic energy of particles in relativity. Consider a simple case: the particle is accelerated from rest under a constant force F in the positive direction of the x-axis, find the work done by the force F in the process of accelerating the particle to the velocity ν. As in classical mechanics, work done by a force can be defined as the product of force and distance in relativity. Therefore, in this case we get

$$A = \int_0^\nu F\mathrm{d}x = \int_0^\nu \frac{\mathrm{d}(m\nu)}{\mathrm{d}t}dx = \int_0^\nu \nu\mathrm{d}(m\nu)$$

In order to find this integral, we square both sides of the mass-velocity equation (1 – 37), and then transform it into

$$m^2(c^2 - \nu^2) = m_0^2c^2$$

Take differential on both sides of the above equation, and get

$$2mc^2\mathrm{d}m - 2m\nu\mathrm{d}(m\nu) = 0$$

Simplify again and get

$$c^2\mathrm{d}m = \nu\mathrm{d}(m\nu)$$

Bring the above formula into the previous integral formula for calculating work, we get

$$A = \int_0^\nu \nu\mathrm{d}(m\nu) = \int_0^\nu c^2\mathrm{d}m = mc^2 - m_0c^2$$

where m_0 is the rest mass of the particle, and m is the relativistic mass of the particle when it is accelerated to velocity ν. As in classical mechanics, the kinetic energy theorem, which states that the kinetic energy of a particle is equal to the work done to accelerate it from rest to a certain velocity, is also valid in relativity, so the kinetic energy of the particle is

$$E_\mathrm{k} = mc^2 - m_0c^2 \tag{1-40}$$

This is the expression of the relativistic kinetic energy. It can be seen that when $\nu \to c, m \to \infty, E_\mathrm{k} \to \infty, A \to \infty$, which means that it will take infinite work to accelerate an object to the speed of light, that is, the limit of the speed of any object is the speed of light. In addition, in the case of $\nu \ll c$, the kinetic energy of an object can be reduced to

$$E_\mathrm{k} = m_0c^2\left(\frac{1}{\sqrt{1 - u^2/c^2}} - 1\right) = m_0c^2\left[\left(1 + \frac{1}{2}\frac{\nu^2}{c^2} + \frac{3}{8}\frac{\nu^4}{c^4} + \cdots\right) - 1\right] \approx m_0c^2 \cdot \frac{1}{2}\frac{\nu^2}{c^2} = \frac{1}{2}m_0\nu^2$$

The result is the same as the expression of the kinetic energy in classical mechanics.

2. The Relativistic Energy

In formula (1 – 40), the kinetic energy is expressed as the difference between the two terms, where m_0c^2 is related to the rest mass and is called the rest energy, which is expressed as E_0; mc^2 is related to the relativistic mass, which is called the relativistic energy or the total energy, expressed by E

$$E = mc^2 \tag{1-41}$$

This is Einstein's Mass-Energy relation, which is the most significant result of special relativity and one of the most famous equations in physics. We know that mass and energy are the two basic attributes of matter, mass is the measure of the inertia or the gravity interaction, and energy is the measure of the transformation of matter motion. They are linked with each other by the factor c^2 in the Mass-Energy relation, which indicates that a certain mass is equivalent to a certain amount of energy, and a certain amount of energy is also equivalent to a certain amount of mass. Mass and energy form a new unity.

The rest energy m_0c^2, which is the energy of a particle as measured in its rest frame, is a new physical concept not found in Newtonian mechanics. It shows that an object has energy even when it is at rest. The rest energy includes all forms of energy such as kinetic energy and potential energy of all microscopic particles in the object. It is the sum of the internal energy of the object.

According to formula(1 - 41), the rest energy of an electron with rest mass of 9.11×10^{-31} kg is 0.511MeV, and the rest energy of a proton with rest mass of 1.67×10^{-27} kg is 937MeV, which is much larger than the chemical energy of particles(Magnitude of several eV). It can be seen that even macroscopic particles with small masses contain huge energy. If we can release and use this kind of energy, our energy will be inexhaustible. It can be said that Einstein's Mass-Energy relation led us to the era of atomic energy.

3. Principle of Conservation of the Relativistic Energy

Energy conservation is a universal law of physics, so the principle of energy conservation is also valid in special relativity, but its content is different from that of classical mechanics. In the theory of relativity, due to Mass-Energy relation(1 - 41), the principle of energy conservation can be written as

$$\sum_i E_i = \sum_i m_i c^2 = \text{常量} \tag{1-42}$$

which is equivalent to principle of mass conservation

$$\sum_i m_i = \text{常量} \tag{1-43}$$

However, it should be made clear that mass conservation refers to the conservation of the relativistic mass, not the conservation of the rest mass. In history, the principle of energy conservation and the principle of mass conservation were discovered as two independent laws, there is no relationship between them, but they are completely unified in the theory of special relativity.

4. The Mass Defect

According to formula(1 - 40), the total energy of a particle is $E = m_0c^2 + E_k$, that is, the total energy is equal to the rest energy plus the kinetic energy. If a reaction occurs in a closed system, the rest mass of all particles in the system is reduced from m_{01} before to m_{02} after the reaction, and the corresponding kinetic energy of particles is changed from E_{k1} to E_{k2}, according to the principle of energy conservation(1 - 42)

$$m_{01}c^2 + E_{k1} = m_{02}c^2 + E_{k2}$$

then we get

$$E_{k2} - E_{k1} = (m_{01} - m_{02})c^2$$

It can be seen that the kinetic energy of all particles in the system before and after the reaction has increased $\Delta E = E_{k2} - E_{k1}$, because the rest mass before and after the reaction has decreased $\Delta m_0 = m_{01} - m_{02}$. This means that the reduction of the rest mass will lead to the increase of the kinetic energy, that is, the rest energy is released in the form of the kinetic energy. The reduction of the rest mass is called the mass defect Δm_0. The kinetic energy released corresponding to the mass defect Δm_0 is

$$\Delta E = \Delta m_0 c^2 \qquad (1-44)$$

in which the kinetic energy released is equal to the mass loss Δm_0 times the square of the speed of light c^2. It can be seen that a very considerable amount of energy can be released, even if there is a small mass defect. Note that there is no contradiction between the mass defect and the mass conservation. The mass defect refers to the reduction of the rest mass, the corresponding rest energy is released in the form of kinetic energy due to the mass defect, so the relativistic mass of particles increases and is conserved. The principle of mass conservation refers to the conservation of the relativistic mass.

In 1938, the German scientist O. Hahn discovered the phenomenon of the fission of uranium nuclei in experiments, and there was a mass defect in the process of the nuclear fission. Since then, the significance of Einstein's Mass-Energy relation has emerged, and human beings quickly entered the era of atomic energy. Besides, the mass defect also occurs in the process of nuclear fusion, for example, hydrogen nuclei (^1_1H) combine to form a helium nucleus (^4_2He) in the sun. Due to Einstein's Mass-Energy relation, tremendous amounts of energy can be released through the nuclear fission and fusion.

In addition, in the presence of antimatter, if the matter and antimatter meet and annihilate, the rest mass will disappear, and the rest energy corresponding to the rest mass can be released completely. If we can make full use of this energy, it will be a much more efficient energy than nuclear reaction. The antiparticle of electron is called positron, in addition to having opposite charges, the positron has the same properties as the electron. When the electron and positron combine, they are converted into two photons. Because the rest mass of the photon is zero, all the energy corresponding to the rest mass of the electron and positron is converted into the kinetic energy of the photon.

Example 1 - 11: The energy of solar radiation is generated by a series reactions of nuclear fusion, and the typical reaction of nuclear fusion is as follow

$$4\,^1_1H \rightarrow \,^4_2He + 2\,^0_1e$$

The rest masses of a single proton (^1_1H), helium nucleus (^4_2He) and positron (^0_1e) are $m_p = 1.672,6 \times 10^{-27}$ kg, $m_{He} = 6.642,5 \times 10^{-27}$ kg and $m_e = 0.000,9 \times 10^{-27}$ kg respectively, find:

(1) How much energy is released through the reaction? In what form is the energy release?

(2) What is the energy release efficiency of this reaction?

(3) How much energy can be released by consuming 1kg proton?

Solution(1)The released energy is

$$\Delta E = \Delta m_0 c^2 = (4m_p - m_{He} - 2m_e)c^2$$
$$= (4 \times 1.672,6 - 6.642,5 - 2 \times 0.000,9) \times 10^{-27}\text{kg} \times (3 \times 10^8 \text{m/s})^2$$
$$= 4.15 \times 10^{-12}\text{J} = 25.9\text{MeV}$$

The energy released appears as the kinetic energy of helium nucleus and positrons, such huge kinetic energy indicates that the motion of particles is very violent, resulting in extremely high temperature inside the sun(up to 10^8K).

(2)The energy release efficiency is

$$\eta = \frac{\Delta E}{4m_p c^2} = \frac{4.15 \times 10^{-12}\text{J}}{4 \times 1.672,6 \times 10^{-27}\text{kg} \times (3 \times 10^8 \text{m/s})^2} = 0.69\%$$

(3)The energy that can be released by consuming 1kg proton is

$$\frac{\Delta E}{4m_p} = \frac{4.15 \times 10^{-12}\text{J}}{4 \times 1.672,6 \times 10^{-27}\text{kg}} = 6.20 \times 10^{14}\text{J/kg}$$

This is equivalent to 2.12×10^7 times of the chemical energy released by the complete combustion of 1kg high-quality coal.

1.5.3 Energy-Momentum Relationships

Converting $m = \dfrac{m_0}{\sqrt{1 - u^2/c^2}}$ to $m^2c^2 = m^2\nu^2 + m_0^2c^2$, multiplying both sides by c^2, and using $E = mc^2$ and $p = m\nu$ to obtain

$$E^2 = p^2c^2 + m_0^2c^4 \qquad (1-45)$$

This is the relativistic energy-momentum relation. It is not difficult to see that if E, pc and m_0c^2 are considered as the length of three sides of a triangle, they just form a right triangle as shown in Figure 1 - 20. The relativistic kinetic energy $E_k = E - m_0c^2$ can be expressed in this figure as the difference between the oblique side E and the right side m_0c^2.

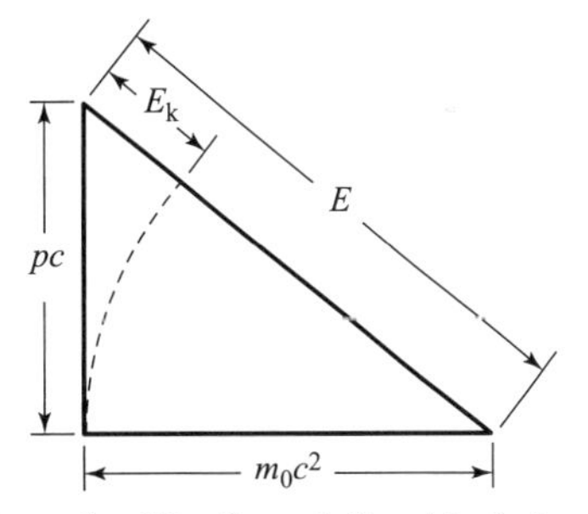

Figure 1 - 20 the relationship between relativistic energy and momentum

Photon has the rest mass $m_0 = 0$, the rest energy $E_0 = 0$, but it has the momentum p, the relativistic mass m and the relativistic energy E. From equation(1 - 45) we have $E = pc$, and from $E = h\nu = hc/\lambda$ we get $p = h/\lambda$, where ν is the frequency of light wave, and h is the Planck constant. Because photon has the relativistic mass, light passing by a large star will be bent by the gravitational force of the star, which has been observed by A. S. Eddington in the observation of the solar eclipse in 1919, which verified that the theory of special relativity is true. Because photon has the momentum, the light on the surface of an object will also produce light pressure.

In closed systems, both the principle of energy conservation and the principle of momentum conservation hold in the relativistic dynamics. Compared with Newtonian mechanics, special relativity reflects the laws of nature more objectively and truly. The theory of special relativity is not a negation of Newtonian mechanics, but a generalization of Newtonian mechanics. Newtonian mechanics is good approximation of special relativity as long as $v \ll c$. For hundreds of years, Newtonian mechanics has withstood the test of experiments in its scope of application; Similarly, special relativity has also been confirmed in experimental observations of modern physics such as cosmology and nuclear physics.

Example 1 – 12: It is known that particle A and particle B have a rest mass of m_0, particle A is at rest, and particle B moves towards particle A with the kinetic energy of $2m_0c^2$, two particles are integrated to one particle after the collision. If there is no release of the energy during the collision, find the rest mass of the particle after collision.

Solution: The energy of particles A and B before collision is respectively

$$E_A = m_0c^2, E_B = m_0c^2 + E_{Bk} = 3m_0c^2$$

According to the principle of energy conservation, the energy of the particle after collision is

$$E_A + E_B = 4\ m_0c^2 = Mc^2$$

we have that the relativistic mass of the particle is $M = 4m_0$, so its rest mass is

$$M_0 = M\sqrt{1 - \frac{v^2}{c^2}} = 4\ m_0\sqrt{1 - \frac{v^2}{c^2}}$$

In order to calculate the velocity v of the particle, we Apply the principle of momentum conservation to the process of the collision, we get

$$0 + p_B = Mv$$

we get $v = \frac{p_B}{M}$. In order to obtain the momentum p_B of the particle B, using the relativistic energy-momentum relation(1 – 45) for particle B

$$E_B^2 = p_B^2c^2 + m_0^2c^4$$

we get $p_B^2 = 8m_0^2c^2$, $v^2 = \frac{p_B^2}{M^2} = \frac{8m_0^2c^2}{16m_0^2} = \frac{1}{2}c^2$, So the rest mass of the particle after collision is

$$M_0 = 4m_0\sqrt{1 - u^2/c^2} = 2\sqrt{2}m_0$$

*1.6 Introduction to General Relativity

1.6.1 Basic Principles of General Relativity

After establishing the theory of special relativity, Einstein found that there were two principled defects in the special relativity: one defect was that after abandoning the concept of the absolute space-time, inertial frames had become impossible to be defined, and in the strict sense inertial frames did not exist in the universe. The second defect is that the law of gravitation has not been revised, the law of gravitation does not remain invariant under the Lorentz Transformation, and it can not be included in the framework of special relativity. In addition, the view of absolute space-time and action at a distance associated with gravitation also contradict Einstein's view of spacetime and the concept of field. In order to eliminate these two defects, Einstein put forward two basic postulates: the principle of general relativity and the equivalence principle, and established the theory of general relativity applicable to any reference frame in 1915. Let's look at two basic postulates of general relativity.

1. The principle of general relativity: The laws of physics are the same in all reference frames

All reference frames mentioned here include both inertial and noninertial frames. In order to eliminate the first defect of special relativity mentioned above, Einstein canceled the special status of inertial frames in the theory and extended the principle of relativity to noninertial frames, thus obtaining the first basic postulates of general relativity.

2. Equivalence Principle: the inertia force and the gravity are equivalent

After Einstein extended the principle of relativity to noninertial frames, he found that inertia force exists in noninertial frames. Through the analysis of the inertial force, Einstein found the connection between the inertial force and the gravity. Austrian scientist E. Mach once pointed out that the inertia of an object is not an inherent attribute of the object itself, but is generated by the action of countless huge celestial bodies in the universe, and the inertial force is essentially a kind of gravity. This thought was later called mach's principle by Einstein. It can be seen that mach's thought has great inspiration and influence on Einstein's proposal of equivalence principle. In addition, although the inertia and the gravity are two completely different properties of objects, the inertial mass and the gravitational mass describing them are equal to each other. On the basis of these thoughts, Einstein put forward the famous equivalence principle, which solved the problems of noninertial frame and gravity together. Einstein also illustrated equivalence principle through the thought experiment of elevator.

In the thought experiment of Einstein's elevator, the elevators on the surface of the earth and in the space without gravity were divided into two cases for comparison. The first case is shown in Figure 1 – 21 (a) and (b), there are two elevators, one is at rest on the earth, another is moving upward with acceleration $a = g$ (the gravitational acceleration on the earth) in space. Under these two conditions, the observers in two elevators can sense their own weight. If they weigh, they can measure their own weight as mg. However, because the elevators are closed, the observers cannot observe the outside, so the observers cannot judge whether they are at rest on the earth or accelerating in space. Obviously, the elevator accelerating up in space is a noninertial frame. It can be seen from the comparsion that the rest reference frame in gravitational fields cannot be distinguished from noninertial reference frame, and the space with gravitational fields is not an inertial frame.

The second case is shown in Figure 1 – 21 (c) and (d), there are also two elevators, one ist the elevator making free fall on the earth, another is at rest or does uniform linear motion in space. Under these two conditions, the observers in two elevators are in weightlessness. If they weigh themself, the measured result is 0. In this case, the observers in the elevators can also not judge whether they make free fall on the earth, or are at rest or do uniform linear motion in space. Since the elevator that is at rest or does uniform linear motion in space can be regarded as an inertial frame, it means that the elevator that falls freely on the earth can also be regarded as an inertial frame. In this way, if a free-

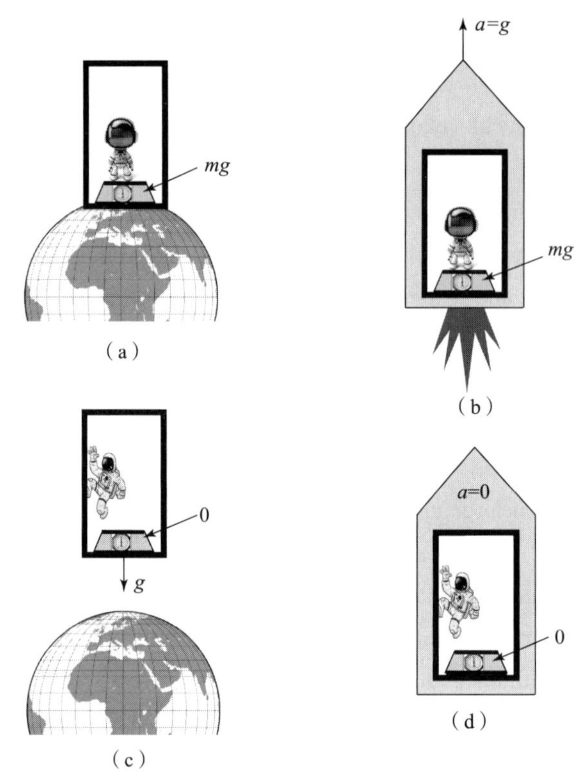

Figure 1 – 21 Schematic diagram of Equivalence Principle: Einstein's elevator

(a) At rest on the earth; (b) Accelerating rise in space; (c) Free fall on the earth; (d) at rest in space

falling frame is introduced in any local area where gravitational fields exist, gravitational fields can be eliminated and the free-falling frame becomes a "local inertial frame", in which the formulas of special relativity are valid.

Through the thought experiment of the elevator, we can see the connection between the gravitational field and non-inertial frame, and we can also see the way to solve the problem of gravitational field and noninertial frame at the same time. In this way, Einstein put forward equivalence principle and established the theory of general relativity on this basis. Among the two basic principles of general relativity, the principle of general relativity has no restriction on the specific content of physical laws, and the physical content of general relativity can be obtained through the equivalence principle.

Based on the equivalence principle, Einstein predicted the deflection of light in gravitational fields and gravitational redshift of light. Here we will first discuss the deflection of light in gravitational fields. In the case of Figure 1 – 21(a) and (b), suppose that a beam of light shoots into the elevators from the left side to the right side. Because the elevator in space [Figure 1 – 21(b)] is accelerating to rise, the light will reach the lower position on the right side of the elevator. For the observer in this elevator, the path of the light is a downward curved parabola. Because the observers in elevators can not distinguish whether the elevator is accelerating in space or is at rest on the earth, it can be concluded that the path of the light in the elevator which is at rest on the earth is the same downward curved parabola. This shows that light rays will bend in gravitational fields. Einstein attributed the bending of light rays to the bending of the space, and pointed out that the gravity can bend the space, in which the path of light rays also bends.

Because we live in a flat three-dimensional space, it is difficult for us to understand what a curved three-dimensional space is like. In order to help us understand the relationship between curved space and flat space, we take curved two-dimensional space and flat two-dimensional space as examples to illustrate. The surface of the earth is a sphere. If a person does not know that the surface of the earth is a sphere, he will feel confused when he moves eastward along a straight line and finally returns to the starting position. From the perspective of high-dimensional space, the truth is actually very simple. Instead of walking along a straight line, he walked through a circle along the latitude of the earth. Here, the earth's surface is a curved two-dimensional space, while the plane of the latitude line is a flat two-dimensional space. This shows that a straight line in a curved two-dimensional space is a curve in a flat two-dimensional space. Similarly, a straight line in a curved three-dimensional space is a curve in a flat three-dimensional space.

In areas where matter is concentrated in the universe, such as in areas near stars, strong gravity causes the curvature of three-dimensional space. When the light passes nearby, although the light travels along a straight line, which is the shortest path in the curved three-dimensional space, the path of the light will bend when observed in the flat three-dimensional space. According to the prediction of general relativity, the space has an extremely strong curvature near a black hole containing a large amount of matter.

In addition, general relativity points out that not only the space will bend, but also the time will bend in gravitational fields. It means that the clock close to the sun will go slower than the clock far away from the sun. This effect is called the gravitational time delation. The bending of time and space is closely linked and inseparable. This kind of curved spacetime is not suitable to be described by the familiar Euclidean geometry, and the Riemannian geometry describing curved space has therefore become a powerful mathematical tool for Einstein to establish general relativity.

Einstein's field equations were established in the Riemannian space in 1915, which marked the establishment of the theory of general relativity. Einstein's field equations show that the matter and the spacetime are unified. Matter tells spacetime how to curve, spacetime tells matter how to move. In this way, Einstein's general relativity unifies the noninertial frame, the gravitational field and the curved space, inherits the reasonable contents of special relativity, extends relativistic physics to noninertial frame, and solves the problem of the gravity.

Now the main application of general relativity is in the field of cosmology and astrophysics. The mathematical description of black holes is obtained by solving Einstein's field equations. The theory of the gravity based on general relativity is the best of all relevant theories at present.

1.6.2 Several Experimental Verifications of General Relativity

At the beginning of the establishment of general relativity, Einstein put forward three experimental test plans, which were later implemented one by one. These three experiments are the precession of Mercury's perihelion, the curvature of light in gravitational fields, and the gravitational redshift of spectral lines.

1. The precession of Mercury's perihelion

Through astronomical observation, the precession of Mercury's perihelion as shown in Figure 1 – 22 is recorded to be about 5600″ every 100 years, but the precession of 5557″ can be explained by Newton's theory, while the difference of 43″ cannot be explained. After establishing general relativity, Einstein made use of the theory that the gravity of the sun bends the space, believed that the orbit of Mercury was the geodesic line in the curved space, and calculated the precession of Mercury's perihelion, which well explained the 43″ which has been unable to be explained by classical theory.

2. curvature of light in the gravitational field

After Einstein proposed equivalence principle, he predicted the curvature of light in the gravitational field. This curvature is not caused by the "force" of gravity, but by the effect of gravity curving space. According to the theory of gravity curving space in general relativity, the starlight emitted from a star should deflect 1.75″ when passing near the sun as shown in Figure 1 – 23. In 1919, the Royal Society of England and the Royal Astronomical Society jointly sent two expedition teams, led by Professor Eddington and Professor Crommelin respectively, to Principe Island in the Gulf of Guinea in Africa and Sobular in Brazil for observation. The two teams found that the starlight passing near the sun deflect 1.61″ ± 0.30″ and 1.98″ ± 0.12″ respectively, which is basically consistent with the results predicted by general relativity. This observation not only confirms the correctness of general relativity, but also makes people realize the significance of Einstein's general relativity.

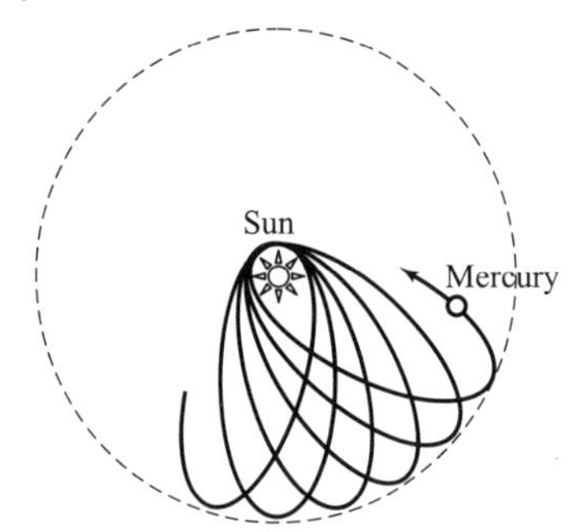

Figure 1 – 22 The precession of Mercury's perihelion

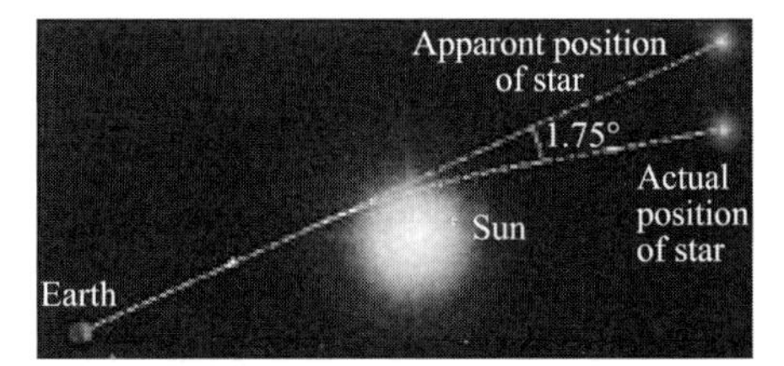

Figure 1 – 23 curvature of light in the gravitational field

In addition, the gravitational lensing effect such as Einstein Ring and Einstein Cross observed in modern astronomy are also a phenomenon that light is curved in the huge gravitational field near a massive object.

3. The Gravitational Redshift of light

After putting forward equivalence principle, Einstein predicted that the frequency of light would decrease when it was far away from the celestial body with huge mass as shown in Figure 1 – 24. This effect is called the gravitational redshift of light. Einstein also gave the formula of gravitational redshift.

This effect can be shown in the starlight spectrum obtained from many astronomical observations. However, due to the existence of Doppler frequency shift at the same time, it is difficult to distinguish the two effects through the observation, so it cannot quantitatively prove the correctness of Einstein's gravitational redshift formula for light.

In 1959, Pound and Rebka of the United States designed an experiment to measure the gravitational redshift of light on the ground. They placed the radioactive source 57Co on the top layer (22.6 meters high) of the Jefferson Tower in Harvard University. The gamma rays emitted by 57Co were directed to the bottom of the tower, and 57Fe was used to receive gamma rays at the bottom of the tower. Using the extremely sensitive Mossbauer effect, they accurately measured the change in the frequency of gamma rays, and the result was consistent with the prediction of general relativity.

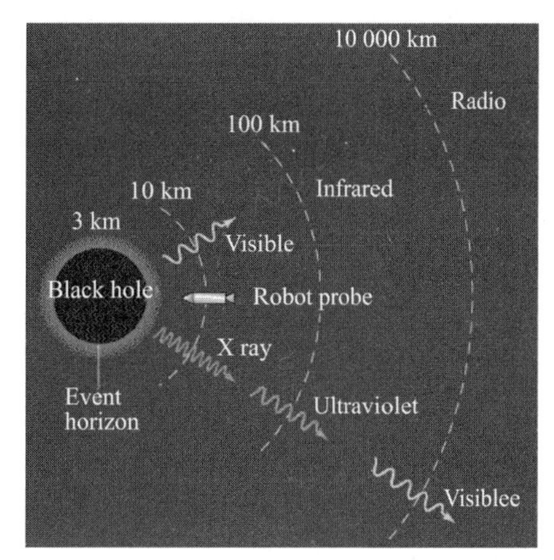

Figure 1 – 24 Gravitational Redshift of light

4. Delay of radar echo

According to general relativity, the gravity of the sun can bend the space near the sun, so it takes longer when light or electromagnetic waves pass near the sun.

In order to test the correctness of general relativity that gravity bends space, Shapiro, an American physicist, put forward the experiment of measuring the time of radar echo in 1964: an electromagnetic pulse is emitted from the earth to a planet, and the signal is reflected back to the earth by the planet, and the round-trip time of signal is measured in two cases. When the path of the signal is far from the sun, the influence of the sun is negligible. When the signal passes near the sun, the round-trip time of the signal will be longer than that of the signal travelling far away from the sun. In this way, we can check whether there is the time delation due to the curvature of the space by massive object. Scientists have used Mercury, Venus and Mars to carry out radar echo experiments, proving that the time delation of radar echo caused by the sun will reach about 200ms, which is consistent with the prediction of general relativity.

5. Black hole and the gravitational wave

Shortly after Einstein established general relativity, the German astronomer Schwarzschild obtained the first exact solution of Einstein's field equations. Schwarzschild's solution describes the curvature of outer spacetime of a spherically symmetric celestial body. The strong gravitational field of the celestial body not only makes space curved, but also time curved. In the solution, a critical radius r_g is also obtained. Schwarzschild predicted: for any celestial body, within the radius less than r_g from its center, its spacetime is so curved that neither light nor matter can escape from it. This "incredible celestial body" was named "black hole" by American physicist Wheeler. This radius r_g was later called the Schwarzschild radius, which gives the radius of the black hole now commonly considered. The sphere formed by the Schwarzschild radius is the event horizon of the black hole.

Modern astronomy believes that black holes are formed by gravitational collapse of stars with sufficient mass after the fuel of nuclear fusion is exhausted. The mass of a black hole is very large, but its volume is very small. The gravitational field generated by it is so strong that any matter and radiation(including the light) can no longer escape after entering the event horizon of the black hole. The event horizon of the black hole completely separates the internal and external space of the black hole, so the black hole cannot be observed directly. The existence of the black hole and its mass can only be inferred indirectly from its impacts on other objects.

In addition, according to general relativity, the gravity of celestial bodies will curve the spacetime. If celestial bodies accelerate, the curved spacetime caused by the gravity can spread outward in the form of waves, and its travelling speed is equal to the speed of light. This is the gravitational wave predicted by Einstein on the basis of general relativity. The gravitational wave can be understood as a kind of spacetime wave travelling at the speed of light, just like the ripples produced when a stone is thrown into the water. The gravitational wave can be regarded as the "ripples of the spacetime" in the universe.

After Einstein predicted the gravitational wave, people began to try to detect gravitational waves. It is generally believed that strong gravitational waves can only be generated in the process of acceleration, collision and merger of massive celestial bodies or black holes in the universe. Since these wave sources are very far away from the earth, when gravitational waves spread to the earth, they have become very weak. Therefore, gravitational waves are also very difficult to be detected experimentally.

In 1974, using radio telescopes physicists Taylor and Hulse discovered a twin-star system consisting of two neutron stars with a mass roughly equal to that of the sun. According to general relativity, the binary system will lose energy by radiating gravitational waves outward. Its orbital period will be reduced by 76.5ms per year, and its orbital semi-major axis will be reduced by 3.5 meters per year. It is estimated that the merger will occur after about 300 million years. Since 1974, Taylor and Hulse have observed the orbit of this binary system for a long time, and the observed values are in good agreement with the values predicted by general relativity, which indirectly proves the existence of the gravitational wave.

In the 1970s, American physicist Weiss and others began to try to detect the gravitational wave using laser interferometry, whose working principle is similar to that of the Michelson interferometer. The laser interferometry includes two mutually perpendicular interference arms. At the intersection of the two arms, a laser beam is divided into two beams, which enter into the two arms and reflect back and forth in the arms. Finally, the two beams meet again to form interference fringes. If gravitational waves pass through the two arms, the spacetime along will be curved to different degrees. The optical path difference will change, and the interference fringe will also change. Because the gravitational wave is very weak, this requires the detection instrument to be able to sense the change of 10^{-18} meters(equivalent to one thousandth of the proton diameter) at a distance of 1000 meters. By the 1990s, the technical conditions for achieving this accuracy are gradually met. In 1991, with the support of the National Science Foundation (NSF) of the United States, Massachusetts Institute of

Technology (MIT) and California Institute of Technology began to build the "Laser Interferometer Gravitational-Wave Observatory" (LIGO) as shown in Figure 1 – 25, which has the high sensitivity required to detect gravitational wave.

On September 14, 2015, LIGO's two twin gravitational wave detectors with a distance of 3 000 km detected the signal of gravitational waves at the same time. As shown in Figure 1 – 26, its initial frequency was 35 Hz, then rapidly increased to 250 Hz, and finally became disordered and disappeared. The whole process lasted only a quarter of a second. After calculation and analysis, the detected gravitational wave is determined to be generated in the final stage of merger of two black holes 1.3 billion light-years away. The initial mass of the two black holes is 29 times and 36 times the mass of the sun respectively. After merging, they become a black hole with 62 times the mass of the sun. The mass defect is converted into energy and released outward in the form of gravitational waves. After a long journey of 1.3 billion years, these energies finally reach the earth and are detected by two twin gravitational-wave detectors of LIGO in the United States.

Figure 1 – 25　Laser Interferometer Gravitational-Wave Observatory (LIGO)

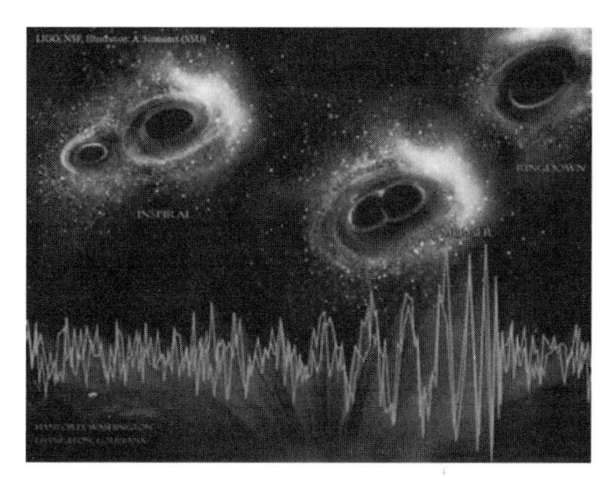

Figure 1 – 26　The detected gravitational-wave generated by the merger of two black holes

On February 11, 2016, the US LIGO researchers announced the results of this observation. This discovery of LIGO has verified Einstein's prediction about gravitational waves made 100 years ago, and the correctness of general relativity has been proved again.

A. Einstein (1879—1955) was born in Germany and settled in the United States. He was one of the greatest physicists in the 20th century. He founded special relativity in 1905 and general relativity in 1915. In 1905, he proposed the hypothesis of photons to explain the photoelectric effect. In 1921, he was awarded the Nobel Prize in physics. He also made many contributions to the establishment of quantum mechanics. In addition, Einstein was also one of the founders of statistical mechanics and modern cosmology.

Summary

1. Basic postulates of special relativity:

Einstein's principle of relativity: The laws of physics are the same in all inertial reference frames, that is, all inertial frames are equivalent, and there is no special absolute inertial frame.

The principle of constant speed of light: in all inertial frames, the propagation speed of light in vacuum is equal to c, regardless of how the light source and observer move.

2. Time dilation.

$$\Delta t = \frac{\Delta t'}{\sqrt{1 - u^2/c^2}} \quad (\Delta t' \text{ is the proper time})$$

3. Length contraction:

$$\Delta x = \Delta x' \sqrt{1 - u^2/c^2} \quad (\Delta x' \text{ is the proper length})$$

4. Relativity of Simultaneity:

If two events occur at different places in the frame S' at the same time, they do not occur at the same time in the frame S. The event behind the relative motion direction occurs first, and the time difference is proportional to the relative motion velocity u and the proper distance $\Delta x'$ of the two events in the frame S', which is

$$\Delta t = t_2 - t_1 = \frac{\Delta x' u/c^2}{\sqrt{1 - u^2/c^2}}$$

5. The Lorentz Transformation:

$$x' = \frac{x - ut}{\sqrt{1 - u^2/c^2}}, y' = y, z' = z, t' = \frac{t - \frac{u}{c^2}x}{\sqrt{1 - u^2/c^2}}$$

6. The relativistic velocity transformation: $v_x' = \frac{v_x - u}{1 - \frac{uv_x}{c^2}}, v_y' = \frac{v_y \sqrt{1 - u^2/c^2}}{1 - \frac{uv_x}{c^2}}, v_z' = \frac{v_z \sqrt{1 - u^2/c^2}}{1 - \frac{uv_x}{c^2}}$

7. The relativistic mass:

$$m = \frac{m_0}{\sqrt{1 - v^2/c^2}} \quad (m_0 \text{ is static mass}, m \text{ is relativistic mass})$$

8. The relativistic momentum:

$$\boldsymbol{p} = m\boldsymbol{v} = \frac{m_0 \boldsymbol{v}}{\sqrt{1 - u^2/c^2}}$$

9. The mass-energy relation:

$$E = mc^2$$

10. The relativistic kinetic energy:

$$E_k = mc^2 - m_0c^2$$

11. The energy-momentum relationships:

$$E^2 = p^2c^2 + m_0^2c^4$$

Questions

1 - 1 Illustrate the difficulties encountered by classical mechanics, and point out the causes of these difficulties.

1 - 2 How to understand the view that the time and the space are indivisible in the relativistic four-dimensional spacetime?

1 - 3 "Two events which occur simultaneously in one inertial frame must not occur simultaneously when observed in another inertial frame." According to the relativistic spacetime view, is this statement correct?

1 - 4 What is the proper time? How to understand that the proper time of a rest frame is the shortest when comparing with the same time interval measured by other frames?

1 - 5 The motion path of a particle is a circle on the xy plane when observed in the inertial frame S, what is its motion path like when observed in the inertial frame S', which moves along the x-axis at constant speed u with respect to the frame S?

1 - 6 Does time delation and length contraction mean that moving clocks really run slow and moving objects become really shorter? How to undersant the relativity of time delation and length contraction?

1 - 7 Will the sequence of two causal events change in differents inertial frames? How about the result for two events without causality? In science fiction films and novels, we often see the plot of time leap. Try to use the relativistic spacetime effect and the law of causality to discuss the possibility of these plots.

1 - 8 The proper lengths of a train and a tunnel are the same. The train moves to the tunnel at speed 0.8c. The front end of the tunnel is closed. The train drives into the tunnel from the back end of the tunnel. When the train tail just enters the back end of the tunnel, the front end of the tunnel immediately opens. Try to discuss the following two discriptions about whether the train can pass through the tunnel:

(1) In the tunnel reference frame: the length of the moving train shrinks due to the length contraction effect, the length of the train is shorter than that of the tunnel. If the front end of the tunnel opens at the time when the tail of the train enters the back end, the train will pass through the tunnel.

(2) In the train reference frame: the length of the moving tunnel shrinks due to the length contraction effect, the length of the tunnel is shorter than that of the train. When the train enters the back end of the tunnel, the head of the train has collided with the closed front end of the tunnel, so the train cannot pass through the tunnel.

1 - 9 Some people think that photons are at rest in the reference frame of photons, is that right?

1 - 10 What is the difference and connection between the momentum theorem of relativity and the momentum theorem of Newtonian mechanics?

1 - 11 From the point of energy, how to understand that the speed of physical particles cannot reach the speed of light?

1 - 12 Einstein put forward his famous Mass-Energy relation in special relativity. Try to discuss under what conditions the huge energy contained in the mass can be released.

Problems

1 - 1 A passenger in a spaceship traveling at constant speed 0.9998c raised his hand for 1s as measured by the ship's clock. If measured in the earth frame, how long does it take for the passenger to raise his hand? If a person on the earth raises his hand for 1s as measured in the earth's rest frame, how long is it measured in the spaceship frame?

1 - 2 Two events occurred successively at the same place of the frame S with a time interval of 2s. The corresponding time interval observed in the frame S' is 3s. Find the space interval of these two events in the frame S.

1 - 3 Two spaceships fly past each other with relative speed 0.98c. The observer in spaceship 1 measures that the length of spaceship 2 is 2/5 of the its own length. Find: (1) the ratio of the proper length of the spaceship 1 to that of the spaceship 2. (2) When measured in the spaceship 2, what is the ratio of the length of spaceship 1 to its own length?

1 - 4 In the frame S, a rod with length l is placed at rest and its included angle with the x-axis is θ. Find the length and the angle with the x' - axis when measured in the frame S'. It is known that the frame S' moves in the x direction at constant speed u with respect to the frame S.

1 - 5 Energetic particles like protons in cosmic rays collide with atoms or molecules in the upper atmosphere and produce particles like muons. The mean lifetime of a muon in its rest frame is 2×10^{-6}s. If a muon is moving toward the ground at speed $0.998c$ at an altitude of 6000m above sea level,

(1) how far can it travel before decay when observed on the ground? Can the muon reach the ground?

(2) If observed in muon's frame, how far away is the ground from itself when it is just produced? Can the muon reach the ground before decay?

1 - 6 The observer in the inertial frame S measured that a flash occurs at $x = 100$km, $y = 10$km, $z = 1$km, $t = 5\times10^{-4}$s. If the inertial frame S' moves at speed $u = -0.8c$ relative to the frame S along the x-axis, calculate the spacetime coordinates x', y', z', t' of the flash measured in the frame S'.

1 - 7 The spacetime coordinates of two events measured in the inertial frame S are: $x_1 = 6\times10^4$m, $y_1 = 0$, $z_1 = 0$, $t_1 = 2\times10^{-4}$s and $x_2 = 12\times10^4$m, $y_2 = 0$, $z_2 = 0$, $t_2 = 1\times10^{-4}$s. If these two

events occur at the same time measured in the inertial frame S'. what is the speed of the frame S' relative to the frame S? What is the space interval between the two events measured in the frame S'?

1 - 8 The spaceship with a proper length of L' moves at constant velocity u relative to the ground. A small ball is moving at velocity v' from the end to the head of the spaceship as observed by the astronauts. Find: (1) the time required for the ball to travel through the spaceship measured by the astronauts; (2) the time required for the ball to move measured by the observers on the ground.

1 - 9 A transmitting station sends radio signals to two receiving stations E and W on the east and west sides of it. The distance from the transmitting station to the two receiving stations is both d. An aircraft flies from west to east at constant speed v. When observed in the frame of the aircraft, which receiving station receives the signal at first, and what is the time interval between the two receivings?

1 - 10 A meterstick moves along its length direction at speed $0.6c$ relative to an observer. How long does it take for the meterstick to pass the observer as measured in the frame of the observer?

1 - 11 Altair star is about 16 light-years away from the earth. If a spaceship is flying to Altair from the earth at constant speed $0.97c$, how long will it take for the spaceship to reach Altair as measured on the spaceship?

1 - 12 Assume that spaceship A and B fly eastward relative to the ground at speeds $0.6c$ and $0.8c$ respectively. Two events occurred at a certain place on the ground. The time interval measured in the frame of the spaceship A is 5s. What is the corresponding time interval measured in the frame of the spaceship B?

1 - 13 A spaceship flied past an observation station on the ground. When the head of the spaceship passed the observation station, a flash was emitted from the head. When the end of the spaceship passed the observation station, a flash was emitted from the end. The time interval between two flashes measured by the observer in the observation station is 75ns. The proper length of the spaceship is 30 m. Find (a) what is the speed of the spaceship relative to the ground? (b) What is the time interval between two flashes measured in the frame of the spaceship?

1 - 14 A spaceship with a proper length of 36m is flying from the earth to other planets. The observers in the frame of the earth observe that the spaceship is 27m long and an astronaut on the spaceship has done exercises for 20 minutes. How long has the astronaut done exercises as measured by his clock?

* 1 - 15 A spaceship flies away from the earth at speed $u = 0.6c$. the spaceship sent a radio signal, which would be reflected by the earth, and the spaceship received the return signal 40s later. If observed in the spaceship frame and the earth frame respectively, how far is the spaceship from the earth at the moments of sending, reflecting and receiving the signal?

1 - 16 The distance between place A and place B is 120km. A train departed the place A at 0:00:00, a flash is emitted at 0:00.0003 in the place B. Find the time interval between the two

events when observed in the frame of a spaceship, which is flying at speed 0.8c from the place A to the place B. Which event happened at first?

1-17 Observers on the earth found that a spaceship travelling east at speed 0.6c would collide in 5s with a comet travelling west at speed 0.8c. For the observers on the spaceship, (1) how fast does the comet approach them? (2) how much time was left for them to avoid collision?

1-18 An electron has the energy of 2.0 MeV, what are the electron's kinetic energy, momentum, speed and relativistic mass? The rest energy of the electron is known to be about 0.51 MeV.

1-19 A moving muon has the total energy of 3000 MeV and the rest energy of 100 MeV. Its lifetime is 2×10^{-6}s in its rest frame. Find the distance it moves in the frame of lab.

1-20 The rest mass of each particle in the thermonuclear reaction ${}_1^2\mathrm{H}+{}_1^3\mathrm{H}\rightarrow{}_2^4\mathrm{He}+{}_0^1\mathrm{n}$ is: deuterium $m_\mathrm{D}=3.3437\times10^{-27}$ kg, tritium $m_\mathrm{T}=5.0049\times10^{-27}$ kg, helium $m_\mathrm{He}=6.6425\times10^{-27}$ kg, neutron $m_\mathrm{n}=1.6750\times10^{-27}$ kg, how much energy is released through this reaction by consuming 1 kg of reactant?

1-21 The strongest cosmic ray has the energy of 50J. If this ray is produced by a proton, what is the difference in speed between the proton and light in vacuum?

1-22 A stationary nucleus shoots two protons in opposite directions at the same time, both of which have a velocity of 0.5c as observed in the laboratory. (1) What is the momentum and the energy of each proton as observed in the laboratory. (2) What is the momentum and the energy of a proton relative to another proton. The results can be expressed by the rest mass of proton m_0 and the speed of light c.

*1-23 An atom with the rest mass of m_0 emits a photon with energy $h\nu$ and recoils. Find the rest mass of the atom after the photon is emitted.

1-24 A rest box has a mass of m and a volume of V. If the box moves relative to the observer at a velocity ν along an edge direction, what is the density of the box as measured by the observer?

1-25 The two particles A and B with the rest mass of m_0 move in opposite direction at the velocity ν, they collide with each other and form a large particle with out energy releasing. Find the rest mass of this large particle.

Chapter 2

Wave-particle Duality of Microscopic Particles

Here is the quantum Kingdom, which is located in the micro world-the atomic and subatomic fields. Its language, legal system and local customs are different from the daily world we are familiar with-the macro world, full of surprises. Now that we are here, we should do as the Romans do and go through the terminology, experimental basis and theoretical framework of quantum physics.

From the late 19th century to the early 20th century, with the progress of science and technology, when people's research tentacles entered the scale of microscopic particles, physicists have successively found a series of new experimental phenomena that cannot be correctly explained by classical physics, such as blackbody radiation, photoelectric effect, Compton effect, atomic linear spectrum and so on. This made classical physics, which had been quite perfect at that time, in a very difficult situation, forcing scientists to jump out of the traditional physical framework and find new solutions, which led to the birth of quantum physics.

Quantum physics first made a breakthrough from the problem of blackbody radiation. In 1900, in order to solve the difficulties encountered by classical theory in explaining the experimental laws of blackbody radiation, M. Plank (1858—1947) first proposed the concept of energon, that is, the concept of energy quantization. This is a great impact on the classical physical theory, because in the classical theory, the continuity of energy is regarded as a natural thing. Einstein's hypothesis of the wave-particle duality of light, and the subsequent hypothesis of wave-particle duality of physical particles by Louis de Broglie(1892—1960), made people realize that all microscopic particles have wave-particle duality. On this basis, in 1926, Erwin Schrödinger, (1887—1961) proposed a non-relativistic equation to describe the law of motion of microscopic particles. In 1928, P. A. M. Dirac (1902—1984) proposed a relativistic equation. They are the fundamental equations of quantum mechanics. After the joint efforts of many physicists, quantum mechanics was finally established in the 1930s. This is the theory about the microscopic world, and together with the theory of relativity, have become the theoretical basis of modern physics.

Quantum physics is the basis for people to know and understand the microscopic world. Its research results and research methods have penetrated into all fields of modern science and technology, and play a more and more important role in the scientific research and technological development of chemistry, biology, information, computer, laser, energy and new materials and etc.

In this chapter we will roughly follow the wheel of history and lead you to appreciate the scenery of quantum physics, through the law of blackbody radiation, the phenomenon of photoelectric effect,

Compton effect, the atomic spectrum of hydrogen, and the wave property of particles and so on, which are incredible and strange things for classical physics. We focuses on introducing some basic concepts, laws and methods in quantum physics proposed successively by Planck, Einstein, N. Bohr (1885—1962), de Broglie, M. Born (1982—1970), W. Heisenberg (1901—1976) and other physicists, so as to provide readers with necessary basic knowledge of modern physics, from which you can experience and appreciate the harmony and beauty of the microscopic world.

2.1 Black Body Radiation and Planck's Energon Hypothesis

2.1.1 Black Body Radiation

First, let's introduce the process of dispersing the second dark cloud once mentioned by Kelvin. In the 19th century, due to the research and development in the fields of metallurgy, high temperature measurement technology and astronomy, people began to study thermal radiation. As shown in Figure 2 - 1, when the heated object reaches a certain temperature, it will emit red light. With the increase of temperature, the color of light will gradually change from red to yellow and then to blue and white. The phenomenon that the thermal motion of molecules (containing charged particles) causes objects to transfer energy outward in the form of electromagnetic waves is called thermal radiation. That the color of light changes regularly with temperature, indicating that the energy distribution of electromagnetic waves of thermal radiation according to frequency (or wavelength) is related to temperature. The higher the temperature, the greater the total energy radiates and the more high frequency (short wavelength) components. Based on this fact, experienced steel workers can estimate the temperature of molten steel according to its color. As the same, one can determine the surface temperature of a star from its color. Generally speaking, the surface temperature of blue stars is above 25,000 K, that of white stars is between 11,500 - 7,700 K, that of yellow stars is between 6,000 - 5,000 K, and that of red stars is between 3,600 - 2,600 K. In 2006, British researchers found that honeybees can also look for warmer flowers by recognizing their color. As the saying goes, one heat makes one light. Any object, regardless of temperature, its thermal radiation does not stop for a moment, and will radiate energy in the form of electromagnetic waves. However, when the temperature is low, the radiation is weak, and the radiated electromagnetic waves mainly concentrate in the invisible infrared or far-infrared band, which cannot be seen by the naked eye. For example, the human body is a natural infrared radiation source, but the infrared radiation characteristics of the human body are different from those of the surrounding environment. The infrared life detector is to take advantage of the differences between them to separate the target to be searched from the background by imaging, as shown in Figure 2 - 2. The infrared radiation range of normal human skin is 3 - 50 μm, of which 8 - 14 μm accounts for about 46% of the total human body radiation energy, and the central wavelength (peak wavelength) where the energy is more concentrated is 9.4 μm. This wavelength is an important technical parameter in the design of human infrared detector.

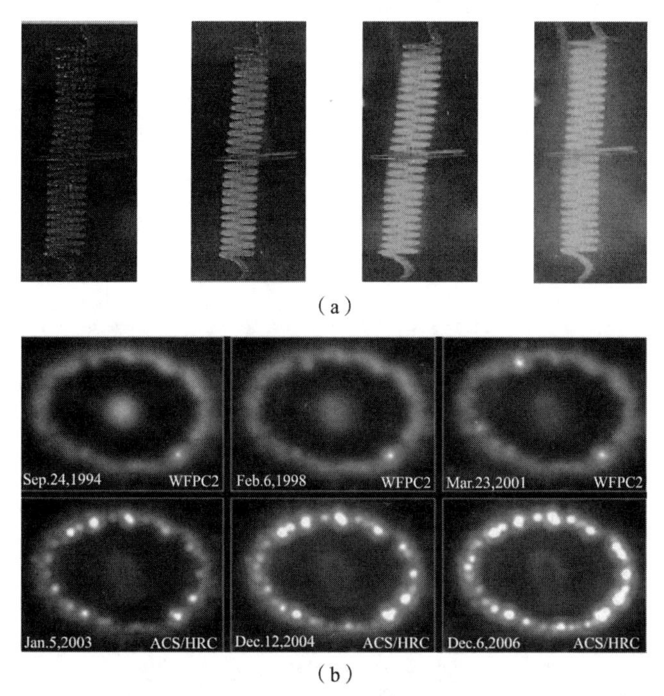

Figure 2 - 1 (a) As the temperature of a metal wire increases, its color changes from dark red to bright red, and then from near yellow to white, finally to dazzling blue and white. (b) The supernova 1987A photographed by Hubble Space Telescope from 1994 to 2006 reveals the process of shock waves crashing into the gas ring, heating the gas inside and making them glow and brighten.

Figure 2 - 2 Images taken by infrared thermal imaging camera

While an object radiates heat outward, it also absorbs thermal radiation from other objects around it. When the energy radiated by the object per unit time is equal to the energy absorbed by external radiation, the thermal radiation process of the object reaches equilibrium, which is called equilibrium thermal radiation. At this time, it has a certain temperature. Next, we discuss the equilibrium thermal radiation through the blackbody model.

The so-called blackbody refers to an ideal object that can absorb all the incident electromagnetic waves of any frequency at any temperature, and when heating it, its ability to radiate out electromagnetic waves of any frequency is stronger than that of any other object at the same temperature. Of course, absolute blackbody does not exist in nature, and even very black coal cannot completely absorb energy of incident electromagnetic waves. And many objects that look black to us only present strong absorption for visible light. They cannot fully absorb the light such as invisible light in infrared and ultraviolet waveband.

Figure 2 – 3 shows the Namibian chameleon land in the desert. This chameleon can make both sides of the body show different colors, black on one side and white on the other, so that in the morning when it is colder, it uses the black side to absorb heat, and uses the other side of white to prevent heat loss.

Figure 2 – 3 Chameleon cold protection

In 1895, German physicists Wilhelm Wien (1864—1928) and Otto Lummer (1860—1925) pointed out that blackbody can be achieved by a cavity, as shown in Figure 2 – 4. A small hole is opened on a cavity wall made of any opaque material (such as steel, copper, ceramics, etc.). The hole is small enough compared to the cavity, and the small hole can be regarded approximately as a blackbody. This is because the electromagnetic waves of all frequencies incident into the cavity through the small hole will go through multiple reflections of cavity's inner wall. At every reflection the inner wall of the cavity will absorb part of it, and finally almost all of the electromagnetic waves are absorbed by the inner wall of the cavity. And then the electromagnetic waves escaping from the small hole are negligible. For example, assume an electromagnetic wave with energy of E_0 goes into a cavity with white inner wall through a small hole, as shown in Figure 2 – 5. Inside the cavity every time the electromagnetic wave is reflected, its energy is only 10% absorbed by the cavity wall. After 100 times of reflections, the energy of the electromagnetic wave coming out of the small hole is only $(0.90)^{100}E_0 = 2.656 \times 10^{-5} E_0$. So the small hole looks very black. You can also do such an experiment, paint the outer wall of the cavity black with ink. When illuminated with visible light, the small hole will look much darker than the surrounding cavity wall. Usually, in the daytime people see that the windows of buildings appear dark, because after the light outside enters the room through the window, by multiple reflections and absorptions it consumes a lot of energy. Thus the light going out of the window and

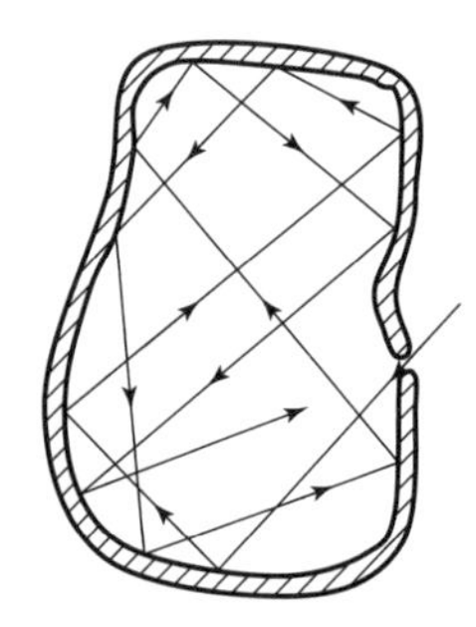

Figure 2 – 4 Blackbody model

reaching the eyes of people outside the house has been very weak. The window here is equivalent to a small hole in the cavity. Similarly, the pupil of an animal's eye often looks dark. If the cavity is heated to a certain temperature, the cavity wall will continue to radiate electromagnetic waves. Then a radiation field will be formed in the cavity. After a certain time, the radiation field in the cavity and the cavity wall will reach thermal equilibrium, that is, the energy radiated from cavity wall into the cavity per unit time is equal to the energy absorbed from the cavity. At this time, the property of equilibrium thermal radiation in the cavity only depends on the temperature. It has nothing to do with the shape and material of the cavity, and it can be represented by the radiation property of the small hole on the cavity at the same temperature. The radiation of the small hole is almost exactly the same as that of the absolute blackbody at the same temperature, that is, blackbody radiation. For example, during metal smelting or ceramic firing, people often open a small hole on the smelting furnace or kiln to measure the temperature in the furnace, as shown in Figure 2 -6.

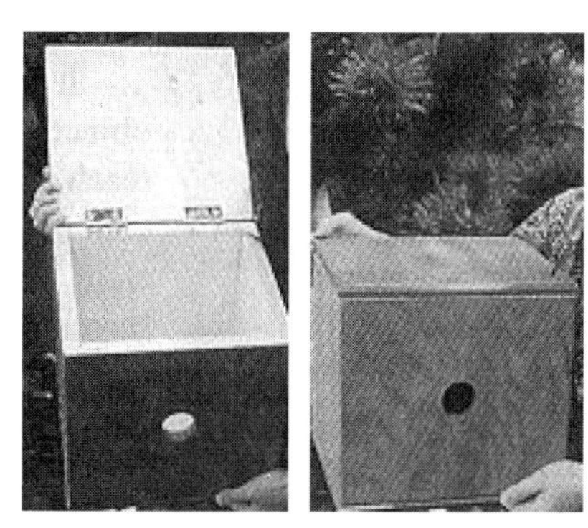

Figure 2 -5 Small hole on the cavity with white inner wall

$T = 300$ K

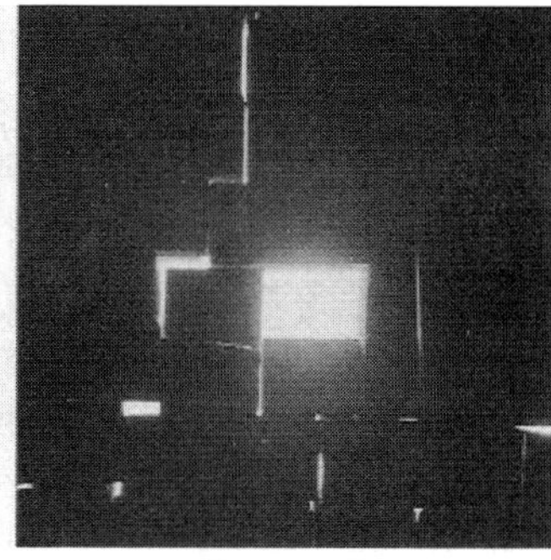

$T = 1\,400$ K

Figure 2 -6 Blackbody at different temperatures

In this way, people can use the above-mentioned blackbody to do experiment in studying monochromatic radiant exitance M_λ of equilibrium thermal radiation, that is, the energy of electromagnetic waves with wavelengths in the unit wavelength range near λ emitted from the unit surface area of the object per unit time. In SI, its unit is W · m^{-3}. The distribution of M_λ by wavelength is mainly related to temperature, and the experi-mental results are shown in Figure 2 -7. As can be seen from the figure that for the distribution curve at a certain temperature, there is a maximum value, and the corresponding wavelength at the dotted line is λ_m,

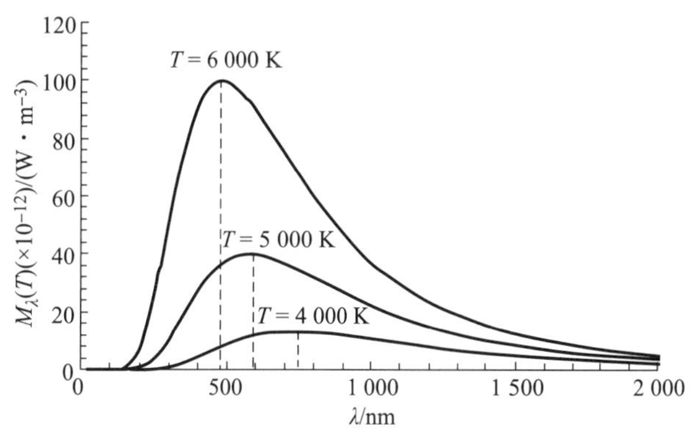

Figure 2 -7 Distribution curves of monochromatic radiant exitance M_λ of blackbody radiation at three temperatures as a function of wavelength

which is called the peak wavelength. For example, when $T = 5,000$ K, $\lambda_m = 580$ nm, it is in the yellow light region. When the thermodynamic temperature increases, the peak wavelength λ_m moves towards the short-wave direction. Conversely, it moves towards the long-wave direction, and it satisfies

$$\lambda_m T = b \tag{2-1}$$

This is Wien displacement law, which was put forward by Wien (Figure 2-8) in 1893 according to thermodynamic theory and Wien won the Nobel Prize in Physics in 1911. In equation (2-1), $b = 2.897,768\,5(51) \times 10^{-3}$ m · K is called Wien's constant. Therefore, as long as the peak wavelength λ_m is measured, the temperature of the object can be estimated, and an optical pyrometer can be made by using Wien displacement law. For example, the sun is approximately a blackbody, and the observation of its radiation shows that the peak wavelength λ_m is about 500 nm, then the solar surface temperature can be calculated as about 5,800 K from equation (2-1). Similarly, the peak wavelength of the human radiation spectrum can be estimated as 9.4 μm.

Figure 2-8 Wien, 1911 Nobel laureate in physics

In Figure 2-7, the area enclosed by the $M_\lambda - \lambda$ curve and the horizontal axis is the total radiation exitance $M(T)$ of the entire electromagnetic spectrum at this temperature, that is, the total energy of electromagnetic waves of various wavelengths radiated from the unit area of blackbody with temperature T in unit time. In SI, its unit is W · m^{-2}. $M(T)$ can also be given by the integral of radiance exitance M_λ over all wavelengths, namely

$$M(T) = \int_0^\infty M_\lambda(T)\,d\lambda \tag{2-2}$$

In 1879, the Slovenian physicist Joseph Stefan (1835—1893) made inductive summary on the experimental data and obtained the relation between radiation exitance $M(T)$ and thermodynamic temperature T as

$$M(T) = \sigma T^4 \tag{2-3}$$

Later, the Austrian physicist L. E. Boltzmann (1844—1906) derived the same result from the thermodynamic theory in 1884, that is, the radiation exitance of the blackbody is directly proportional to the fourth power of the thermodynamic temperature of blackbody, which is called Stefan-Boltzmann law. In equation (2-3), $\sigma = 5.670,367(13) \times 10^{-8}$ W · m^{-2} · K^{-4} is called Stefan Boltzmann constant.

Stefan-Boltzmann law and Wien displacement law are widely used in modern science and technology. They are the physical basis for technologies such as high temperature measurement, remote sensing and infrared tracking.

A Short Story

In 1964, American radio astronomers A. A. Penzias (1933—) and R. W. Wilson (1936—), working at Bell Telephone Company in the United States unexpectedly received a kind of microwave signal noise uniformly distributed in space when studying the signal sent back from the satellite. After analysis, they believe that this noise certainly didn't come from artificial satellite, nor can it come from any galaxy. Because when turning the antenna, the noise intensity remained constant. After they took the antenna apart and reassembled it, they still received that kind of unexplainable noise. Therefore, so it wasn't the noise of antenna or receiver. They called this noise existing in the whole cosmic background as cosmic microwave background (CMB). Later, a lot of experiments were carried out to measure the radiation energy spectrum distribution. It was found that the peak wavelength λ_m appeared near 1.0 mm, corresponding to the $M_\lambda - \lambda$ curve of blackbody radiation at the thermodynamic temperature of 2.7 K. This confirms the prediction of the big bang theory, that is, due to the explosion about 13.7 billion years ago, there is the residual thermal radiation with a temperature of about 2.7 K in the universe today. Because the discovery of the cosmic microwave background radiation is of great significance in cosmology, Penzias and Wilson (Figure 2 - 9) won half of the 1978 Nobel Prize in Physics. The Swedish Academy of Sciences said in its award decision that this discovery allows us to obtain information on cosmic processes that took place during the creation of the universe a long time ago.

Figure 2 - 9 Wilson and Penzias, 1979 Nobel laureates in physics

Then, John C. Mather (1946—) from Goddard Flight Center, NASA, and George F. Smoot (1945—) from the University of California, Berkeley (Figure 2 - 10) made analysis and calculation according to the results measured by the Microwave background explorer (COBE) laun-ched in November 1989, and found that the temperature of cosmic microwave background radiation has extremely small differences in different directions. In other words, there is the so-called anisotropy, which won the 2006 Nobel Prize in Physics.

Figure 2 - 10 Mather and Smoot, 2006 Nobel laureates in physics

2.1.2 Planck's Quantum Hypothesis

At the end of the 19th century, in order to explain the phenomenon of blackbody radiation theoretically, many physicists tried to derive the relation of the monochromatic radiant exitance of blackbody radiation with thermodynamic temperature and wavelength by using the methods of classical thermodynamics and statistical mechanics. However, they all failed. Wherein, the most typical classical theoretical formulas for blackbody radiations are Wien formula and Rayleigh-Jeans formula, which are the following equations (2 - 4) and (2 - 5), respectively.

$$M_\lambda = \frac{C_1}{\lambda^5}\exp\left(-\frac{C_2}{\lambda T}\right) \tag{2-4}$$

$$M_\lambda = \frac{2\pi}{\lambda^4}kTc \tag{2-5}$$

Equation (2 - 4) is obtained by Wien in 1896 assuming that the energy distribution of blackbody radiation by wavelength was similar to the Maxwell distribution of ideal gas molecules by velocity at the same temperature, where C_1 and C_2 are constants, which are called the first radiation constant and the second radiation constant respectively. The result of Wien formula is similar to the experiment only in the short waveband, but it is very different from the experiment in the long waveband, as shown by the Wien line in Figure 2 - 11.

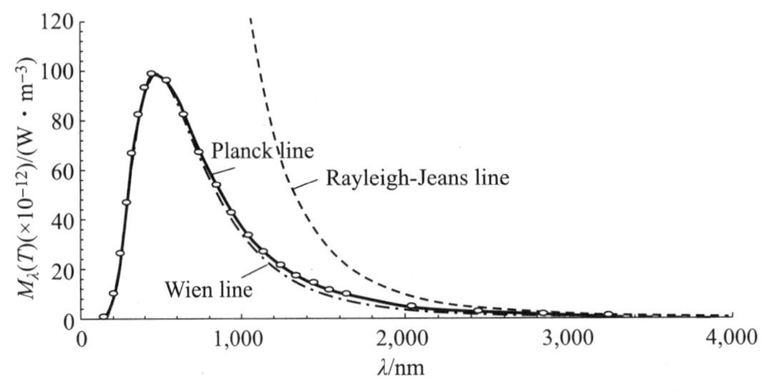

Figure 2 - 11 The results of two classical theoretical formulas of blackbody radiation are compared with the experimental curve at T = 6,000 K

In June 1900, the British physicist Lord Rayleigh (i. e. J. W. Strutt, 1842—1912, Figure 2 - 12) published his research results entitled "Remarks upon the law of complete radiation" in Philosophy Magazine. He applied the energy equipartition theorem according to degree of freedom in statistical physics to blackbody radiation. However, Rayleigh made a wrong factor 8 in his derivation. This error was corrected by the British James Hopwood Jeans (1877—1946) in a letter to the magazine Nature in 1905, that is, the original Rayleigh formula was divided by 8, and the present equation (2 - 5) is obtained, which is called the Rayleigh-Jeans formula.

Figure 2 - 12 Rayleigh, 1904 Nobel laureate in physics

Where k is the Boltzmann constant and c is the speed of light in vacuum. Compared with the experimental results, the Rayleigh-Jeans formula is applicable only in the long waveband, but it is obviously inconsistent with the experiment in the short waveband, as shown by Rayleigh-Jeans line in Figure 2 – 11. In particular, the Rayleigh-Jeans formula indicates that in the short wave region, including ultraviolet, even X – rays and γ – rays, there will be higher and higher monochromatic radiant exitance. This is obviously absurd. Hence, the failure was called "ultraviolet catastrophe" by Dutch physicist P. Ehrenfest (1880—1933).

Figure 2 – 13 Planck, 1918 Nobel laureate in physics

In 1894, the problem of blackbody radiation caused the attention of German physicist Max Planck (Figure 2 – 13). At the end of 1899, Planck learned that the German physicist Heinrich Rubens (1865—1922) et al pointed out the defects of Wien formula in the experimental report published that year. On October 7, 1900, Rubens visited Planck and told him the success and failure of Rayleigh formula. Inspired by this, Planck used mathematical interpolation to derive an empirical formula that approached the Wien formula in the short waveband and approached the Rayleigh-Jeans formula in the long waveband, that is

$$M_\lambda = \frac{2\pi hc^2}{\lambda^5}\frac{1}{e^{\frac{hc}{\lambda kT}} - 1} \tag{2-6}$$

This is Planck blackbody radiation law, also referred to as Planck's law or blackbody radiation law for short. Where $h = 6.626,070,040(81) \times 10^{-34}$ J · s $\approx 6.626 \times 10^{-34}$ J · s is a universal constant, later known as Planck constant. On October 19, 1900, Planck reported his results entitled "On an improvement of Wien's equation for the spectrum" at the meeting of the German Physical Society. When Rubens learned Planck's formula, he checked it with his experimental data and found that Planck's new formula was completely consistent with the experiment in the whole waveband, as shown by Planck line in Figure 2 – 11. Planck was greatly encouraged by the news and decided to look for the physical essence hidden in the formula.

In the next two months, Planck considered the atoms and molecules on the inner wall of the cavity as many charged simple harmonic oscillators, which can radiate and absorb energy and reach equilibrium with the radiation in the cavity. The electromagnetic waves of various frequencies radiated by the blackbody are radiated by these charged simple harmonic oscillators on the inner wall of the cavity. Planck boldly assumed that the energy E of a simple harmonic oscillator with frequency ν will no longer be continuous as specified by classical physics. It will be discontinuous and can take only a series of specific discrete values as follows

$$E = nh\nu \ (n = 1, 2, 3, \cdots) \tag{2-7}$$

That is to say, the energy of a simple harmonic oscillator is quantized. During emitting radiation or absorbing radiation, the energy can only change in a jumping manner by an integer multiple of $\varepsilon = h\nu$. As the smallest unit of energy, $h\nu$ is called quantum of energy. Where h was called the basic quantity by Planck, and is now called Planck constant. The above assumption is called the Planck quantum hypothesis. Because the value of h is very small, it is difficult to detect the discontinuity of energy macroscopically. Based on this assumption, Planck used classical statistical physics method to get the blackbody radiation formula, i. e. equation(2 - 6).

On December 14, 1900, Planck officially announced his hypothesis and the derivation from it entitled "On the theory of the energy distribution law of the normal spectrum" at the meeting of the German Physical Society, and fired the first shot of the quantum revolution. In 1919, the German physicist Sommerfeld first called December 14, 1900 "the birth of quantum theory" in his book "Atomic structure and spectral lines". Later historians of science set this day as the birthday of quantum physics. People call Planck the father of quantum. Planck won the 1918 Nobel Prize in Physics for the concept of energy quantum.

Planck quantum hypothesis misfits classical physics. At that time, the response to it was extremely cold in the physics community. People only recognized Planck's blackbody radiation formula which was consistent with the experiment, but did not accept his hypothesis of energy quantum. Even Planck himself felt it doubtful at that time. In order to be back to the theoretical system of classical physics, he always wanted to use the continuity of energy to solve the problem of blackbody radiation for a period of time, but he didn't succeed. In the literatures from 1900 to 1904, it was hard to find out papers about Planck quantum hypothesis, and Planck's work was almost forgotten. It was not until the advent of Albert Einstein's photon hypothesis and photoelectric effect equation in 1905 that people gradually accepted concept of quantum.

In 1948, at Planck's memorial meeting, Einstein highly praised Planck's work. He commented: "The discovery of the basic quantity became the basis of physics research in the 20th century. Since then it has almost completely determined the discovery of physics. Without this discovery, it would be impossible to establish the theory of molecules, atoms and the energy processes that govern their changes. Moreover, it also shattered the framework of classical mechanics and electrodynamics, and put forward a new task for science: to find a new conceptual basis for all physics."

Max Planck (1858—1947), a German theoretical physicist, has been a professor at Kiel University, Munich University and Berlin University, and has long been engaged in the research of thermodynamics. In 1900, he put forward the hypothesis of energy quantization, and deduced the energy distribution formula of blackbody radiation, which successfully explained the experimental phenomenon and opened the prelude to quantum theory. For this reason, he won the Nobel Prize in Physics in 1918. Although in later time, Planck has been trying to bring his theory into the framework of classical physics, he is still regarded as one of the pioneers of modern physics. Since the 1920s, Planck has become a central figure in the German scientific community, and has close ties with well-known physicists in the world at that time. He was elected as an academician of the Berlin Academy of Sciences in 1894 and permanent secretary from 1912 to 1938, a member of the Royal Society in 1918, a foreign academician of the Soviet Academy of Sciences in 1926. He served as president of Kaiser Wilhelm Society for the Advancement of Science from 1930 to 1935. In order to show respect for Planck, after 1945, the Society was renamed Max Planck Society for the Advancement of

Science. Planck's tomb is in the Gottingen city cemetery. Its mark is a simple rectangular stone tablet with his name engraved on it and a line of words in the bottom corner, $h = 6.63 \times 10^{-27}$ erg · s. The is an affirmation of his greatest contribution in his life – putting forward the quantum hypothesis.

2.2 Photoelectric Effect and Einstein's Photon Theory

2.2.1 Photoelectric Effect

In 1887, the German physicist H. Hertz(1857—1894) found in the experiment of confirming the existence of electromagnetic waves and Maxwell's electromagnetic theory of light that when ultraviolet is incident on the metal, it can make the metal emit charged particles. In 1900, the German physicist P. Lennard(1862—1947), Hertz's assistant, confirmed by measuring the charge-mass ratio of these charged particles that ultraviolet light causes metal to release electrons. Such escape phenomenon of electrons on the metal surface under the irradiation of light is called the photoelectric effect, and the escaped electrons are called photoelectrons.

Figure 2 – 14 shows a schematic diagram of the experimental device of photoelectric effect. Apply variable DC voltage U between anode A and cathode metal C of vacuum photocell, and its value can be given by voltmeter V. When light is incident on the surface of cathode metal C through the quartz window of the photocell, photoelectrons escape from the surface of metal C and accelerate to anode A under the action of the electric field applied between electrodes A and C, forming a current, which is called photocurrent i. This photocurrent i can be measured by galvanometer G.

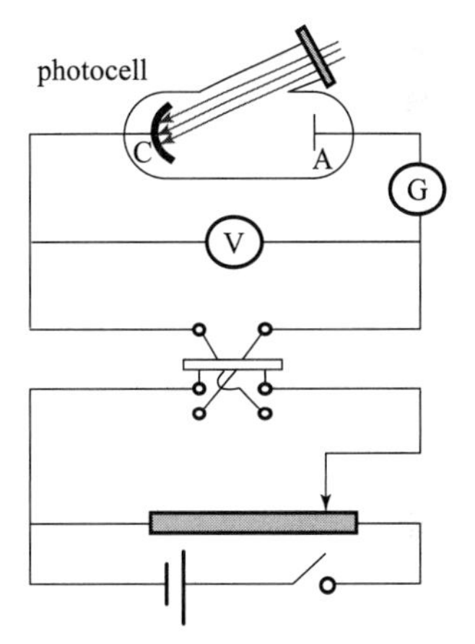

Figure 2 – 14 Experimental device for photoelectric effect

The photoelectric effect has the following experimental laws.

(1) Saturation photocurrent i_m.

Figure 2 – 15 shows the experimental curve of photocurrent i changing with the acceleration voltage U under the irradiation of incident light with three different intensities(I_1, I_2, I_3, and $I_1 < I_2 < I_3$) at a certain frequency. Experiments show that when the intensity I of the incident light is constant, the photocurrent i starts to increase with the increase of the acceleration voltage U, subsequently, when the acceleration voltage increases to a certain value, the photocurrent tends to a saturation value i_m, which is called saturation photocurrent. At this time, photoelectrons escaping from the cathode per unit time all shall reach the anode. As can be found in experiment, when the frequency of incident light is constant, the saturated photocurrent i_m is directly proportional to the intensity I of incident light. In other words, the number of electrons released by a metal plate exposed to light per unit time is directly proportional to the intensity of incident light.

(2) Cut off voltage U_a.

The experimental curves in Figure 2 – 15 also show that when the accelerating voltage U is reduced to zero and becomes negative on the reverse, the photocurrent does not immediately drop to

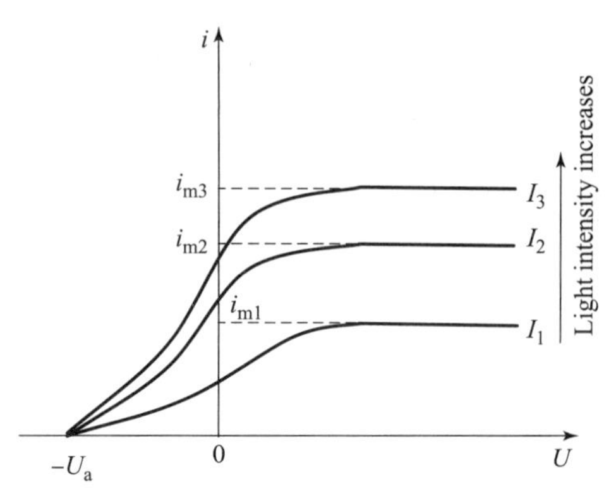

Figure 2 – 15 Experimental curves of photocurrent *i* with voltage *U* when the frequency of incident light is constant

zero. This shows that the photoelectrons escaping from the cathode surface have a certain initial kinetic energy. Although the voltage is zero, which cannot accelerate the photoelectrons, or the reverse voltage is applied to prevent the photoelectrons from moving to anode A, there are still some photoelectrons which can reach the anode A. When the reverse voltage applied between A and C is equal to U_a, the photocurrent can be zero. Here, the absolute value of the applied potential difference U_a is called cut-off voltage. The cut-off voltage U_a makes the photoelectron escaping from cathode C with maximum initial velocity v_m use its initial kinetic energy $m_e v_m^2/2$ all to overcome the obstruction by the external electric field force. The required energy is eU_a, so that the photoelectron just cannot reach anode A. Therefore, the relation between the cut-off voltage U_a and the maximum initial velocity v_m of photoelectrons escaping from the metal surface is as follows:

$$\frac{1}{2}m_e v_m^2 = eU_a \tag{2-8}$$

where m_e is the mass of the electron, e is the absolute value of an electron charge. It can be seen from equation(2 – 8) that the maximum initial kinetic energy of photoelectrons $m_e v_m^2/2$ is directly proportional to the cut-off voltage U_a. As shown in Figure 2 – 15, under the irradiation of incident light with the same frequency and different intensity, the $i - U$ relation curves intersect at one point, where $U = -U_a$. This shows that the cut-off voltage U_a and the maximum initial kinetic energy of the photoelectrons $m_e v_m^2/2$ both have nothing to do with the intensity I of the incident light ν_0.

(3) Cut off frequency ν_0.

Figure 2 – 16 shows the experimental curves of the cut-off voltage U_a of three different materials (Cs, Na, Ca) with the frequency ν of incident light. Experiments show that when the frequency ν of incident light increases, the cut-off voltage U_a will increase linearly, that is

$$U_a = K\nu - U_0 \tag{2-9}$$

where K is a universal constant independent of the properties of cathode metal materials, and it is also the slope of the straight line in Figure 2 – 16. U_0 is a quantity related to metal materials. Substituting equation(2 – 9) into equation(2 – 8), the relation between the maximum initial kinetic

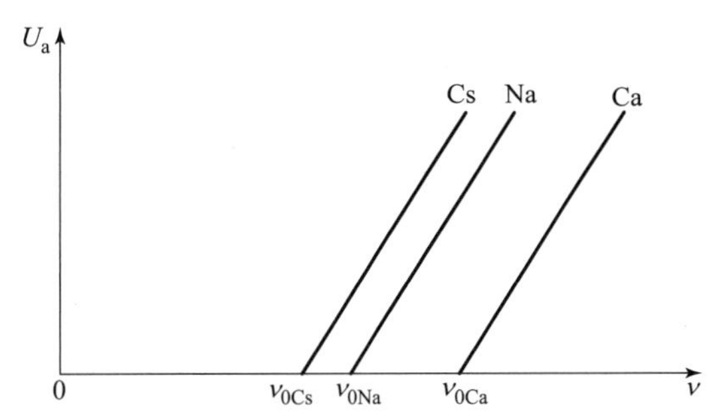

Figure 2 – 16 The relation between cut-off voltage and incident light frequency

energy $mv_m^2/2$ and the frequency ν of incident light can be obtained as

$$\frac{1}{2}m_e v_m^2 = eK\nu - eU_0 \tag{2-10}$$

That is, the maximum initial kinetic energy of photoelectron $m_e v_m^2/2$ increases linearly with the increase of incident light frequency ν.

It can be known from equation(2 – 10)that when the incident light frequency ν decreases to ν_0, that is

$$\nu_0 = \frac{U_0}{K} \tag{2-11}$$

The cut-off voltage and the maximum initial kinetic energy of the photoelectrons will be reduced to zero. Thus electrons cannot escape from the metal surface and there will not happen photoelectric effect. The frequency ν_0 is called the cutoff frequency of photoelectric effect. It can also be figured out by the straight line of the $U_a - \nu$ relation shown in Figure 2 – 16. The frequency corresponding to the intersection of the straight line and the abscissa is the cutoff frequency.

It can be seen from Figure 2 – 16 that different metals have different cut-off frequencies ν_0. In order to let a metal show photoelectric effect, the frequency of incident light must be greater than its corresponding cut-off frequency ν_0. This means that for incident light with a frequency below the cutoff frequency ν_0, photoelectric effect cannot occur even if the light intensity is very strong and the irradiation lasts a long time. Therefore the cut-off frequency ν_0 is also known as red limit frequency. The corresponding wavelength λ_m is called the red limit wavelength. Photoelectric effect can occur only when the wavelength of incident light is less than the red limit wavelength.

(4) Relaxation time.

Experiments also found that the photoelectric effect is instantaneous. No matter how weak the light intensity is, as long as the frequency of the incident light is greater than the cut-off frequency ν_0, photoelectrons almost immediately escape from metal surface, and the lag time is no more than 1 ns.

Some experimental results of photoelectric effect cannot be explained by classical wave theory. Firstly, according to the classical wave theory, the greater the intensity of the incident light is, the greater the energy of the light is, and the greater the maximum initial kinetic energy of the photoelectrons escaping from the metal will be. But in fact, the maximum initial kinetic energy of photoelectrons increases linearly with the frequency of incident light, regardless of the intensity of the incident light. Secondly, according to the classical wave theory, the electrons in the metal can continuously absorb the energy of light. As long as the irradiation lasts long enough, no matter how small its frequency is, it can always obtain enough energy to escape, and there should be no cut-off frequency. But in fact, when $\nu < \nu_0$, no matter how strong the intensity of incident light was, photoelectric effect could not occur. Thirdly, according to the classical wave theory, the energy of light wave is evenly distributed on the wave surface, and an electron in the metal can only absorb a small part of it. It should take a period of time to accumulate enough energy, so that it can break free from the metal surface. Thus it should be impossible that photoelectric effect can occur instantaneously. But in fact, no matter how weak the intensity of the incident light is, photoelectrons will escape from the metal surface immediately as long as $\nu > \nu_0$. Obviously, there is a sharp contradiction between the classical wave theory of light and the photoelectric effect experiment.

2.2.2 Einstein's Photon Theory

In order to explain the phenomenon of photoelectric effect, in March 1905, Einstein, who was only 26 years old, published a paper entitled "On a heuristic viewpoint concerning the production and transformation of light" in Volume 17 of the German "Annals of Physics". In this paper, Einstein further proposed the light quantum hypothesis on the basis of Planck energy quantum hypothesis.

According to Planck energy quantum hypothesis, when electromagnetic wave is absorbed and emitted, it is not continuous, but one by one. Inspired by this, Einstein proposed that when light (electromagnetic wave) is propagating in space, it is also discontinuous and had particle property. Namely, a light beam can be regarded as a particle stream composed of microscopic particles, each of them is called light quantum. In a vacuum, each light quantum moves at the speed of light $c = 3 \times 10^8 \text{ m} \cdot \text{s}^{-1}$. The energy of each light quantum has the same expression as Planck energy quantum. That is, for a light beam with frequency ν, the energy ε of each light quantum is directly proportional to the frequency ν of the light.

$$\varepsilon = h\nu \tag{2-12}$$

where h is the Planck constant. In addition, light quantum has integrity property, that is, each light quantum does not disintegrate in motion, and it can only be absorbed or emitted as a whole. The above content is Einstein's light quantum hypothesis.

In 1926, American physicist G. N. Lewis (1875—1946) changed the name of "light quantum" to "photon", which is still used today.

According to Einstein's photon hypothesis, when a light wave is spreading outward from a point, its energy is not continuously distributed in an increasing volume as believed by the classical theory, but it is composed of a finite number of energy quanta localized in space.

When the light with frequency of ν is incident on the metal, after an electron in the metal absorbs an incident photon, it will get energy $h\nu$. A part of the energy is used for the energy required for the electron to escape from the metal surface, and the rest part is the initial kinetic energy of photoelectron. If the minimum energy required for an electron to escape from the metal surface is A, which is called metal electron work function, then the maximum initial kinetic energy obtained by photoelectron can be known from the conservation of energy as

$$\frac{1}{2}m_e\nu_m^2 = h\nu - A \tag{2-13}$$

This is the Einstein photoelectric effect equation.

The intensity of light I is defined as the energy incident on a unit area in the direction perpendicular to light propagation per unit time. Therefore, the intensity of light with a frequency of ν can be expressed as

$$I = Nh\nu \tag{2-14}$$

where N is the number of photons incident on a unit area in the direction perpendicular to light propagation per unit time.

The photoelectric effect can be satisfactorily explained by using Einstein's photon hypothesis and photoelectric effect equation.

The energy of a photon only depends on the frequency of the light. For example, no matter how weak the violet light is, the energy of a violet photon is always greater than that of a red photon. The different intensity of light with the same color reflects the different number of photons incident on a unit area of metal per unit time. That is, when the frequency ν is the same, according to equation (2-14), the greater the light intensity is, the greater the number of photons incident on a unit area of metal per unit time is, the greater the number of photoelectrons which absorb photons and escape from the metal surface per unit area per unit time will be, and the greater the saturated photocurrent will be. Therefore, it is natural to explain the experimental results that the saturated photocurrent is directly proportional to the intensity of the incident light when the frequency of the incident light is constant.

According to Einstein's photoelectric effect equation (2-13), the maximum initial kinetic energy of photoelectrons has linear relation with the frequency of incident light, but it has nothing to do with the light intensity. Then according to equation (2-8), the linear relation between the cut-off voltage and the frequency of the incident light can be explained.

The existence of the cutoff frequency in the photoelectric effect can also be explained according to Einstein's photoelectric effect equation. The photon energy absorbed by electron should reach a certain value so as to overcome the work function and the electron can escape from metal surface. If the photon energy $h\nu_0$ absorbed by the electron is equal to the work function A, the initial kinetic energy of the electron $m_e v_m^2/2 = 0$, and the electron can just escape from the metal surface, the value of ν_0 is

$$\nu_0 = \frac{A}{h} \tag{2-15}$$

Obviously, if the frequency of the photon is less than ν_0, the photon energy absorbed by electron will be less than the work function A, then no matter how strong the light intensity is and how long the incident time is, any electron cannot escape from the metal surface to become photoelectron, that is, there will not happen. ν_0 is the cut-off frequency. Because the work functions of various metals are different, it can be seen from equation (2 – 15) that the cut-off frequency ν_0 will also be different. Table 2 – 1 shows the work function A and cut-off frequency ν_0 of several metals.

Table 2 – 1 Electronic work function A and cut-off frequency ν_0 of several metals

Metal	Tungsten	Zinc	Calcium	Sodium	Potassium	Rubidium	Cesium
cut-off frequency $\nu_0/10^{14}$ Hz	10. 95	8. 06	7. 73	5. 53	5. 44	5. 15	4. 69
work function A/eV	4. 54	3. 34	3. 20	2. 29	2. 25	2. 13	1. 94

When the light with a frequency higher than the cut-off frequency is incident on a metal, the energy of a photon can be immediately absorbed as a whole by a free electron in the metal, and the time for energy accumulation is almost not required. Therefore, the emission of photoelectrons from the metal surface occurs almost at the same time as the light is incident.

Figure 2 – 17 Millikan, 1923 Nobel laureate in physics

In 1915, in order to verify Einstein's photoelectric effect equation, American physicist R. Millikan (1868—1953, Figure 2 – 17) measured the photoelectric effect with high precision. Comparing equation (2 – 10) and equation (2 – 13), we can get

$$A = eU_0 \tag{2-16}$$

$$h = eK \tag{2-17}$$

where U_0 is also known as escape potential of metal electron. In fact, Millikan originally wanted to prove that Einstein's theory was wrong, but he unexpectedly found he had proved by experiment that every detail of Einstein's equation is valid. He also successfully obtained the Planck constant $h = 6.56 \times 10^{-34}$ J · s by using the slope K of the $U_c - \nu$ straight line shown in Figure 2 – 16 and equation (2 – 17), and the result coincides very well with the value of h measured by the thermal radiation method at that time. The experiment was also a good proof for the correctness of Einstein's photon theory at that time. Thus Einstein's work on photoelectric effect really began to cause people's attention. In 1923, Millikan was awarded the Nobel Prize in Physics for his work on the elementary charge of electricity and on the photoelectric effect.

The photoelectric effect has played an important role in the understanding of the essence of light and the development of quantum theory. Einstein received the 1921 Nobel Prize in Physics for his successful explanation on photoelectric effect.

Later, with further research of photoelectric phenomena, the connotation of the concept of photoelectric effect has changed greatly. Physicists call collectively the photo-change-electricity phenomenon such that light incident on certain substances causes the change of electrical properties of the substances as photoelectric effect. The phenomenon studied around the beginning of the 20th century that incident light led to photoelectrons escaping from the metal surface, is called the external photoelectric effect, to be distinguished from the internal photoelectric effect. The internal photoelectric effect of semiconductor materials is more obvious. When light is incident on some semiconductors, it will be absorbed. Although semiconductors sometimes do not emit photoelectrons, but the conductive carriers are excited inside the semiconductors, so that the conductivity of materials has a significant increase. This is the so-called photoconductive effect. Or due to the charge accumulation caused by the movement of such carriers, there is a certain potential difference between the two sides of the material. That is the so-called photovoltaic effect. These phenomena are collectively referred to as internal photoelectric effect. Optoelectronic devices based on the principle of external photoelectric effect include photocell and photomultiplier, while the internal photoelectric effect is more widely used at present. For example, solar cells and various photodetectors based on photosensitive elements are all devices developed based on the internal photoelectric effect.

Photoelectric devices using photoelectric effect have been widely used in production, scientific research and national defense, and are still opening up new application fields. For example, in audio movies and TV and radio fax technology, photocells or photoelectric cells are used to convert optical signals into electrical signals. Figure 2 – 18 shows a device that uses photoelectric conversion to realize sound playback when showing a movie. The dubbing of the film is to convert the sound signal into optical signal, and record it on the sound track at the edge of the film with light and dark stripes. When showing a movie, the light from a source changes in intensity after passing through the moving sound track, and it is received by the photocell. The photocell converts the intensity-varying light into a correspondingly varying current, which is then amplified by the amplifier. After that, the sound is emitted from the speaker. In photometric measurement and radioactivity measurement, photocells or photoelectric cells are also commonly used to convert light into current and make measurement after amplification. Optical counter, photoelectric tracking, photoelectric protection and other devices are more widely used in production automation. Popular anti-theft automatic alarm bell, smoke detector and elevator door anti-pinch sensor usually use a beam of infrared light and a photocell as the switch. When the beam is blocked, the current passing through the photocell will be

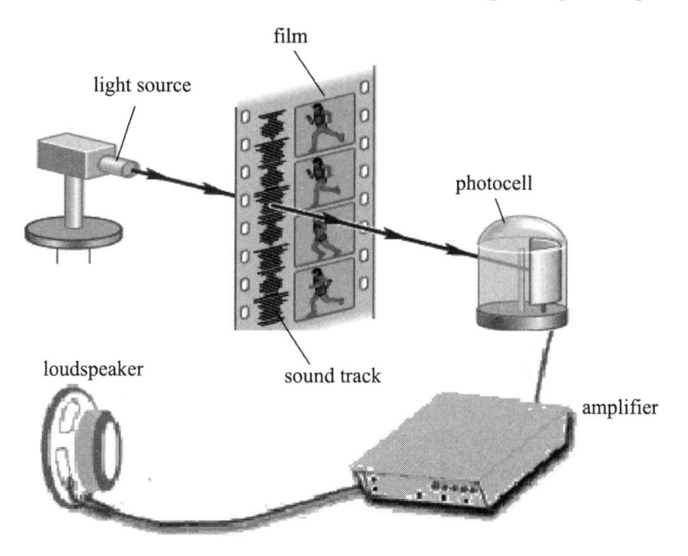

Figure 2 – 18 Sound production system of a movie

interrupted or weakened, and the alarm will be triggered. Or the elevator door will not close, even if it is closing, it will reopen. However, note that the anti-pinch function of the elevator is not always on. Once the elevator door is almost closed, this function will fail. Therefore, don't force entry or exit when the elevator is about to close, as shown in Figure 2 – 19, so as to avoid injury.

Figure 2 – 19 This is a dangerous action. Don't force entry or exit when the elevator is about to close

Example 2 – 1: The monochromatic light with a wavelength of 200 nm is incident on the surface of metal aluminum. The work function of aluminum is known to be 4.2 eV. Find (1) the maximum kinetic energy of photoelectrons, (2) the cut-off voltage, (3) the cut-off wavelength of aluminum.

Solution: (1) According to Einstein's photoelectric effect equation, the maximum kinetic energy of photoelectrons is

$$E_{\mathrm{km}} = h\nu - A = h\frac{c}{\lambda} - A = \left(\frac{6.63 \times 10^{-34} \times 3 \times 10^{8}}{200 \times 10^{-9} \times 1.6 \times 10^{-19}} - 4.2\right)\ \mathrm{eV} = 2.0\ \mathrm{eV}$$

(2) From equation (2 – 8), the cut-off voltage can be obtained as

$$U_{\mathrm{a}} = \frac{E_{\mathrm{km}}}{e} = \frac{2.0}{1}\ \mathrm{V} = 2.0\ \mathrm{V}$$

(3) According to equation (2 – 15), the cut-off wavelength is

$$\lambda_{\mathrm{m}} = \frac{\mathrm{c}}{\nu_0} = \frac{\mathrm{hc}}{\mathrm{A}} = \frac{6.63 \times 10^{-34} \times 3 \times 10^{8}}{4.2 \times 1.6 \times 10^{-19}}\ \mathrm{m} = 2.96 \times 10^{-7}\ \mathrm{m} = 296\ \mathrm{nm}$$

Example 2 – 2: Figure 2 – 20 shows a common axis system in vacuum. The outside is a quartz cylindrical shell with a length of 20 cm, and its inner surface is coated with translucent aluminum film which radius is $r_2 = 1$ cm. In the middle there is a cylindrical sodium rod with a radius of $r_1 = 0.6$ cm and a length of 20 cm. It is known that the red limit wavelength of sodium is $\lambda_m = 540$ nm and that of aluminum is $\lambda'_m = 296$ nm. If the monochromatic ultraviolet light with a wavelength of $\lambda = 300$ nm is used to irradiate the system from the periphery of the cylinder perpendicular to the axis, and the edge effect is ignored. Find (1) the electric potential difference between the sodium rod and the aluminum film in equilibrium, (2) the amount of charge carried by the sodium rod in equilibrium.

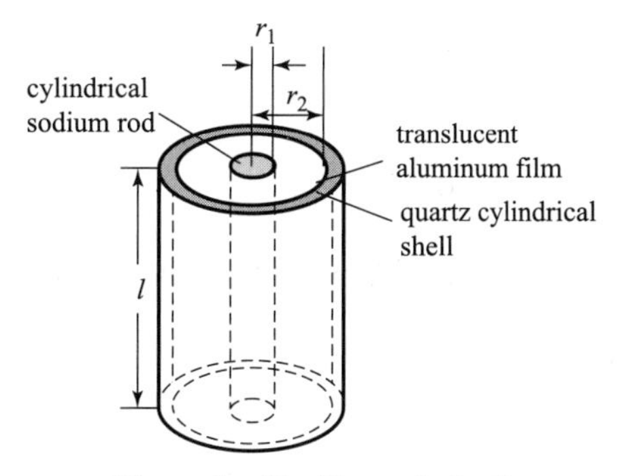

Figure 2 – 20 Example 2 – 2

Solution: (1) Due to $\lambda'_m < \lambda < \lambda_m$, under ultraviolet light irradiation, aluminum does not show photoelectric effect, while sodium shows photoelectric effect. According to Einstein's photoelectric effect equation, the maximum initial kinetic energy of photoelectrons escaping from the sodium rod is

$$\frac{1}{2}m_e v_m^2 = \frac{hc}{\lambda} - \frac{hc}{\lambda_m} \quad ①$$

These photoelectrons gather on the aluminum film, then the surface of the sodium rod and the aluminum film are charged with positive and negative charges, respectively. When the electric potential difference U between them satisfies the relation

$$eU = m_e v_m^2/2 \quad ②$$

That is, the system is in equilibrium when U is

$$U = \frac{hc}{e}\left(\frac{1}{\lambda} - \frac{1}{\lambda_m}\right) = \frac{6.63\times10^{-34}\times3\times10^8}{1.6\times10^{-19}}\left(\frac{1}{300\times10^{-9}} - \frac{1}{540\times10^{-9}}\right)\text{ V} = 1.84\text{ V}$$

(2) If the edge effect is ignored, the relation of the electric field at r between the sodium rod and the aluminum film with the amount of charge Q on the surface of the sodium rod can be obtained from the Gauss theorem as follows,

$$E = \frac{Q}{2\pi\varepsilon_0 lr}$$

Then according to the relation between electric potential difference U and electric field $\boldsymbol{E}$, we have

$$U = \int_{r_1}^{r_2}\boldsymbol{E}\cdot d\boldsymbol{l} = \int_{r_1}^{r_2}\frac{Q}{2\pi\varepsilon_0 lr}dr = \frac{Q}{2\pi\varepsilon_0 l}\ln\frac{r_2}{r_1} \quad ③$$

It can be obtained from equations ①, ② and ③ that

$$Q = \frac{2\pi\varepsilon_0 hcl}{e\ln\left(\frac{r_2}{r_1}\right)}\left(\frac{1}{\lambda} - \frac{1}{\lambda_m}\right)$$

$$= \frac{2\pi \times 8.85 \times 10^{-12} \times 6.63 \times 10^{-34} \times 3 \times 10^{8} \times 0.2}{1.6 \times 10^{-19} \times \ln\left(\frac{1}{0.6}\right)}\left(\frac{1}{300 \times 10^{-9}} - \frac{1}{540 \times 10^{-9}}\right) \text{ C}$$

$$= 4.0 \times 10^{-11} \text{ C}$$

2.2.3 Wave-Particle Duality of Light

In 1917, Einstein further assumed that photon not only has energy but also has momentum. By combining equation (2 – 12) of the energy expression of photon with the relativistic mass-energy relation, $\varepsilon = mc^2$ it can be seen that the relativistic mass of photon is

$$m = \frac{h\nu}{c^2} = \frac{h}{c\lambda} \tag{2-18}$$

Then the relativistic momentum of photon is

$$p = mc = \frac{h\nu}{c} = \frac{h}{\lambda} \tag{2-19}$$

As a result, photon has properties such as energy, mass, momentum and etc. that particles commonly have

The physical quantities that describe the particle nature of light are energy ε, mass m and momentum p. And the physical quantities describing the wave nature of light are frequency ν and wavelength λ . As can be seen from equations (2 – 12), (2 – 18) and (2 – 19), the physical quantities describing the wave nature and particle nature of light are linked by Planck constant.

Photoelectric effect and Einstein's light quantum theory established the particle property of light. People began to realize that light has both wave property and particle property, that is, wave-particle duality of light. In some cases, light stands out for its wave property. In other cases, it highlights its particle property. Generally speaking, the higher the frequency is or the shorter the wavelength is, and the higher the energy the photon has, the more significant its particle property shows. The longer the wavelength is, the lower the energy the photon has, the more significant its wave property shows. It is worth mentioning that under the same conditions, photon either shows particle property or wave property, but generally both cannot be shown at the same time.

Example 2 – 3: If the wavelength of red light is $\lambda_1 = 600$ nm, the wavelength of X – rays is $\lambda_2 = 0.1$ nm. Find the energy and momentum of a photon in them, respectively.

Solution: According to equation (2 – 12), the energy of a red light photon is

$$\varepsilon_1 = h\nu_1 = \frac{hc}{\lambda_1} = \frac{6.63 \times 10^{-34} \times 3 \times 10^{8}}{600 \times 10^{-9}} \text{ J} = 3.31 \times 10^{-19} \text{ J}$$

This is also the minimum energy that red light with a wavelength of 600 nm can absorb and emit in any process. Similarly, the energy of an X – ray photon can be obtained as

$$\varepsilon_2 = h\nu_2 = \frac{hc}{\lambda_2} = \frac{6.63 \times 10^{-34} \times 3 \times 10^{8}}{0.1 \times 10^{-9}} \text{ J} = 1.99 \times 10^{-15} \text{ J}$$

It can be seen that the energy of X – ray photon is greater than that of visible light photon by order of magnitude of 10^4. Therefore, X – rays is a highly-transmissive electromagnetic wave. It can penetrate an object of a certain thickness and display the internal situation of the object on screen, which can be used for medical diagnosis.

Because the greater the energy of X - ray photon is, the greater the damage to the human body will be, so we should be exposed to X - rays as little as possible. As lead can absorb X - rays well, when carrying out medical X - ray diagnosis, as shown in Figure 2 - 21 (a), the human body can protect non radiation sensitive organs by wearing lead apron, lead scarf, lead cap or lead glasses, thereby reducing the effect of X - rays on gonads, thyroid gland, brain or ocular lens. Similarly, the curtain in front of the security detector is a lead curtain, which is also a protective equipment to prevent X - ray leakage. Never lift the curtain of the security detector in a hurry to get your luggage back as shown in Figure 2 - 21 (b).

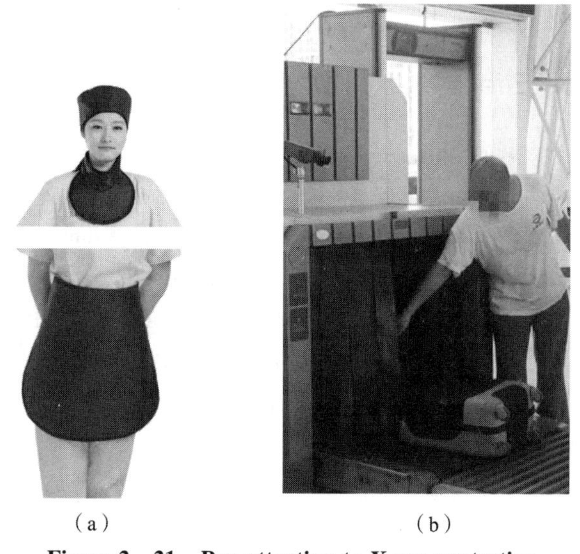

(a) (b)

Figure 2 - 21 Pay attention to X-ray protection

(a) Lead apron, lead scarf and lead cap should be worn during chest X-ray; (b) This "door" curtain cannot be lifted

According to equation (2 - 19), the momentum of the red photon is

$$p_1 = \frac{h}{\lambda_1} = \frac{6.63 \times 10^{-34}}{600 \times 10^{-9}} \text{ kg} \cdot \text{m} \cdot \text{s}^{-1} = 1.1 \times 10^{-27} \text{ kg} \cdot \text{m} \cdot \text{s}^{-1}$$

The momentum of X - ray photon is

$$p_2 = \frac{h}{\lambda_2} = \frac{6.63 \times 10^{-34}}{0.1 \times 10^{-9}} \text{ kg} \cdot \text{m} \cdot \text{s}^{-1} = 6.63 \times 10^{-24} \text{ kg} \cdot \text{m} \cdot \text{s}^{-1}$$

It can be seen that the momentum of X - ray photon is also greater than that of visible light photon by order of magnitude of 10^4.

Figure 2 - 22 Einstein, 1921 Nobel laureate in physics

Albert Einstein (1879—1955, Figure 2 - 22), a world-famous German-American scientist, winner of the 1921 Nobel prize in physics, founder of modern physics. On December 26, 1999, Einstein was selected as "The Great Man of the Century" by the American "Time" magazine. Einstein graduated from the Zurich university of technology in 1900 and became a Swiss citizen. In 1905, he received a doctorate in philosophy from the University of Zurich. He once worked in the Berne patent office, became a professor at Zurich University of technology and Deutsche University in Prague. He returned to Germany in 1913, served as director of the Kaiser Wilhelm Institute of Physics and a professor at University of Berlin, and was elected as an academician of the Prussian Academy of Sciences. In 1933, persecuted by the Nazi regime, he moved to the United States and became a professor at the Institute for Advanced Study in Princeton, engaged in theoretical physics research, and became an American citizen in 1940.

2.3 Compton Effect

2.3.1 Compton Effect

In 1923, American physicist A. H. Compton, (1892—1962, Figure 2 – 23) found that when observing the rays scattered by substances such as graphite, there are rays with longer wavelength than the incident wavelength in addition to the rays with the same wavelength as the incident wavelength. This scattering phenomenon with changed wavelength is called Compton effect. Compton published this discovery and its theoretical explanation in the American "Physics Review" entitled "A quantum theory of the scattering of X – rays by light elements".

Figure 2 – 24 is the schematic diagram of Compton experimental device. A beam of X – rays with wavelength of λ_0 emitted by an X – ray source is incident on a piece of graphite. After scattering by the graphite, the wavelength λ and intensity of the scattered light passing through the collimating system in the direction of scattering angle φ can be measured by a spectrograph composed of crystal and detector.

Figure 2 – 23 Compton, 1927 Nobel laureate in physics, is measuring the scattering of X-rays by crystal

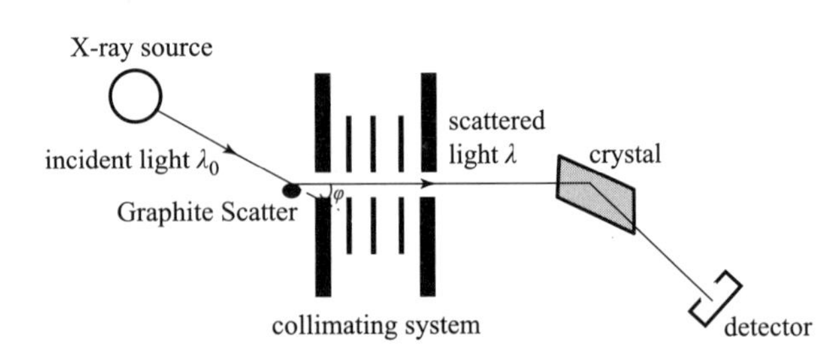

Figure 2 – 24 Schematic diagram of Compton experimental device

Compton used X – rays from molybdenum with wavelength of $\lambda_0 = 0.071$ nm as incident light, and measured the distribution of X – ray wavelength with intensity in various scattering angle directions. The experimental results are shown in Figure 2 – 25. Where the abscissa represents the wavelength of scattered light, and the ordinate represents the intensity of scattered light. The experimental results show that when $\varphi = 0°$, that is, measuring in the direction facing the incident light, only a single wavelength of light is found to be scattered, and this wavelength is the same as the wavelength of the incident light. When $\varphi \neq 0°$, such as $\varphi = 45°$, $90°$ and $135°$, the intensity distribution curve of scattered light has two peaks. This indicates that there are two wavelengths of scattered light. That is, in addition to the spectra with the same wavelength λ_0 as the incident light, there are also spectra with wavelengths $\lambda > \lambda_0$ in the scattered light. Moreover, with the increase of scattering angle φ, the wavelength shift $\Delta\lambda = \lambda - \lambda_0$ also increases accordingly. The experiment indicates that the relation between the wavelength shift $\Delta\lambda = \lambda - \lambda_0$ and the scattering angle φ is

$$\Delta\lambda = \lambda_C(1 - \cos\varphi) \tag{2-20}$$

The above formula is called Compton scattering formula. Where λ_C is a universal constant, which is called the Compton wavelength of the electron, and its magnitude is

$$\lambda_C = \frac{h}{m_e c} = 2.426,310,236\,7(11) \times 10^{-3} \approx 2.43 \times 10^{-3}\ \text{nm} \tag{2-21}$$

where h is Planck constant, m_e is the rest mass of the electron, and c is the speed of light in vacuum.

It can also be seen from Figure 2 – 25 that the intensity of the scattering spectral line with the original wavelength λ_0 decreases with the increase of the scattering angle φ, while the intensity of the scattering spectral line with the wavelength λ increases with the increase of the scattering angle φ.

From 1925 to 1926, the Chinese physicist Y. H. Woo (1897—1977), who was working with Compton at that time, carried out a large number of experiments by using X – rays of silver with a wavelength of $\lambda_0 = 0.056,2$ nm as the incident rays and 15 elements of different weight as scattering materials. The scattered light intensity of various wavelengths was measured in the direction of the scattering angle $\varphi = 120°$, as shown in Figure 2 – 26. Woo Y. H. pointed out that Compton scattering is stronger for substances with small atomic weight, Compton scattering is weak for substances with large atomic weight. And at the same scattering angle, the change of wavelength is the same for all scattering substances.

Classical wave theory cannot explain the above-mentioned wavelength-shifted Compton effect. According to the classical wave theory, under the irradiation of X – rays, the charged particles in matter will absorb energy from the incident light and do forced vibration with the same frequency. The frequency of the radiated electromagnetic wave should also have the same frequency as the incident light. Therefore, no wavelength change occurs. Obviously, there is a sharp contradiction between the classical wave theory of light and Compton effect.

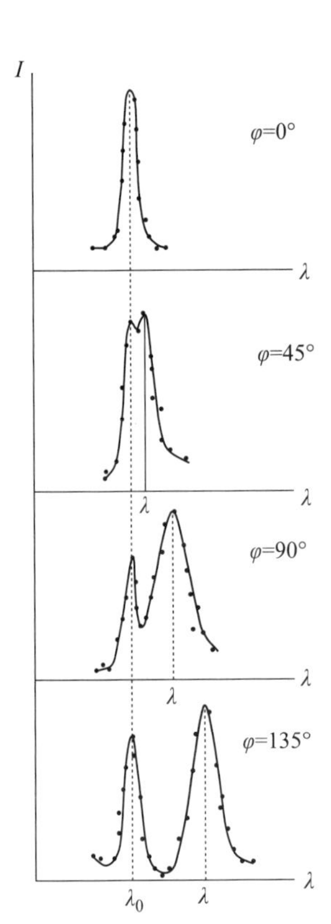

Figure 2 - 25 Experimental results of Compton effect

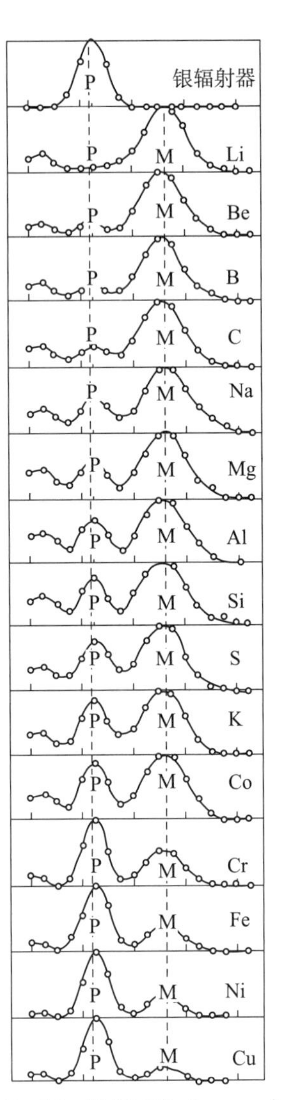

Figure 2 - 26 Y. H. Woo's experimental results of 15 element scatterers

2.3.2 Explanation of Compton Effect by Light Quantum Theory

In 1923, with the help of Einstein's light quantum theory, Compton regarded the above-mentioned scattering effect as the result of elastic collision between photons in X - rays and stationary free electrons, and assumed that the energy and momentum are conserved respectively during the collision. The equation (2 - 20) of the wavelength shift is obtained, and the Compton effect is correctly explained.

As particle, photon has both energy and momentum. The energy of a photon in X - rays is about $10^4 - 10^5$ eV. The outer electrons in the atoms of the scattered material are bound weakly, and the energy required to make an electron escape from the surface of the scatterer is only a few eV, which is much less than the energy of a photon in X - rays. So the binding energy of these electrons can be ignored, and they can be considered approximately as free electrons. And since the average kinetic energy of thermal motion of these electrons is about 10^{-2} eV, which is also much smaller than the energy of photon in X-rays, the kinetic energy of these electrons can be ignored, and approximately they are considered to be in static.

When a photon hits a scatterer and collides with an outer electron in an atom, the electron will absorb some energy, and recoil away from the atom, which is called a recoil electron. Therefore, the energy of the scattered photon is less than that of the incident photon, then the frequency of scattered light becomes smaller and the wavelength becomes longer. As shown in Figure 2-27, if the frequency of the incident photon before collision is ν_0, then its energy is $h\nu_0$. According to equation (2-19), the momentum of photon is $\frac{h\nu_0}{c}\boldsymbol{e}_0$, where $\boldsymbol{e}_0$ is the unit vector in the direction of incident light. The free electron at rest has energy of $m_e c^2$ and zero momentum. After collision, the energy of the scattered photon with scattering angle of φ is $h\nu$, its momentum is $\frac{h\nu}{c}\boldsymbol{e}$, where $\boldsymbol{e}$ is the unit vector in the direction of the scattered light. The mass of the electron with a recoil velocity $\boldsymbol{v}$ is

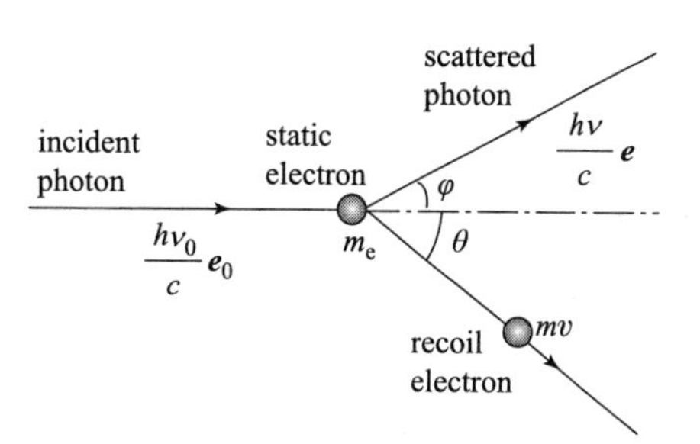

Figure 2-27 Collision of a photon with a static free electron

$$m = \frac{m_e}{\sqrt{1-\left(\frac{v}{c}\right)^2}} \tag{2-22}$$

Its energy is mc^2 and momentum is $m\boldsymbol{v}$. By the law of conservation of energy and the law of conservation of momentum, we can list the equations as follows

$$h\nu_0 + m_e c^2 = h\nu + mc^2 \tag{2-23}$$

$$\frac{h\nu_0}{c}\boldsymbol{e}_0 = \frac{h\nu}{c}\boldsymbol{e} + m\boldsymbol{v} \tag{2-24}$$

Equation (2-24) is a vector formula and can be written into two component formulas

$$\frac{h\nu_0}{c} = \frac{h\nu}{c}\cos\varphi + mv\cos\theta \tag{2-25}$$

$$\frac{h\nu}{c}\sin\varphi = mv\sin\theta \tag{2-26}$$

Eliminate θ from equations (2-25) and (2-26), we get

$$m^2v^2c^2 = h^2(\nu_0^2 + \nu^2 - 2\nu_0\nu\cos\varphi) \tag{2-27}$$

Move the term $h\nu$ in equation (2-23) to the left side of the equation and then square on both sides, we have

$$m^2c^4 = h^2(\nu_0^2 + \nu^2 - 2\nu_0\nu) + m_e^2c^4 + 2hm_ec^2(\nu_0 - \nu) \tag{2-28}$$

Subtract equation (2-27) from equation (2-28) and combine with equation (2-22), we get

$$2h^2\nu_0\nu(1 - \cos\varphi) = 2hm_ec^2(\nu_0 - \nu)$$

Therefore, the relation between the wavelength shift and the scattering angle φ is

$$\Delta\lambda = \lambda - \lambda_0 = \frac{c}{\nu} - \frac{c}{\nu_0} = \frac{h}{m_e c}(1 - \cos\varphi) = \lambda_C(1 - \cos\varphi) \tag{2-29}$$

This is in full agreement with the experimental results of the Compton effect, namely equations (2 - 20) and (2 - 21).

In order to explain that there is also a component of the scattered light with the same wavelength as the incident light, Compton thought that when a lot of photons shoot at scattering material, some photons do not collide with the outer electrons in the atoms that can be regarded as free electrons, but collide with the inner electrons in the atom, which are tightly bound. Because the inner electrons are closely bound with the atoms, this collision can be regarded as the collision between a photon and an entire atom. Since the mass m_a of the atom is far larger than that m_e of the electron, for example, the mass of the lightest hydrogen atom is about 2000 times greater than that of an electron, thus, after the collision, the energy of the photon transmitted to the atom is very little, while the photon hardly changes its energy and only changes the direction of its motion. That is, for this case, the mass m_a of the atom should replace the mass m_e of the electron in equation (2 - 29). Relatively speaking, the mass m_a of the atom is very big, so the wavelength shift is $\Delta\lambda \approx 0$. Therefore, the frequency of the scattered light is almost unchanged, that is, the scattered light also contains wavelength-invariant light.

Because the outer electrons of the atoms in the scatterer can be regarded as free electrons, the amount of wavelength shift is independent of the type of scatterer. Due to the electrons are weakly bound with light atoms, there are more inner electrons in heavy atoms than in light atoms, and the inner electrons are very tightly bound with heavy atoms, so the Compton effect is more significant for substances with small atomic weight. For substances with large atomic weight, the Compton effect is not significant. This is also consistent with the experimental results of Woo Y. H. et al.

In principle, Compton effect can occur when light of any wavelength is scattered by matter. However, for visible light and infrared light with long wavelength, the wavelength shift amount is much smaller than the wavelength of incident light, and it is not easy to observe. For example, for purple light with wavelength of $\lambda = 400$ nm, at the scattering angle $\varphi = \pi$, the wavelength shift amount $\Delta\lambda = 0.004,9$ nm, then $\Delta\lambda/\lambda \approx 10^{-5}$. However, for X - rays with wavelength of $\lambda = 0.05$ nm, $\Delta\lambda/\lambda \approx 10\%$. Therefore, in general, Compton effect is mainly observed for the light such as X - rays and γ rays with very short wavelength.

Instead of absorbing photons as electrons in photoelectric effect, electrons in Compton effect scatter photons. Assuming that free electrons can absorb photons, the equations are listed by the law of conservation of energy and the law of conservation of momentum as

$$h\nu_0 + m_e c^2 = mc^2$$

$$\frac{h\nu_0}{c}\boldsymbol{e}_0 = mv\boldsymbol{e}_0$$

Solving the above two formulas and equation (2 - 22) simultaneously, we get

$$1 - \frac{v}{c} = \sqrt{1 - \frac{v^2}{c^2}}$$

Thus it can be deduced that $v = c$. Therefore, it violates the theory of relativity. It can be seen that the process of free electrons absorbing photons cannot meet both conservation of energy and conservation of momentum. Therefore, free electron cannot absorb photon, it can only scatter photon.

Note that conservation of momentum is not considered in the interpretation of photoelectric effect. This is because in photoelectric effect, the incident light is visible light or ultraviolet light, and its photon has low energy. The connection between electron and the whole atom cannot be ignored, that is, it cannot be regarded as a free electron, and the atom also participates in the momentum exchange. Therefore, the momentum of the photon-electron system is not conserved. However, the atomic mass is large and the energy exchange can be negligible, so it can still be considered that the energy is conserved for the photon-electron system.

Photoelectric effect and Compton effect have become important experimental basis for the particle property of light. The photoelectric effect reveals the relation between energy and frequency of photon, that is, equation (2 - 12), the Compton effect further reveals the relation between momentum and wavelength of photon, that is, equation (2 - 19). It also demonstrates that the law of conservation of energy and the law of conservation of momentum are strictly established during the interaction between microscopic particles. This does not contradict with Einstein's claim that a photon "never split". In Compton effect, the interaction between electron and photon is a "two-step process", including two possible ways. One way is to absorb first and then release, that is, the free electron first absorbs a whole incident photon, and then releases another whole scattered photon, as shown in Figure 2 - 28 (a). The other way is to release first and then absorb, that is, the free electron first releases a scattered photon, and then absorbs an incident photon, as shown in Figure 2 - 28 (b). Whether the order is to absorb first and then release or to release first and then absorb, photon is "released or absorbed as a whole", and the whole process follows the laws of conservation of energy and the laws of conservation of momentum.

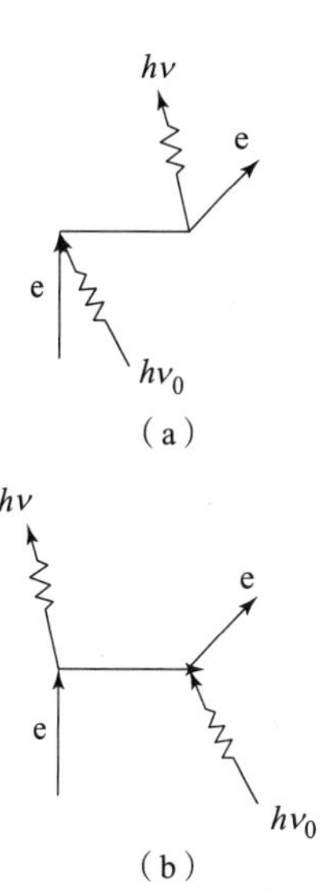

Figure 2 - 28 The "two - step process" of the interaction between electron and photon

(a) Absorb first and then release;
(b) Release first and then absorb

Compton shared the Nobel Prize in Physics in 1927 [with the Scottish physicist C. T. R. Wilson (1869—1959)]. But Compton's success was not plain sailing. He had taken detours for nearly five years. In his early papers, he always believed that the change in the wavelength of scattered light was due to the "mixing of some fluorescent radiation". At first, only the conservation of energy was considered in the calculation. It was later realized that the conservation of momentum should also be used. This reveals the uneven process of emergence and development of modern physics from one side.

Through experiments such as interference and diffraction of light, it has been recognized that light has wave property. Through photoelectric effect and Compton effect, it is recognized that light also has particle property. Thus, in 1923 – 1924, the wave-particle duality of light has been understood and accepted by people. In order to explain all the experimental facts about light, a comprehensive understanding of the nature of light should be that light has both wave property and particle property, that is, light has wave-particle duality.

Example 2 – 4: In a Compton scattering experiment, the wavelength of the incident X – rays is known to be $\lambda_0 = 0.07$ nm, and the scattered light is perpendicular to the incident rays. Find, (1) the kinetic energy E_k of recoil electron, (2) The angle θ at which the moving direction of recoil electron deviates from the incident X – rays.

Solution: (1) According to the Compton scattering formula, the wavelength shift of scattered light is

$$\Delta\lambda = \lambda_C(1 - \cos\varphi) = 2.43 \times 10^{-3} \times 10^{-9}\left(1 - \cos\frac{\pi}{2}\right)\ \text{m} = 0.002\,43\ \text{nm}$$

So the wavelength of scattered light is

$$\lambda = \lambda_0 + \Delta\lambda = (0.07 + 0.002\,43)\ \text{nm} = 0.072\,43\ \text{nm}$$

According to the conservation of energy, the kinetic energy E_k obtained by recoil electron is the energy lost by scattered photon, i. e

$$E_k = h\nu_0 - h\nu = hc\left(\frac{1}{\lambda_0} - \frac{1}{\lambda}\right)$$

$$= 6.63 \times 10^{-34} \times 3 \times 10^8\left(\frac{1}{0.07 \times 10^{-9}} - \frac{1}{0.072\,43 \times 10^{-9}}\right)\ \text{J}$$

$$= 7.89 \times 10^{-7}\ \text{J}$$

(2) According to the vector relation of momentum conservation, as shown in Figure 2 – 29, there is

$$\boldsymbol{p}_0 = \boldsymbol{p} + m\boldsymbol{v}$$

The magnitude of the momentum of recoil electron is

$$mv = \sqrt{p_0^2 + p^2} = \sqrt{\left(\frac{h}{\lambda_0}\right)^2 + \left(\frac{h}{\lambda}\right)^2}$$

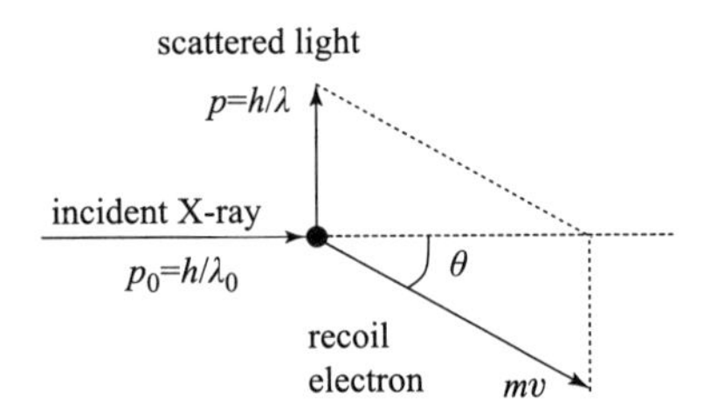

Figure 2 – 29 Example 2 – 4

The angle at which the moving direction of recoil electron deviates from the incident X – rays satisfies

$$\cos\theta = \frac{p_0}{mv} = \frac{h/\lambda_0}{\sqrt{(h/\lambda_0)^2 + (h/\lambda)^2}} = \frac{\lambda}{\sqrt{\lambda_0^2 + \lambda^2}} = \frac{0.072\,43}{\sqrt{0.07^2 + 0.072\,43^2}} = 0.779$$

It can be solved that $\theta = 44.03°$.

Woo Y. H. (Figure 2 – 30), a famous Chinese physicist, educator, and one of the founders of the Chinese Physics Society. In 1921, Woo Y. H. went to the United States to enter the University of Chicago. He studied physics with Compton and participated in the research of Compton effect. He obtained X - ray scattering spectra of 15 kinds of substances, which made up for the lack that Compton had only one substance spectrum and became one of the most favorable experimental evidence of Compton effect. By this, Compton won the Nobel Prize in Physics in 1927, so Compton effect is also known as "Compton-Woo Y. H" effect. In 1926, Woo Y. H. returned to China after receiving his doctorate degree and taught successively at Shanghai Jiaotong University, Nanjing Central University and Tsinghua University, etc, for more than 20 years. With rigorous scientific style, he cultivated many excellent students. In 1950, he served as vice president of the Chinese Academy of Sciences, director of the Institute of Modern Physics, and former chairman of the Chinese Physical Society.

Figure 2 - 30 Woo Y. H., the Chinese famous physicist

2.4 Hydrogen Atom Spectrum and Bohr's Theory of Hydrogen Atom

2.4.1 Hydrogen Atom Spectrum

In 1853, the Swedish Anders Jöns Angström (1814—1874) first observed the red line with a wavelength of 656.3 nm in the hydrogen atom spectrum from the spectrum of gas discharge, i. e. H_α line. The unit of wavelength Å was named after him. Someone took the year 1853 as the beginning of scientific spectrum. Later, several other hydrogen spectral lines in visible light region were found, which are green line (H_β line), blue line (H_γ line) and purple line (H_δ line) with wavelengths of 486.1 nm, 434.1 nm and 410.2 nm, respectively, as shown in Figure 2 - 31. These wavelength values seemed to be messy. Finding out the rule and its internal relation with the atomic structure was an important research topic for spectrologists and physicists at that time.

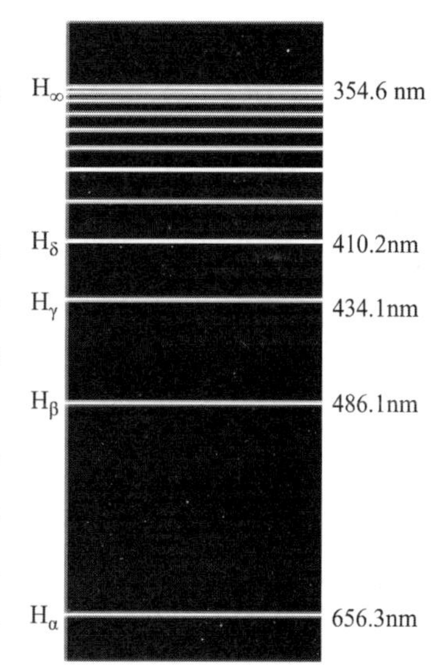

Figure 2 - 31 Balmer series of hydrogen atomic spectrum

In 1885, a middle school teacher in Switzerland, J. J. Balmer (1825—1898), found that the wavelengths of four visible spectral lines in hydrogen spectrum can be summarized as the formula

$$\lambda = B\frac{n^2}{n^2 - 2^2}; n = 3, 4, 5, \cdots \qquad (2-30)$$

where $B = 365.47$ nm, it is a constant determined by Balmer from experimental data.

In 1890, the Swiss physicist J. R. Rydberg (1854—1919) used the reciprocal of the wavelength $\tilde{\nu} = 1/\lambda$ to replace the wavelength in Balmer formula, and obtained

$$\tilde{\nu} = R_{\infty}\left(\frac{1}{2^2} - \frac{1}{n^2}\right);n = 3,4,5,\cdots \quad (2-31)$$

where $R_{\infty} = 1.097 \times 10^7\ \text{m}^{-1}$, it is called the Rydberg constant. $\tilde{\nu}$ is called wave number, which is equal to the number of wavelengths per unit length. Equation (2 – 31) is called Balmer formula, and its corresponding spectrum is called Balmer series.

In 1914, inspired by Balmer's research, American physicist T. Lyman (1874—1954) found the formula for describing ultraviolet series of hydrogen spectrum. It only changed the denominator 2 to 1 in Balmer formula, that is

$$\tilde{\nu} = R_{\infty}\left(\frac{1}{1^2} - \frac{1}{n^2}\right);n = 2,3,4,\cdots \quad (2-32)$$

The spectral series represented by equation (2 – 32) is called Lyman series.

In 1908, the German physicist F. Paschen (1865—1947) found the formula describing one of the infrared series of hydrogen spectrum. It only changed the denominator 2 to 3 in Balmer formula, that is

$$\tilde{\nu} = R_{\infty}\left(\frac{1}{3^2} - \frac{1}{n^2}\right);n = 4,5,6,\cdots \quad (2-33)$$

The spectral series represented by equation (2 – 33) is called Paschen series.

In 1922, the American physicist F. S. Brackett (1896—1972) gave the formula describing the far-infrared series of hydrogen spectrum. It only changed the denominator 2 to 4 in Balmer formula, that is

$$\tilde{\nu} = R_{\infty}\left(\frac{1}{4^2} - \frac{1}{n^2}\right);n = 5,6,7,\cdots \quad (2-34)$$

The spectral series represented by equation (2 – 34) is called Brackett series.

In 1924, the American physicist H. A. Pfund (1879—1949) gave the formula describing the far-infrared series of the hydrogen spectrum. It only changed the denominator 2 to 5 in Balmer formula, that is

$$\tilde{\nu} = R_{\infty}\left(\frac{1}{5^2} - \frac{1}{n^2}\right);n = 6,7,8,\cdots \quad (2-35)$$

The spectral series represented by equation (2 – 35) is called Pfund series.

When the denominator 2 in equation (2 – 31) is changed to other positive integer m, the generalized Balmer formula (or Rydberg formula) is obtained as

$$\tilde{\nu} = R_{\infty}\left(\frac{1}{m^2} - \frac{1}{n^2}\right);m = 1,2,3,\cdots;\ n = m+1,\ m+2,\ m+3,\cdots \quad (2-36)$$

In 1908, E. Rutherford found in the experiment of α particles scattering on gold foil that a considerable number of α particles were scattered by gold foil at a large angle. Rutherford's experiment strongly suggests that the positively charged part of the atom seems to be a "nucleus". In 1911, Rutherford proposed the nuclear model of atomic structure that an atom with atomic number Z has a nucleus with positive charge Ze, and its radius is about 10^{-15} m. The mass of the nucleus accounts for the vast majority of atomic mass. Z electrons are distributed outside the nucleus and rotate around the nucleus with a radius of about 10^{-10} m. Rutherford used this solar system like model to explain a series of experiments on α particle scattering. At the same time, he also formally proposed the concept of atomic nucleus, or first discovered the atomic nucleus.

However, classical physics encountered insurmountable difficulties in explaining the stability of atoms and their linear spectra.

According to Rutherford's nuclear model of atom, the motion of electrons around the nucleus under the Coulomb force of the nucleus is movement with acceleration. It keeps radiating electromagnetic waves outward, that is, energy is continuously emitted in the form of radiation. As the energy decreases, the radius of electron orbit will become smaller and smaller, and it will soon fall to the nucleus. This process takes less than 10^{-12} s. Therefore, there can be no stable hydrogen atom. In other words, according to the classical theory, atoms are "short-lived". That is, the Rutherford's nuclear structure is not a stable system, which cannot explain the stability of the orbital motion of electrons in atom.

Because the frequency of electromagnetic wave radiated by the electron should be equal to the frequency at which it rotates around the nucleus, the frequency ν of the radiated electromagnetic wave is proportional to the radius $r^{-3/2}$ of the electron's rotation around the nucleus. In the process of electron falling to the nucleus, the continuous reduction of orbital radius will inevitably lead to the continuous change of the frequency or wavelength of the radiated electromagnetic wave. The atomic spectrum should be a continuous spectrum. Therefore, the classical theory cannot explain the separated atomic linear spectrum.

2.4.2 Bohr's Theory of Hydrogen Atom

In 1913, based on Rutherford's nuclear structure model and the analysis of spectroscopy data, the Danish physicist Niels Bohr published a long paper entitled "on the constitution of atoms and molecules" in the British Journal of philosophy in three times. In this paper, Bohr broke through the bondage of classical concepts, combined the quantum concept of light with Balmer formula, and put forward the atom theory containing the following three basic assumptions.

(1) Steady state assumption. An atomic system can and can only often be in a series of states corresponding to discrete energy values $E_1, E_2, \cdots$. In these states, although the electron moves around the nucleus, it does not radiate electromagnetic wave. These states are called the steady states of the atomic system. That is to say, the electrons in an atom move around the nucleus only in some certain circular orbits, as shown in Figure 2-32. Electrons do not radiate energy as they run in these orbits, so any energy change of the system can only be caused by the transition between these states.

(2) Frequency condition. Only when electron transits from a higher-energy orbit to a lower-energy orbit, it emits radiation. Conversely, it absorbs radiation. In other words, the atom will absorb or emit a photon with a frequency of ν only when there is a transition between two steady states, and there is the following relation called frequency condition,

$$h\nu = |E_f - E_i| \tag{2-37}$$

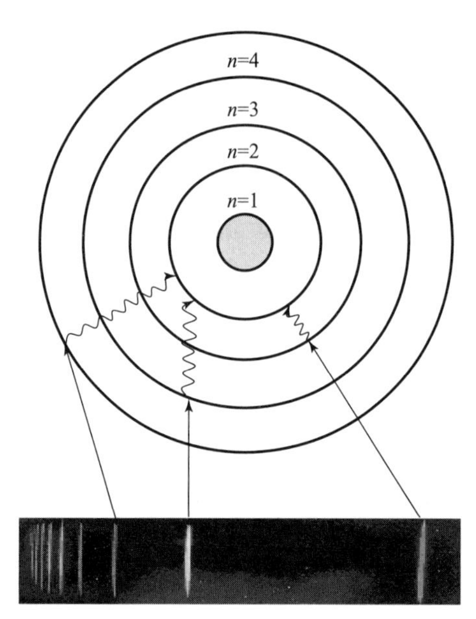

Figure 2 – 32 Bohr's theory of hydrogen atom

where E_i and E_f are the energy values of the initial and the final steady states respectively.

(3) Quantization assumption of angular momentum. When the electron moves around the nucleus with velocity v on a circle of radius r, only those orbits where the magnitude of angular momentum L of the electron is equal to an integral multiple of $\hbar$ are stable, that is,

$$L = m_e vr = n\hbar; n = 1,2,3,\cdots \tag{2-38}$$

where m_e is the mass of electron, n is a positive integer, and $\hbar = h/2\pi = 1.054,571,800(13) \times 10^{-34}$ J · s≈1.05 × 10^{-34} J · s is called the reduced Planck constant.

From these basic assumptions, Bohr derived the energy formula for the hydrogen atom. Considering that the electron in hydrogen atom make circular motion around the nucleus, let the orbital radius of the electron around the nucleus be r, then there is Newton's equation of motion

$$\frac{1}{4\pi\varepsilon_0}\frac{e^2}{r^2} = m_e\frac{v^2}{r} \tag{2-39}$$

where e is the absolute value of electron charge and v is the electron velocity. By eliminating v from equation (2 – 38) of the quantized condition and equation (2 – 39), it can be solved that the electron orbital radius of allowable steady state in hydrogen atom is

$$r_n = \frac{4\pi\varepsilon_0\hbar^2}{m_e e^2} \cdot n^2; n = 1,2,3,\cdots \tag{2-40}$$

The above equation shows that in hydrogen atom the orbital radius of electron around the nucleus in steady state is also quantized. When $n = 1$, the orbital radius is minimum, which is

$$a_0 = \frac{4\pi\varepsilon_0\hbar^2}{m_e e^2} = 0.529,177,210,67(12) \times 10^{-10}\ m \approx 0.052,9\ \text{nm} \tag{2-41}$$

It is called the Bohr radius, which is the radius of the first circular orbit in Bohr atom theory.

When the electron moves on a certain steady state orbit, the total energy of hydrogen atomic system, that is, the steady state energy is:

$$E = E_k + E_p = \frac{1}{2}m_e v^2 - \frac{e^2}{4\pi\varepsilon_0 r} = -\frac{e^2}{8\pi\varepsilon_0 r} \tag{2-42}$$

Equation (2-39) has been used in the above equation. Substitute equation (2-40) into it to obtain the energy formula of hydrogen atom

$$E_n = -\frac{m_e e^4}{2(4\pi\varepsilon_0)^2\hbar^2}\cdot\frac{1}{n^2}; n = 1,2,3,\cdots \tag{2-43}$$

where n can only take a series of positive integers, which is called principal quantum number. This formula shows that the energy of hydrogen atom can only take discrete values, which is the quantization of energy. The possible value of each energy given by equation (2-43) is called an energy level. The energy levels of hydrogen atom can be represented by energy level diagram shown in Figure 2-33.

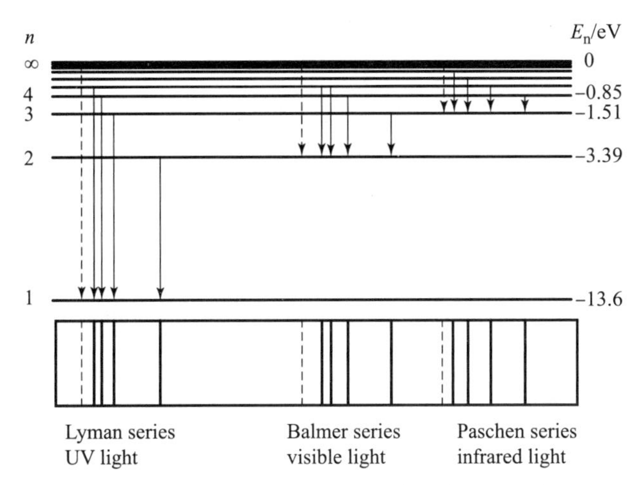

Figure 2-33 Energy levels and spectral series diagram of hydrogen atom

When $n = 1$, the energy of hydrogen atom is the lowest, which is called the ground state energy, and its value is

$$E_1 = -\frac{e^2}{8\pi\varepsilon_0 a_0} = -\frac{m_e e^4}{2(4\pi\varepsilon_0)^2\hbar^2} \approx -13.6 \text{ eV} \tag{2-44}$$

The energy level state of $n > 1$ is called the excited state. For hydrogen atom, the energy of any excited state is $1/n^2$ of its ground state energy, i. e

$$E_n = \frac{E_1}{n^2} \approx \frac{-13.6}{n^2} \text{ eV}; n = 2,3,4,\cdots \tag{2-45}$$

It is easy to calculate that the energy levels of each excited state at $n = 2,3,4$ have the energy values of $E_2 = -3.4$ eV, $E_3 = -1.5$ eV, $E_4 = -0.85$ eV, etc. In the region of $E < 0$, the electron is attracted by the nucleus, i. e. proton, and it is in a bound state. The energy is discrete, that is quantized. The energy level interval of hydrogen atom decreases rapidly with the increase of n. When

n is large, the energy level interval is very small, and the energy level can be regarded as continuous. When $n \to \infty, E_\infty \to 0$, it shows that electron is no longer bound around the nucleus, but out of the nucleus, the atom is said to be ionized.

The energy difference between E_∞ and E_1 of the ground state is called ionization energy E_i. The ionization energy E_i of hydrogen atom is

$$E_i = E_\infty - E_1 = -E_1 = 13.6 \text{ eV}$$

If the electron gains enough energy to be in the region where $E > 0$, it means that the electron has completely separated from the attraction of the nucleus and becomes a free electron, that is, the hydrogen atom is in an ionized state. The energy of free electron at this time also has continuous values.

If we make slight revision on the energy level formula of hydrogen atom, it can also apply to hydrogen-like ions, such as the helium ion He^+. There are two electrons outside the nucleus of He atom. When it is ionized and loses an electron, its structure is similar to that of hydrogen atom, but the nuclear charge is $+2e$. If Z represents the nuclear charge number of hydrogen-like ion, then the energy level formula of hydrogen-like ion is

$$E_n = -\frac{m_e e^4}{2(4\pi\varepsilon_0)^2\hbar^2} \cdot \frac{Z^2}{n^2}; n = 1, 2, 3, \cdots \tag{2-46}$$

According to Bohr atom theory, when a hydrogen atom transits between different energy levels, it will emits or absorbs photon with energy of $h\nu$. Under normal circumstances, the hydrogen atom is in the ground state with lowest energy. However, when enough energy is provided by the outside, usually a hydrogen atom will absorb a photon to obtain energy $h\nu$ and transit to an excited state. When transitioning from a higher energy level to a lower energy level, it will emit light of corresponding frequency. From equation (2 - 37) of the Bohr frequency condition, when a hydrogen atom transits from a high energy level with energy E_n to a low energy level with energy E_m, the frequency of photon emitted is

$$\nu = \frac{E_n - E_m}{h}$$

Substitute equation (2 - 44) of the hydrogen atom energy into the above formula, we can get

$$\nu = \left(\frac{1}{4\pi\varepsilon_0}\right)^2 \frac{m_e e^4}{4\pi\hbar^3}\left(\frac{1}{m^2} - \frac{1}{n^2}\right)$$

The corresponding wavenumber of the hydrogen atom spectrum is

$$\tilde{\nu} = \frac{1}{\lambda} = \frac{\nu}{c} = \frac{1}{hc}(E_n - E_m) = \frac{m_e e^4}{4\pi (4\pi\varepsilon_0)^2\hbar^3 c}\left(\frac{1}{m^2} - \frac{1}{n^2}\right) \tag{2-47}$$

This formula is exactly the same as the generalized Balmer formula (equation (2 - 36)). By comparing the two formulas the Rydberg constant can be obtained

$$R_\infty = \left(\frac{1}{4\pi\varepsilon_0}\right)^2 \frac{m_e e^4}{4\pi\hbar^3 c} \tag{2-48}$$

Substituting the values of the basic constants into the above equation, we get

$$R_\infty = 1.097,373,156,850,8(65) \times 10^7 \text{ m}^{-1}$$

The wavenumbers calculated by this formula are in good agreement with those of various spectral lines of the hydrogen spectrum measured experimentally.

Hydrogen atoms emit lights of different frequencies to form different spectral lines, forming a spectral line system. According to Bohr atom theory, it is further clarified that each spectral line of the hydrogen spectrum corresponds to the light emitted by the transition between different energy levels. As shown in Figure 2 – 33, the lights emitted by the transitions of hydrogen atom from higher energy levels back to the ground state where $n = 1$ form the Lyman series, and the frequencies of these lights are in the ultraviolet region. The lights emitted by the transitions of hydrogen atom from higher energy levels back to the state where $n = 2$ form the Balmer series, and the frequencies of these lights are in the visible light region. The lights emitted by the transitions of hydrogen atom from higher energy levels back to the energy level of $n = 3$ form the Paschen series, and the frequencies of these lights are in the infrared region. The lights emitted by the transitions of hydrogen atom from higher energy levels back to $n = 4$ form the Brackett series, the frequencies of these lights are in the far infrared region, and so on.

Example 2 – 5: The hydrogen atoms where in ground state are bombarded with electrons with an energy of 12.5 eV. Find the possible wavelengths of the spectral lines emitted when an excited hydrogen atom transits to the low energy levels.

Solution: The energy of the excited state level is $E_n = E_1/n^2$, so when a hydrogen atom transits from the ground state to the second excited state, the required energy, that is, the excitation energy is

$$E_3 - E_1 = -1.51\ \text{eV} + 13.6\ \text{eV} = 12.09\ \text{eV}$$

When a hydrogen atom transits from the ground state to the third excited state, the required excitation energy is

$$E_4 - E_1 = -0.85\ \text{eV} + 13.6\ \text{eV} = 12.75\ \text{eV}$$

It can be seen that an electron with an energy of 12.5 eV can excite a hydrogen atom from the ground state to energy level of E_3 ($n = 3$). Thus, there are three possibilities for the transition from the second excited state to low energy level: E_3 to E_2, E_2 to E_1 and E_3 to E_1. Therefore, three spectral lines can be emitted. One belongs to the Balmer series, and its wavelength is

$$\lambda_{32} = \frac{ch}{E_3 - E_2} = \frac{3 \times 10^8 \times 6.63 \times 10^{-34}}{[-13.6/3^2 - (-13.6/2^2)] \times 1.6 \times 10^{-19}}\ \text{nm} = 658\ \text{nm}$$

The other two belong to the Lyman series, and their wavelengths are

$$\lambda_{31} = \frac{ch}{E_3 - E_1} = \frac{3 \times 10^8 \times 6.63 \times 10^{-34}}{[-13.6/3^2 - (-13.6)] \times 1.6 \times 10^{-19}}\ \text{nm} = 103\ \text{nm}$$

$$\lambda_{21} = \frac{ch}{E_2 - E_1} = \frac{3 \times 10^8 \times 6.63 \times 10^{-34}}{[-13.6/2^2 - (-13.6)] \times 1.6 \times 10^{-19}}\ \text{nm} = 122\ \text{nm}$$

Example 2 – 6: An electron with a kinetic energy of 20 eV collides with a hydrogen atom in the ground state, causing the hydrogen atom to be excited. When the hydrogen atom transits back to the ground state, it emits a spectrum of 121.6 nm. Find the velocity of the electron after the collision.

Solution: The frequency can be obtained from the wavelength of the radiation spectrum, there is

$$\nu = \frac{c}{\lambda} = \frac{3.0 \times 10^{8}}{121.6 \times 10^{-9}} \text{ Hz} = 2.467 \times 10^{15} \text{ Hz}$$

Since the energy of hydrogen atom gained by the collision is equal to the energy lost by its radiation, we have

$$E = h\nu = 6.626 \times 10^{-34} \times 2.467 \times 10^{15} \text{ J} = 1.635 \times 10^{-18} \text{ J} = 10.216 \text{ eV}$$

The kinetic energy of the electron after the collision can be obtained as

$$E_{\text{k}} = 20 \text{ eV} - 10.214 \text{ eV} = 9.786 \text{ eV} = 1.566 \times 10^{-18} \text{ J}$$

Then the velocity of the electron after the collision is

$$v = \sqrt{\frac{2E_{\text{k}}}{m}} = \sqrt{\frac{2 \times 1.566 \times 10^{-18}}{9.109 \times 10^{-31}}} \text{ m} \cdot \text{s}^{-1} = 1.854 \times 10^{6} \text{ m} \cdot \text{s}^{-1}$$

In 1921, Bohr delivered a long speech on "Atomic structure of each element and their physical and chemical properties". He expounded the new development of spectrum and atomic structure theory, interpreted the formation of the periodic table of elements, and explained the atomic structure of various elements starting from hydrogen in the periodic table. Meanwhile he made a prophecy on the properties of element 72 in the periodic table. In 1922, the element hafnium was discovered, which confirmed Bohr's prediction. In 1922, Bohr won the Nobel Prize in Physics for his research on atomic structure and atomic radioactivity.

Bohr atomic theory successfully explains the linear spectrum of hydrogen atom on the basis of classical theory adding some new quantum assumptions. As an early quantum theory, it played an important leading role in the development of quantum mechanics. Einstein later praised the theory as "the highest musical verve in the realm of thought". However, Bohr theory has defect. It is far from reflecting the essence of the microscopic world. For example, it cannot explain the spectrum of multi-electron atom, nor can it do anything about the intensity and width of spectral lines. The reason for this defect is that Bohr atom model is a mixture of concepts of Newtonian mechanics and quantization conditions.

The correct theory of atomic structure should be based on new quantum mechanics. Although some basic concepts of Bohr theory, such as "steady state", "energy level", "energy level transition determines the radiation frequency", etc. are still important basic concepts in quantum mechanics, but from the perspective of quantum mechanics, the orbit in the classical sense is no longer applicable to the microscopic atomic world.

To sum up, from the end of the 19th century to the beginning of the 20th century, a series of important physical phenomena, such as blackbody radiation, photoelectric effect, Compton effect, and atomic spectral line system, exposed the limitations of classical physics and highlighted the contradiction between classical physics and the laws of the microscopic world. Those inspired scientists to explore and discover the laws of the microscopic world. Planck energy quantum hypothesis, Einstein photon quantum theory and Bohr atom theory laid the foundation for discovering the laws of the microscopic world. In the 1930s, quantum mechanics was established.

Bohr and the Copenhagen Spirit

In 1921, the Institute of Theoretical Physics in University of Copenhagen was established at the initiative of Bohr (Figure 2 - 34). Bohr led the institute for 40 years and attracted a large number of outstanding physicists at home and abroad around him with his high prestige, including the two most talented disciples, Heisenberg and Pauli. During the rise of quantum mechanics, this institute once became the most important and most active academic center in the world, founded the famous Copenhagen School in quantum mechanics, and Bohr also became the leader of Copenhagen School. They not only created the basic theory of quantum mechanics, but also gave reasonable explanations, so that quantum mechanics had many new applications, such as atomic radiation, chemical bond, crystal structure, and metal state, etc. What is even more valuable is that while Bohr and his colleagues were creating and developing science, they also created the "Copenhagen Spirit". This is a unique, strong academic atmosphere of equal and free discussion and close cooperation with each other. Until today, many people still say that the "Copenhagen Spirit" is unique in the international physics community. Bohr was once asked, "why can you attract so many scientists to work at the Institute of Theoretical Physics in Copenhagen?" He replied, "because I'm not afraid to admit my lack of knowledge in front of young people, and I'm not afraid to admit that I'm a fool." This attitude of frankness and seeking truth from facts was the reason why the Institute of Theoretical Physics in Copenhagen led by him at that time was always full of vitality and prosperity. Actually, people's understanding of atomic physics, that is, the understanding of the so-called quantum theory of atomic system, began in the early 20th century and completed in the 1920s. However, from beginning to end, Bohr's spirit full of high creativity, acuity and criticalness always guided the direction of his career and made it deeper until the final completion.

Figure 2 - 34 Bohr, 1922 Nobel laureate in physics

Einstein and Bohr once had a long and fierce debate around the interpretation for the theoretical basis of quantum mechanics, but they were always a pair of good friends who respect each other. Bohr spoke highly of this debate and thought of it as "the source of his many new ideas". While Einstein highly praised Bohr, "As a scientific thinker, Bohr's amazing attraction lies in his rare combination of boldness and prudence. Few people have such an intuitive understanding on secret things and meanwhile have such a strong critical ability. He not only has full knowledge of details, but also always firmly focuses on the fundamentals. He is undoubtedly one of the greatest discoverers of our time in the field of science."

2.5 The Wave Property of Particle and Born's Statistical Interpretation

2.5.1 De Broglie Wave

In 1924, Louis de Broglie, a young doctoral student at the University of Paris in France (Figure 2 – 35), inspired by the wave-particle duality of light, in his doctoral thesis entitled "Research on Quantum Theory" boldly put forward a hypothesis: physical particles such as electron, proton and so on also have wave property. He believed that just as the understanding of light was one-sided in the past, the understanding of physical particles may also be one-sided. Duality is not just for light, physical particles also have wave-particle duality.

Figure 2 – 35 de Broglie, 1929 Nobel laureate in physics

With his keen mind, de Broglie applied the description of the wave-particle duality of light to physical particles. A physical particle with mass m and moving with velocity v has both particle property described by energy E and momentum $\boldsymbol{p}$, and wave property described by frequency ν and wavelength λ. For the physical particle, the relations between energy E and frequency ν, momentum $\boldsymbol{p}$ and wavelength λ are the same as the relations $\varepsilon = h\nu$ and $p = h/\lambda$ for the photon, and the frequency ν and the wavelength λ of the wave associated with the physical particle are respectively

$$\nu = \frac{E}{h} = \frac{mc^2}{h} \tag{2-49}$$

$$\lambda = \frac{h}{p} = \frac{h}{mv} \tag{2-50}$$

where $m = \dfrac{m_0}{\sqrt{1-(v/c)^2}}$, m_0 is the rest mass of the physical particle. Equations (2 – 49) and (2 – 50) are called de Broglie relations, λ is called de Broglie wavelength. The wave associated with physical particle is called de Broglie wave, also known as matter wave.

De Broglie then connected matter wave with standing wave, and more naturally derived the orbital quantization condition in Bohr atom theory that was once puzzling. As shown in Figure 2 – 36, he believed that the electron in a steady state of hydrogen atom moves in a circle around the nucleus, and the corresponding electron wave propagates around the nucleus. After one period of propagation the wave should be smoothly connected, which is equivalent to that the electron wave forms a

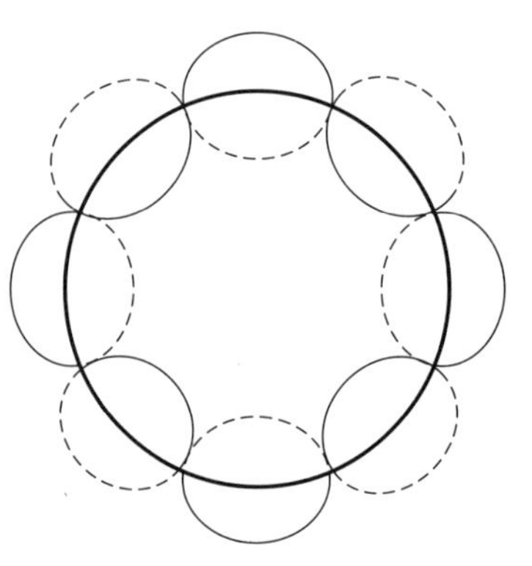

Figure 2 – 36 Derivation of Bohr's quantization condition

stable standing wave on this circumference. Only the standing wave is a stable vibration state without radiating energy, which corresponds to the steady state of the atom. Therefore, the orbit of the electron around the nucleus is limited, that is, the circumference of the electron orbit should be equal to an integral multiple of the electron's wavelength, so that a stable standing wave can be formed. Let r be the radius of the electron's stable orbit, then we have

$$2\pi r = n\lambda; n = 1, 2, 3, \cdots\cdots$$

From equation (2 – 50) of the de Broglie relation, the angular momentum of the electron moving around the nucleus can be obtained

$$rmv = n\frac{h}{2\pi}$$

This is exactly Bohr's quantization condition for electron orbital angular momentum as equation (2 – 38).

When the velocity v of the particle is much less than the speed of light c, the relativistic effect does not need to be considered, and the de Broglie wavelength is

$$\lambda = \frac{h}{m_0 v} \qquad (2-51)$$

In fact, Einstein played an important role in bringing de Broglie's work to the attention of the academic community at that time. P. Langevin (1872—1946, French physicist), de Broglie's supervisor, handed his thesis to Einstein for review before his defense. Einstein fully affirmed de Broglie's work and praised him for "he has lifted a corner of the great veil", "Read it, even though it might look crazy, it is absolutely solid." After Einstein's recommendation, de Broglie's theory of matter wave soon attracted people's attention.

The de Broglie hypothesis had no any experimental basis when it was proposed, but when de Broglie defended his doctoral thesis on November 27, 1924, Jean Baptiste Perrin (1870—1942, 1926 Nobel laureate in physics), the chairman of the defense committee, proposed how the matter wave can be confirmed by experiment. De Broglie gave such an answer that the diffraction experiment of crystals on electrons can be used to verify the existence of matter waves. In fact, he had already thought so. In his paper entitled "light quantum, diffraction and interference" published in the Bulletin of the French Academy of Sciences on September 24, 1923, he made a prophesy that electron beam via a very small hole may have diffraction phenomenon.

2.5.2 Experimental Verification of de Broglie Wave

(1) Davidson-Germer experiment

In 1927, scientists at Bell Laboratory in the United States, C. J. Davidson (1881—1958, left in Figure 2 – 36) and L. H. Germer (1896—1971, right in Figure 2 – 37), observed electron diffraction phenomenon similar to X – ray diffraction through the experiment of electron beam scattering on the surface of nickel single crystal, which first confirmed the wave property of electron.

Figure 2 – 38 is a schematic diagram of the Davison-Germer experimental device. An electron beam emitted from electron gun is accelerated by the voltage U of 40 eV – 600 eV, and then it passes

Figure 2 – 37　Davidson(left),1937 Nobel laureate in physics, and Germer(right)

through a set of slits and becomes a very thin electron beam to be projected on a selected crystal plane of nickel cubic single crystal. After being scattered by the crystal plane, it enters the electron detector, and the intensity of electron beam is measured by the galvanometer G. In the experiment, instead of using the surface of the cubic crystal as the incident surface of the electron beam, a corner of the cubic crystal is symmetrically cut off to form a triangular plane, which is used as the incident surface of the electron beam. The electron beam is directed right in front of this face with a predetermined speed, while the crystal itself rotates with the incident electron beam as the axis. So that the intensity of the scattered electron beam in any direction in front of the crystal surface can be measured. Or keep the scattering angle φ unchanged and measure the intensity of scattered electron beam under different accelerating voltages.

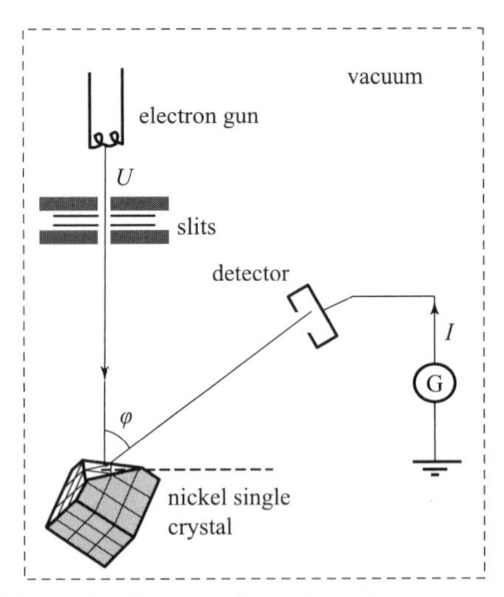

Figure 2 – 38　Schematic diagram of Davisson-Germer experimental device

It is found in the experiment that when the accelerating voltage U keeps constant, the scattered electron beam is especially strong in some directions, as shown in Figure 2 – 39. In the direction of a certain scattering angle φ, when the accelerating voltage U increases monotonically, the current in the electron detector does not increase monotonically, but obviously it shows a regular selectivity, that is, only when the accelerating voltage U takes a certain value, the intensity of scatted electron beam has maximum value, as shown in Figure 2 – 40.

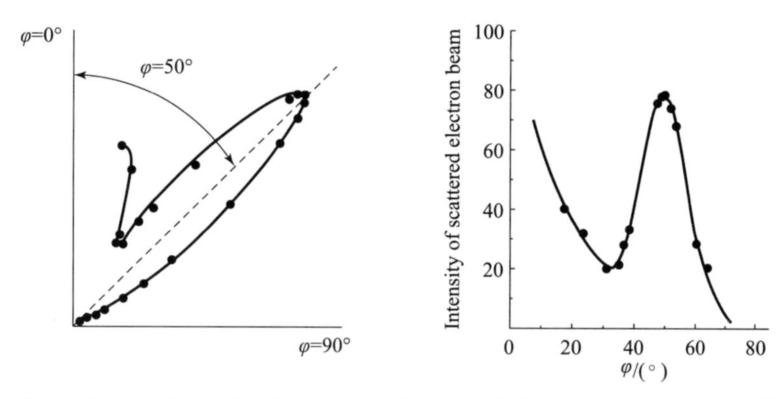

Figure 2 – 39 Intensity distribution of scattered electron beam at U = 54 V

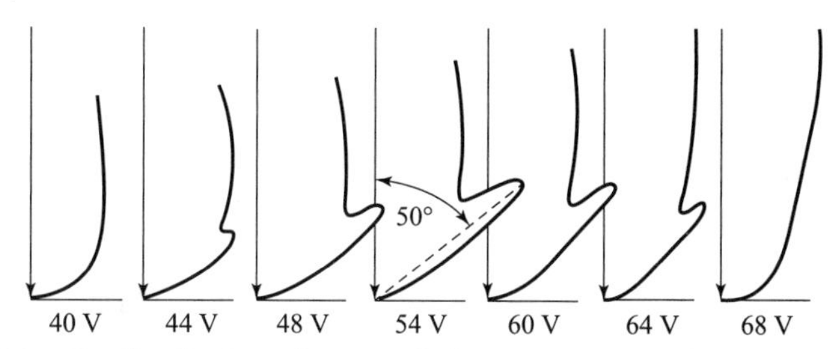

Figure 2 – 40 Intensity distribution of scattered electron beam at different accelerating voltages

The characteristics of the intensity distribution of scattered electron beam can be explained as follows by de Broglie relation and diffraction theory.

When the electron velocity v, which is accelerated by voltage U, is much less than the speed of light c, the kinetic energy of the electron is

$$\frac{1}{2}m_e v^2 = eU \tag{2-52}$$

where m_e is the rest mass of electron. According to equations (2 – 51) and (2 – 52), the de Broglie wavelength of the electron is

$$\lambda = \frac{h}{\sqrt{2m_e e}}\frac{1}{\sqrt{U}} = \frac{1.225}{\sqrt{U}}\ \text{nm} \tag{2-53}$$

For example, when U = 54 V, the de Broglie wavelength of electrons is $\lambda = 1.67 \times 10^{-10}$ m, this is the same order of magnitude as the wavelength of X – rays.

As shown in Figure 2 – 41, the wave path difference of two scattered rays along the φ direction scattered by the nickel crystal plane with an atomic spacing of $d = 2.15 \times 10^{-10}$ m is

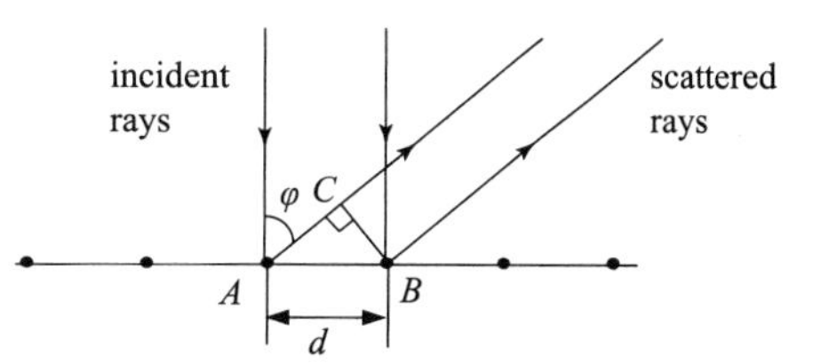

Figure 2 – 41 Analysis of electron beam diffraction on crystal

$$AC = d\sin\varphi$$

The condition for coherent enhancement is

$$d\sin\varphi = k\lambda \tag{2-54}$$

where $k = 1, 2, 3, \cdots\cdots$. This is the Bragg equation for X - ray diffraction on a crystal.

As can be seen from Figure 2 - 39, when the energy of the incident electron is 54 eV, the scattered electron beam has largest intensity in the direction of $\varphi = 50°$. This phenomenon is similar to the case where X - rays are diffracted by a single crystal. From equation (2 - 54), and taking $k = 1$, the wavelength of electron $\lambda = 1.65 \times 10^{-10}$ m can be obtained. It differs very little from the theoretical value calculated by equation (2 - 53).

Substitute equation (2 - 53) into equation (2 - 54), there is

$$\sqrt{U} = k\frac{1.225 \times 10^{-9}}{d\sin\varphi} \tag{2-55}$$

That is, when the accelerating voltage U meets the above formula, the intensity I of scattered electron beam measured by the detector has maximum value. The calculated values of accelerating voltage U are consistent with the experimental results.

The above analysis indicates that electrons are like X - rays to own wave property, and it also verifies the correctness of de Broglie wavelength formula.

(2) George Thomson's experiment

Also in 1927, the British physicist G. P. Thomson (1892—1975, Figure 2 - 42), the son of the electron discoverer J. J. Thomson (1856—1940, 1906 Nobel laureate in physics), did the experiment of high-energy electron beam transmitting polycrystalline thin film, he also observed the phenomenon of electron diffraction similar to X - ray diffraction.

Figure 2 - 42 George Thomson, 1937 Nobel laureate in physics

Figure 2 - 43 is a schematic diagram of George Thomson's experimental principle. An electron beam with energy of 1,000 - 8,000 eV is vertically incident on a very thin sheet of gold, or platinum, or aluminum sheet, etc. Then the diffraction pattern is photographed with a photographic plate. Because polycrystal is composed of a large number of randomly oriented tiny single crystals, it is possible to satisfy Bragg equation along the planes of various orientations, so diffraction can be observed simultaneously from all directions. As shown in the left picture of Figure 2 - 44, the diffraction pattern is a series of concentric circles, which is similar to the diffraction pattern obtained by X - rays (the right picture of Figure 2 - 44). According to the diameters of these diffraction rings, the de Broglie wavelength of the incident electron can be calculated. The experimental result is perfectly consistent with that given by the de Broglie relation. This fully proves that electron has wave property. It once again convincingly shows the correctness of de Broglie's theory.

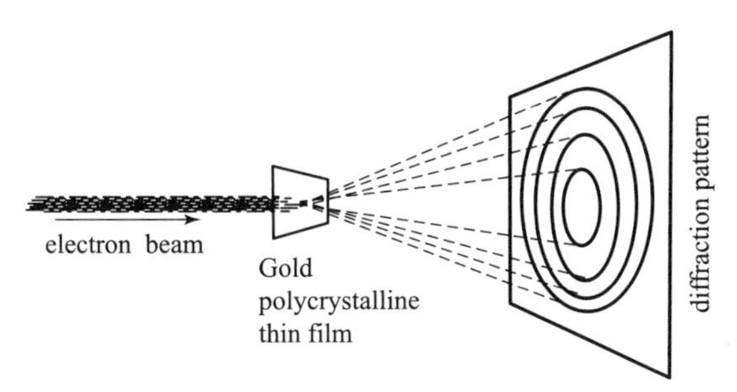

Figure 2 – 43 Schematic diagram of George Thomson's experimental principle

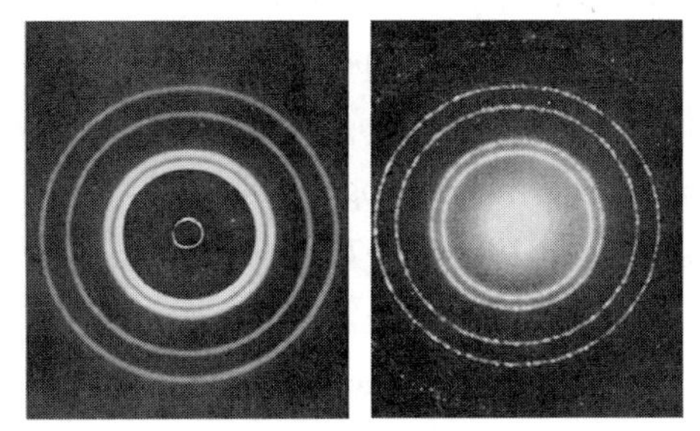

Figure 2 – 44 Electron diffraction pattern (left picture) and X – ray diffraction pattern (right picture) obtained with the same aluminum film

Ten years later, Davidson and Thomson shared the 1937 Nobel Prize in Physics for their achievements in electron diffraction experiments.

(3) Jönsson experiment

In 1961, German scholar C. Jönsson directly performed electron diffraction experiments of single-slit, double-slits, etc. with slits formed on a copper foil. He first carved 5 slits on the copper foil with a slit length of 50 μm, a slit width of 0. 3 μm, and a spacing of 1 μm, and then did experiments with 1 – 5 slits respectively. After being accelerated by a voltage of 50 kV, that is, electrons with a wavelength of about 0. 005 nm were incident vertically, and diffraction patterns similar to that of light were obtained, as shown in Figure 2 – 45.

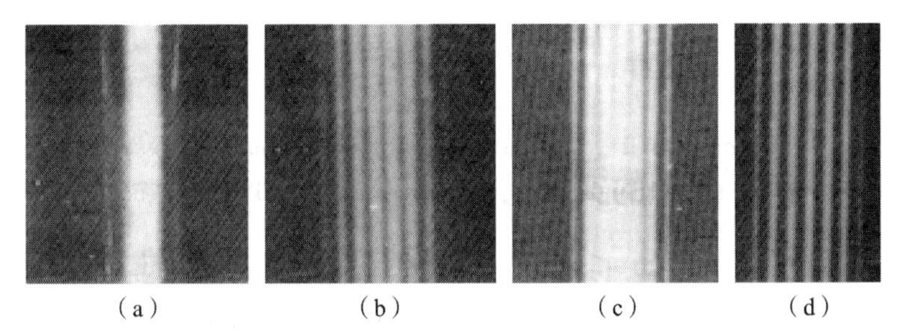

(a) (b) (c) (d)

Figure 2 – 45 Electronic Diffraction patterns of single slit, double slit, etc

(a) single slit; (b) double slits; (c) triple slits; (d) four silts

(4) Crommie experiment

In 1993, American scientist M. F. Crommie et al used scanning tunneling microscopy technology to arrange 48 iron atoms on a copper surface as a circular quantum fence with a radius of 7. 13 nm, and the moving electrons inside the fence formed concentric circular standing wave, as shown in Figure 2 – 46. This is the first intuitive observation of electron standing wave pattern in the world. It also intuitively confirmed

Figure 2 – 46 Quantum fence

the wave property of electron.

After the wave property of electron was confirmed, people used diffraction experiments to further confirmed that microscopic particles such as atom, molecule, neutron and proton all have wave property, and the de Broglie relation is also correct for them. For example, in 1930, German physicists I. Estermann(1900—1973) and O. Stern(1888—1969) produced diffraction with rays of helium atom and hydrogen molecular respectively, which confirmed for the first time that the de Broglie relation can also apply to atom and molecule. In 1988, Austrian physicist A. Zeilinger(1945 –) et al did the double slit experiment of neutrons and obtained the intensity distribution curve reflecting the diffraction phenomenon, as shown in Figure 2 – 47. Therefore, all microscopic particles have wave property. There is no doubt about the existence of matter waves, and the de Broglie relation has become the basic expression to reveal the wave-particle duality of microscopic particles. De Broglie also won the Nobel Prize in Physics in 1929, and he was the first scholar to win the Nobel Prize by his doctoral dissertation.

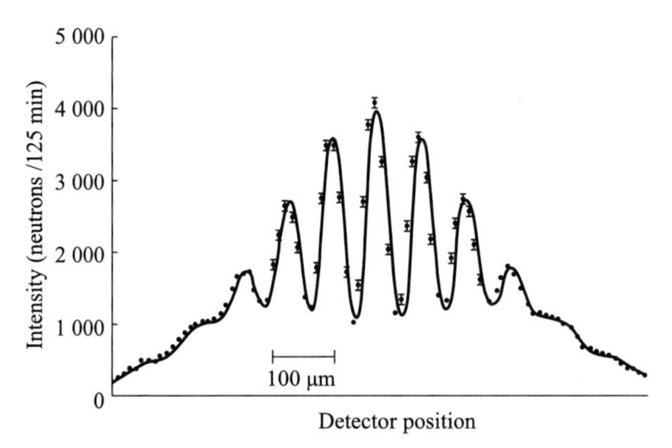

Figure 2 – 47 The intensity distribution curve of neutron double slit experiment done by Zeilinger et al. (**A. Zeilinger et al, Reviews of Modern Physics, Vol. 60, 1988**)

As can be known from wave optics, because the resolution power of the microscope is inversely proportional to the wavelength, the maximum resolving distance of optical microscope is greater than 0.2 μm, and the maximum magnification is only about 1 000 times. Since the discovery of the wave nature of electron, as the de Broglie wavelength of electron beam is much shorter than that of light, and it is very convenient to change the wavelength of electron wave. Thus electron microscope, electron diffraction technology and neutron diffraction technology developed according to the wave property of microscopic particles have become effective means to detect and analyze the microstructure of materials.

Example 2 – 7: Calculate the de Broglie wavelength of a bullet with mass of $m = 0.01$ kg and velocity of $v = 300\ \text{m} \cdot \text{s}^{-1}$.

Solution: Due to $v \ll c$, there is no need to consider relativistic effect. By equation (2 – 51), it can be obtained that

$$\lambda = \frac{h}{m_0 v} = \frac{6.63 \times 10^{-34}}{0.01 \times 300} \text{ m} = 2.21 \times 10^{-34} \text{ m}$$

The result shows that since Planck constant is a very small quantity, the de Broglie wavelength of a macro object is too small to be measured in the experiment. Therefore, under normal circumstances, the wave property of macro objects is hard to show, and they only show particle property.

2.5.3 Born's Statistical Interpretation

(1) Probability wave.

In classical physics, "particle" and "wave" are two distinct concepts. The classical "particle" refers to the substance that can be localized in a small area of space, which shows certain "granular" properties such as mass and charge when interacting with matter. And its movement has the characteristics of a certain orbit, that is, a certain position and velocity at any time. The classical "wave" refers to the periodic change of the spatial distribution of a real physical quantity, with frequency and wavelength, that is, with the periodicity of space and time. And it exhibits the phenomena such as interference and diffraction reflecting the superposition of coherence. Obviously, according to the concept of classical physics, because waves and particles are two different research objects, they have very different performances. Matter is either particle-like or wave-like, or either.

However, modern physics has shown that microscopic particles have wave-particle duality. How can we correctly understand this?

In the early days of the establishment of quantum mechanics, people once regarded the wave-particle duality of micros particles as some kind of superposition of classical particles and classical waves. For example, an electron wave is a wave packet that represents the entity of an electron, and the electron itself is a wave of matter dispersed in space. There is also the view that the wave property of electron is the interaction between many electrons. However, these views eventually must be abandoned because they cannot satisfactorily explain the experimental phenomena. For example, if the electron is a wave packet, the diffraction wave generated when the electron hits the crystal surface will propagate in different directions, and only "a part of the electron" can be observed in different directions of space. This is contradictory to the fact that the electrons measured in the experiment are always one by one. Secondly, in the non-relativistic case, the matter wave packet of a free particle must diffuse, or the electron will become more and more "fat" with time. This is also contradictory to the existing experiments.

In order to understand the wave-particle duality of micros particles, we might as well analyze the electron double-slit diffraction experiment.

In 1974, Italian physicist Pier Giorgio Merli (1943—2008) et al ejected electrons successfully one by one in the physics laboratory at the University of Milan and shot them in turn toward the double slit. With the accumulation of the incident electrons, the image as shown in Figure 2 – 48 was obtained on the detection screen. Where, figure (a) is the image that only one electron reaches the screen, figure (b) is the image formed when more than a dozen electrons arrive on the screen successively, and figure (c) shows the image formed when dozens of electrons arrive on the screen successively. As can be seen from these three figures, every time the electron hits the screen, it is always a bright spot, so the observed electron always arrives at the receiving point "as a whole", which shows the particle property of the electron. And the positions of these bright spots on the screen seem to be irregular, which shows that the whereabouts of electrons are uncertain every time, and where an electron arrives is purely a probabilistic event. However, as time goes by, more and more electrons hit the screen, and the accumulation of electrons on the screen is in turn as shown in figures (d), (e) and (f), which gradually shows fringes, and becomes more and more clear. Finally, the diffraction fringes presented are the same as those formed by a large number of electrons passing through the double slits in a short time (Figure 2 – 45 (b)). This clearly reflects the wave property of a single electron, which is not only the phenomenon that occurs when many electrons gather together in space. Because the second electron is ejected after the previous electron has reached the screen, the diffraction fringes on the screen are not the result of interaction between electrons, but the collective contribution of the wave property of single electron. In addition, the diffraction fringes that appear over time indicate the positions where electrons hit the screen have a certain probability distribution. The probability of electrons falling on bright fringes is high, while the probability of falling on dark fringes is low, that is, the probability of falling on each point is not equal. That is to say, although the whereabout of individual electron is probabilistic, its probability still has certain rules under certain conditions (such as double slits). These are the core of the concept of probability wave proposed by physicist Max Born (German of British nationality, Figure 2 – 49) in 1926.

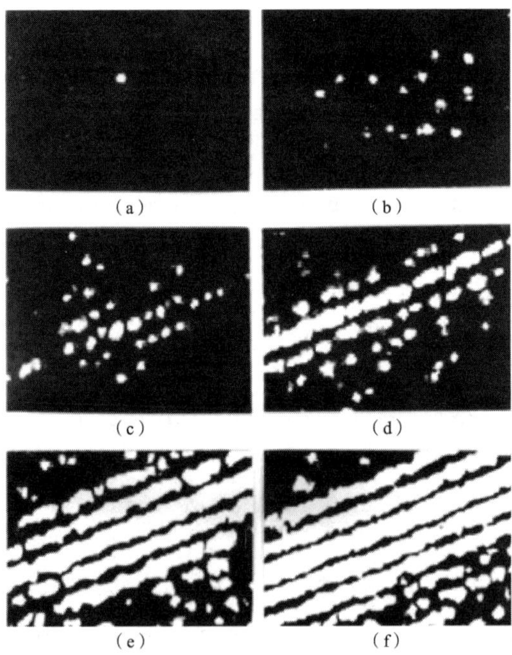

Figure 2 – 48 The electron double-slit diffraction experiment

Figure 2 – 49 Born, 1954 Nobel laureate in physics

Born believed that the de Broglie wave is not like the classical wave which represents the wave of real physical quantity, but it is a probability wave which describes the probability distribution of particles in space. That is, the de Broglie wave is a probability wave. Due to the wave property of microscopic particles, where a particle appears in space is a random behavior. The Particle's movement is not like classical particle moving along the orbit. When and where it appears depend on probability. In the experiment, the double slit diffraction pattern obtained by a large number of electrons shooting at the same time is the same as that single electron shooting for multiple times, indicating that the movement of microscopic particle follows the statistical law. When a single electron is incident multiple times, the bright spots on the screen are randomly distributed at the beginning. As the number of electrons increases, a diffraction pattern is gradually formed, which is the statistical result of an electron repeating the same experiment many times. When a strong electron flow is incident, the diffraction pattern will soon appear on the screen, which is the statistical result of many electrons in the same experiment, and it is also a probability distribution shown by a large number of events. According to Born's concept of probability wave, at the location of the bright fringes in the electron double-slit diffraction pattern the intensity of de Broglie wave is large, and there are more electrons arriving, or in other words, the probability of electrons arriving at the bright fringes is large. On the contrary, at the dark fringes the intensity of de Broglie wave is small, and there are fewer electrons arriving, or in other words, the probability of electrons arriving at the dark fringes is small. This is Born's explanation for the physical meaning of de Broglie wave, that is, the probability density of microscopic particles appearing somewhere is directly proportional to the intensity of de Broglie wave. That is to say, the probability of microscopic particles appearing at each place has obvious physical meaning.

Born's thought of probability wave unifies the particle property and wave property of microscopic particle. This statistical interpretation makes people have a deeper understanding for the wave-particle duality. For other microscopic particles, because they also have wave-particle duality, the matter waves associated with them are also probability waves. That is, the position of a single micros particle is uncertain, but for a large number of microscopic particles, the probability distribution leads to definite macro result, such as diffraction fringes. If a slit is closed, that is to do the electron single-slit diffraction experiment, the resulting image is peculiar one owned by the single slit, as shown in Figure 2-45(a).

The American physicist Richard Phillips Feynman (1918—1988, Figure 2-50) once said: "All of quantum mechanics can be gleaned from carefully thinking through the implications of this single experiment." In the first chapter of Feynman Lectures on Physics III published in 1963, an ideal experiment once designed by Feynman to compare bullets, water wave and electrons passing through double slits respectively was introduced, so as to illustrate the difference between microscopic particles, classical particles and classical wave.

Figure 2-50 Feynman, 1965 Nobel laureate in physics

As shown in Figure 2 – 51 (a) , when a machine gun fires bullets continuously to double slits and the bullets hit the target behind. If only one slit(slit 1 or slit 2) is open and N bullets are fired, the bullet density distribution on the target is curve P_1 or P_2. When both slits are open, $2N$ bullets are fired, and the final bullet density distribution is curve P_3. Obviously, $P_3 = P_1 + P_2$, then the density distribution of bullets on the target after passing through the double slits is equal to the direct addition of the distributions when the two single slits are open separately. There is no interference phenomenon here, which reflects the characteristics of classical particles.

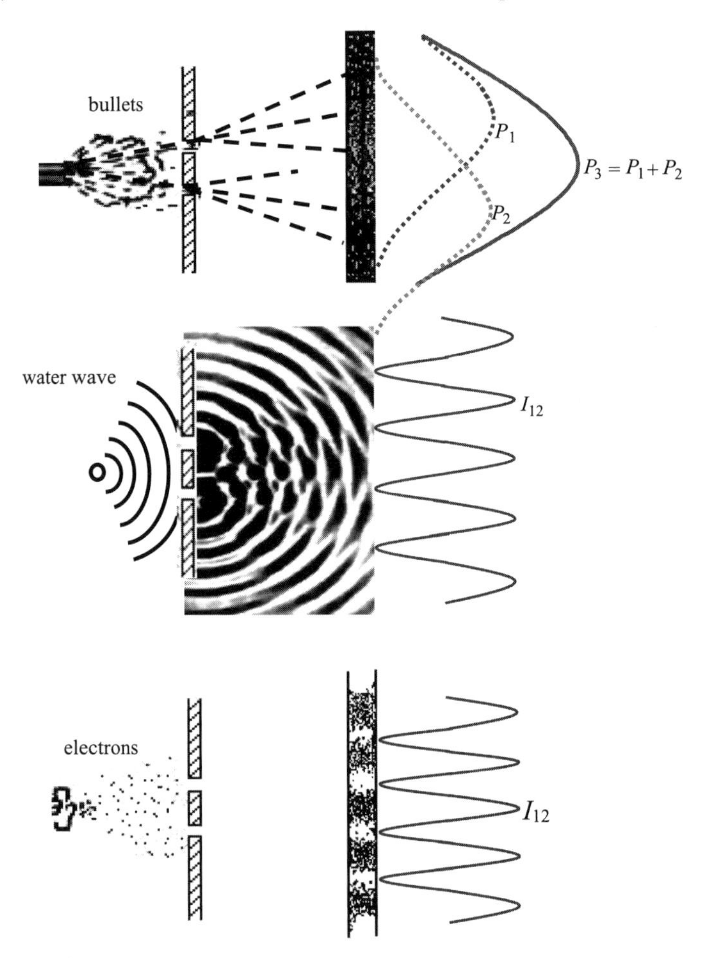

Figure 2 – 51 Schematic diagram of the experimental setup comparing bullets, water wave and electrons passing through double slits

As shown in Figure 2 – 51 (b), when the water wave passes through the double slits, it is divided into two coherent sub wave sources, which are coherently superimposed in space, so it presents a double slit interference pattern. The intensity of the wave on the observation screen is represented by curve I_{12}.

As shown in Figure 2 – 51 (c), this is an electronic double slit experiments. On the one hand, the electrons falling on the detection screen present particle property as bullets. On the other hand, the number distribution curve I_{12} of electrons on the screen presents the double slit interference phenomenon similar to water wave. When only one slit is open, the microscopic particles form a single slit diffraction pattern. When double slits are open at the same time, there are two possibilities for the movement of microscopic particles: maybe passing through slit 1, maybe passing through slit 2. This is different from the possible state of motion in space when only one slit is open, and we should use different probability wave to describe it. In order to obtain the double slit diffraction pattern, in anthropomorphic statement, when electrons pass through this slit, they seem to "know" that another slit is also open, so they act according to the probability under the condition of double slits. That is, when microscopic particles such as electrons and photons pass through the slit, they behave like waves rather than particles, and when they reach the screen, they behave as indivisible particles.

However, in the double slit interference experiment of electrons, if a detector is placed at one of the slits to detect which slit the electrons pass through, the interference pattern on the screen will disappear. In order to make interference happen, obviously, we should not know which slit the particles pass through actually.

Someone once used a scene similar to Figure 2 – 52 to describe this strange behavior of microscopic particles. A skier wearing a pair of sleds is sliding towards a tree. She is not seen until she slipped over the tree. The track she slipped split in two as she passed the tree. There is a slide to the left of the tree and another slide to the right of the tree. You didn't see how the skier completed this miraculous action. Without being observed by others, other people can think that in front of the tree the skier bypasses the tree from both sides of the tree at the same time, goes to the back of the tree, and the skier becomes a so-called wave. But once the skier is observed by the observer, the observer is sure which side the skier bypassed the tree, and the skier is a particle. It depends on whether the skier's information is leaked out to the observer.

Figure 2 – 52 The strange behavior of a skier

This change is due to the complementarity of quantum mechanics, because the motion of matter has the dual properties of particle and wave. But in the same experiment, the two are mutually exclusive. In the double slit interference experiment, measuring which slit a particle passes through is equivalent to emphasizing the particle property of wave-particle duality. The wave property complementary to the particle property is excluded, and the interference fringes no longer exist. This phenomenon of disappearance of coherence due to measurement or other influences is called quantum decoherence. As far as quantum measurement is concerned, it is called wave packet collapse. This is beyond the scope of this course.

In September 2002, double-slit electron diffraction experiment was rated as the No. 1 of the "top ten most beautiful physical experiments in the history of science" by the readers of the American magazine "Physics World".

Through the above comparative experiments, we can see that microscopic particles such as electrons are neither classical particles nor classical waves. The particle property presented by an electron only has the so-called "granularity" or "integrity", that is, it always appears with certain properties such as mass and charge, but it has nothing to do with the concept of "particles have exact orbits". The wave property presented by an electron is just the most essential thing in wave property, which is "the superposition of waves", such as interference and diffraction, and it is not related to the wave of some physical quantity in space. Therefore, the wave-particle duality unifies the "granularity" of microscopic particles and the "superposition" of waves. Moreover, microscopic particles show particle property under some conditions and they show wave property under other conditions. Although these two properties are in the same body, generally they are not presented at the same time.

For example, what do you see in Figure 2 – 53? Is it an old woman? Or a girl? If you look at it from a certain angle, you will see an old woman in the picture. If you look at it from another angle, you will see a girl in the picture. Both images will not appear in your vision at the same time. Similarly, when we observe an electron, generally, it can only show one property, either particle or wave. However, as for the overall concept of electron, it has wave-particle duality. It can show the particle side and can also show the wave side. It all depends on how we observe it.

Figure 2 – 53 Old woman or girl?

(2) Wave function.

In quantum mechanics, in order to reflect the wave-particle duality of a microscopic particle, its motion state can be described by wave function, denoted with $\Psi(\boldsymbol{r},t)$, which is a function of time and space. In general, the wave function of microscopic particle is a complex function. In 1926, Born gave a statistical interpretation of the wave function on the basis of the concept of probability wave. The content is that the wave function $\Psi(\boldsymbol{r},t)$ itself has no direct physical meaning, but the square of its modulus $|\Psi(\boldsymbol{r},t)|^2 = \Psi^*(\boldsymbol{r},t)\Psi(\boldsymbol{r},t)$ represents the probability of finding particle in unit volume near the spatial coordinate r at time t, that is, the probability density of particle occurring. Where $\Psi^*(\boldsymbol{r},t)$ is the complex conjugate of $\Psi(\boldsymbol{r},t)$, and $\Psi^*(\boldsymbol{r},t)$ can be obtained by changing the imaginary number i in $\Psi(\boldsymbol{r},t)$ to $-$i. In a small volume element $\mathrm{d}V$ in space, the wave function can be regarded as invariant, so the probability of finding the particle in the volume element $\mathrm{d}V = \mathrm{d}x\mathrm{d}y\mathrm{d}z$ is proportional to

$$|\Psi|^2\mathrm{d}V = \Psi\Psi^*\mathrm{d}V \tag{2-56}$$

Since the intensity of any wave is proportional to the square of the amplitude of the corresponding wave function, the de Broglie wave associated with microscopic particles is a probability wave. As it can be seen from the analysis of the electron double-slit diffraction experiment that the intensity of de Broglie wave is proportional to the probability of microscopic particle somewhere, so the wave function $\Psi^*(\boldsymbol{r},t)$ is also called the probability amplitude.

According to Born's statistical interpretation, since $|\Psi|^2$ represents the probability density, and the probability should be single valued, finite and continuous, the wave function itself is required to satisfy the condition of being single valued, finite and continuous, which is called the standard condition for the wave function.

In addition, because the particle must appear at some point in space, the total probability of finding the particle in the whole space at any time t should be 1. Therefore, there is a normalizing condition for the wave function

$$\int_{-\infty}^{\infty} |\Psi(\boldsymbol{r},t)|^2 \mathrm{d}V = 1 \tag{2-57}$$

where the volume integral covers the whole space.

For matter wave (probability wave), the wave function $\Psi(\boldsymbol{r},t)$ and the wave function $C\Psi(\boldsymbol{r},t)$ (C is a complex constant) represent the same state of microscopic particle, because the probability distribution of microscopic particle in space only depends on the relative intensity of the wave at each point in space, but does not depend on the absolute magnitude of the intensity. It is not difficult to see that for the $C\Psi(\boldsymbol{r},t)$ state, the ratio of the probability densities of microscopic particle at any two points $\boldsymbol{r}_1$ and $\boldsymbol{r}_2$ in space, that is, the relative probability density is

$$\frac{|C\Psi(\boldsymbol{r}_1,t)|^2}{|C\Psi(\boldsymbol{r}_2,t)|^2} = \frac{|\Psi(\boldsymbol{r}_1,t)|^2}{|\Psi(\boldsymbol{r}_2,t)|^2}$$

It is exactly the same as the relative probability density of the $\Psi(r,t)$ state. Therefore, the wave function can always differ by an arbitrary complex constant factor C, which is different from waves in classical physics. If the amplitude of a classical wave is doubled, the corresponding wave energy will be quadrupled the original, thus it represents different wave state. Just because of this, the classical wave does not involve "normalized" at all. For the probability wave, normalization or not does not affect the relative probability density of microscopic particle, that is, it does not affect its probability distribution. Usually for convenience, an appropriate constant factor is chosen so that the wave function satisfies the normalization condition.

If the wave function $\Phi(\boldsymbol{r},t)$ has not been normalized, then we can always make

$$\Psi(\boldsymbol{r},t) = A\Phi(\boldsymbol{r},t) \tag{2-58}$$

where A is an undetermined normalization constant. By the normalization condition given by equation (2-57), we have

$$\int_{-\infty}^{\infty} |\Psi(\boldsymbol{r},t)|^2 \mathrm{d}V = |A|^2 \int_{-\infty}^{\infty} |\Phi(\boldsymbol{r},t)|^2 \mathrm{d}V = 1$$

As can be obtained from this

$$A = \left[\int_{-\infty}^{\infty} |\Phi(\boldsymbol{r},t)|^2 \mathrm{d}V\right]^{-1/2} \tag{2-59}$$

Substituting the above formula into equation (2 - 58), the normalized wave function $\Psi(\boldsymbol{r},t)$ is obtained. Of course, $\Psi(\boldsymbol{r},t)$ and $\Phi(\boldsymbol{r},t)$ represent the same probability wave.

The meaning of equation (2 - 57) can be interpreted as that the probability of the particle appearing in the whole space must be equal to 1. After normalization, the square of the wave function modulus $|\Psi(\boldsymbol{r},t)|^2$ is called the probability density of particle distribution, and $|\Psi(\boldsymbol{r},t)|^2 \mathrm{d}V$ represents the probability of finding the particle in the volume $\mathrm{d}V$ near point $\boldsymbol{r}$ in space at time t.

In quantum mechanics, the wave function describing the motion state of microscopic particle must also satisfy the state superposition principle. Its content is that if $\Psi_1(\boldsymbol{r},t)$, $\Psi_2(\boldsymbol{r},t)$, $\Psi_3(\boldsymbol{r},t)$, $\cdots$ represent a series of different possible states of the system, then their linear combination $\Psi(\boldsymbol{r},t) = C_1\Psi_1(\boldsymbol{r},t) + C_2\Psi_2(\boldsymbol{r},t) + C_3\Psi_3(\boldsymbol{r},t) + \cdots$ is also a possible state of the system, where $C_1, C_2, C_3, \cdots$ are arbitrary complex constants. The state superposition principle is a fundamental principle of quantum mechanics.

The state superposition principle also has the following meaning. When a particle is in the linear superposition state $\Psi(\boldsymbol{r},t)$ of state $\Psi_1(\boldsymbol{r},t)$ and state $\Psi_2(\boldsymbol{r},t)$, the particle is in both state $\Psi_1(\boldsymbol{r},t)$ and state $\Psi_2(\boldsymbol{r},t)$.

For example, in the electron double-slit diffraction experiment, if the wave functions $\Psi_1(\boldsymbol{r},t)$ and $\Psi_2(\boldsymbol{r},t)$ are used to represent the states of electron passing through slit 1 and slit 2 respectively, then when the double slits are all open, an electron is in the state $\Psi_1(\boldsymbol{r},t)$ and the state $\Psi_2(\boldsymbol{r},t)$ at the same time, and the state of the electron induced simultaneously by the double slits is the linear superposition state of $\Psi_1(\boldsymbol{r},t)$ and $\Psi_2(\boldsymbol{r},t)$, that is, $\Psi(\boldsymbol{r},t) = \Psi_1(\boldsymbol{r},t) + \Psi_2(\boldsymbol{r},t)$. According to Born's statistical interpretation of the wave function, the probability distribution of finding electron on the screen is $|\Psi(\boldsymbol{r},t)|^2$, we have

$$|\Psi(\boldsymbol{r},t)|^2 = |\Psi_1 + \Psi_2|^2 = |\Psi_1|^2 + |\Psi_2|^2 + (\Psi_1^* \Psi_2 + \Psi_1 \Psi_2^*) \qquad (2-60)$$

where $|\Psi_1|^2$ represents the probability density of electron appearing on the screen when there is only slit 1 open, and $|\Psi_2|^2$ represents the probability density of electron appearing on the screen when there is only slit 2 open. When two slits are open at the same time, in addition to $|\Psi_1|^2$ and $|\Psi_2|^2$, there is an interference term $\Psi_1^* \Psi_2 + \Psi_1^* \Psi_2$, which is the reason for the double-slit diffraction pattern.

Although the superposition principle of states in quantum mechanics is the same mathematically as that of classical wave, there is a fundamental difference in physical essence. The superposition of quantum states refers to the superposition of two states of a particle, and its interference is also the interference between the two states, by no means two particles interfere with each other. So no matter if the experiment is done under the incident condition of strong particle flowing, or weak particle flowing, or let the particles enter one by one repeatedly, the diffraction fringe results obtained are the same.

Born's statistical interpretation unifies the wave property and the particle property of microscopic particle. It can satisfactorily explain the experimental phenomena encountered. Therefore, it was quickly recognized by the majority of physicists and became a basic hypothesis of quantum mechanics. Born shared the Nobel Prize in Physics in 1954 (with the German physicist W. Bothe

(1891—1957)) for his fundamental research in quantum mechanics, especially his statistical interpretation of wave function.

Example 2 - 8: Normalize the wave function $f(x) = e^{-\alpha^2x^2/2}$.

Solution: Assume the normalization factor is A, then the wave function is

$$\psi(x) = Ae^{-\alpha^2x^2/2}$$

Substitute it into the normalization condition given by equation (2 - 57), we have

$$\int_{-\infty}^{+\infty} A^2 e^{-\alpha^2x^2} dx = 1$$

Solve the integral, it can be obtained that

$$A = \left(\frac{\alpha}{\pi^{1/2}}\right)^{1/2}$$

Then the normalized wave function is

$$\psi(x) = \left(\frac{\alpha}{\pi^{1/2}}\right)^{1/2} e^{-\alpha^2x^2/2}$$

2.5.4 Wave Function of Free Particle

For a free particle, because it is not affected by external force, it moves in a straight line at a uniform speed, and its momentum and energy are both constant. Therefore, according to the de Broglie hypothesis, the frequency and wavelength of the de Broglie wave associated with the free particle are also constant. In wave theory, any wave whose frequency and wavelength are both determined values is a planar simple harmonic wave, which is the simplest kind of wave.

In classical wave theory, the wave function of a planar simple harmonic wave with angular frequency ω and angular wavenumber k propagating along the positive direction of x-axis is given by

$$y(x,t) = A\cos(\omega t - kx) \tag{2-61}$$

Its complex form is

$$\tilde{y}(x,t) = Ae^{-i(\omega t - kx)} \tag{2-62}$$

In quantum mechanics, if a free particle moves along the x - axis with momentum p_x and energy E, its corresponding de Broglie wave can be represented by frequency $\nu = E/h$ or angular frequency $\omega = 2\pi\nu = E/\hbar$ and wavelength $\lambda = h/p_x$ or angular wavenumber $k = 2\pi/\lambda = p_x/\hbar$, where $\hbar = h/2\pi = 1.054\ 571\ 800(13) \times 10^{-34}\ \mathrm{J \cdot s} \approx 1.05 \times 10^{-34}\ \mathrm{J \cdot s}$ is called the reduced Planck constant. The wave function of the de Broglie wave can be written as a complex expression similar to the classical equation (2 - 62) of planar simple harmonic wave, as long as the parameters ω and k that describe the wave property are expressed as the parameters p_x and E describing the particle property. To distinguished from classical wave theory, $\tilde{y}(x,t)$ will be changed to $\Psi(x,t)$, because for the de Broglie wave (matter wave or probability wave), the wave function is no longer a real physical quantity. Then the wave function of a free particle can be written in the form

$$\Psi(x,t) = Ae^{\frac{i}{\hbar}(p_x x - Et)} \tag{2-63}$$

where A represents the normalization constant. This is the wave function of matter wave of the free particle with momentum p_x and energy E propagating along the x direction.

For a free particle moving with momentum $\boldsymbol{p} = p_x\boldsymbol{i} + p_y\boldsymbol{j} + p_z\boldsymbol{k}$ and energy E in any direction in three-dimensional space, its wave function is

$$\Psi(\boldsymbol{r},t) = \Psi_0 e^{-\frac{i}{\hbar}(Et-\boldsymbol{p}\cdot\boldsymbol{r})} \tag{2-64}$$

where $\boldsymbol{p} \cdot \boldsymbol{r} = p_x x + p_y y + p_z z$. In the non-relativistic case, the energy of the free particle is

$$E = \frac{p_x^2 + p_y^2 + p_z^2}{2m} \tag{2-65}$$

where m is the mass of the particle. Because free particle is not affected by external force field, its kinetic energy is the total energy E.

If the particle is subjected to an external force in a force field and is not a free particle, then the matter wave of the particle is no longer a planar simple harmonic wave, and its wave function is more complicated.

Schrodinger Cat

There is controversial debate for the statistical interpretation of Bonn. In 1935, Austrian physicist Owen Schrodinger proposed a paradox. As shown in Figure 2 – 54, there is a cat and a sealed glass container containing highly toxic substance such as cyanide in a closed box. A hammer is hung directly above the glass container. The hammer is controlled by gears operated by a relay, and the relay is controlled by Geiger counter switch. There is also a trace of radioactive material in the box. Its half-life guarantees that within two hours there will be an atom decaying and releasing a particle. Its radiation triggers the counter switch, and the hammer will fall through the action of relay, smash the glass container below, release cyanide gas inside, and the cat will die immediately.

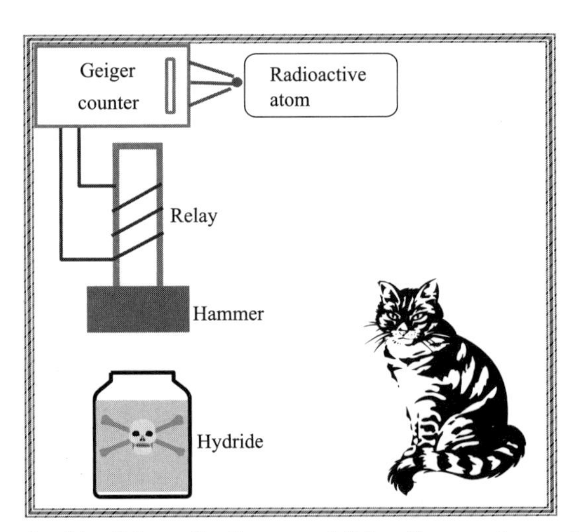

Figure 2 – 54 Schematic diagram of Schrodinger cat experiment

If the experiment is only performed for an hour, for radioactive material in the box during this period there may be an atom decaying or there may be no such decay. According to quantum theory, the cat is both dead and alive. It is in the superposition state of neither dead nor alive. For a macro animal cat, it is obviously absurd, because the cat is either dead or alive. However, quantum mechanics believes that a microscopic particle can be in a superposition state of "neither dead nor alive", that is, "dead and alive". This is the origin of the Schrödinger cat paradox, which has puzzled people for a long time.

However, since the 1990s, scientists have done a series of "electron cat", "photon cat", "atom cat" and "ion cat" experiments. They use microscopic "kittens", such as electron, photon, atom or ion, to prove that these "kittens" are indeed in the superposition state of "dead and alive". For example, in 1991, C. R. Stroud and J. A. Yeazell at the University of Rochester did experiment with the highly excited state of potassium atom and found that an electron formed two separate wave packets at different spatial positions in the range of 0.5 μm ~ 10^{-7} m away from the nucleus, which showed that one part of an electron was indeed in a certain position and the other part was in another position. We regard one spatial position as "dead state" and the other spatial position as "alive state", then the electron cat is in a superposition state of "dead and alive". In 1996, D. J. Wineland (1944 –) at the National Institute of Standards and Technology in the United States confirmed that a beryllium ion has two states of spin-up and spin-down in two spatial locations separated by a distance of 80 nm. This is a "dead and alive" "ion cat". In 1997, S. Haroche and etc. (1944 –) at the Ecole Normale Super Paris (ENS) sent highly excited atom into the cavity, so that a photon generated two different vibration states, which confirmed the "dead and alive" "photon cat". Haroche and Wineland also won the 2012 Nobel Prize in Physics for their ground-breaking research in the field of quantum mechanics that "enable measuring and manipulation of individual quantum systems". This method of measuring quantum particles without destroying them makes it possible for humans to build a new type of super-fast computer that is much faster than any computer in the world today. The statement issued by the Royal Swedish Academy of Sciences that the review committee belongs to, claimed: "Perhaps the quantum computer will change our everyday lives in this century in the same radical way as the classical computer did in the last century."

Although so far all experiments have confirmed that the statistical interpretation is correct, this debate on the fundamental issues of quantum mechanics has not only promoted the development of quantum mechanics, but also laid the foundation for the birth of some emerging disciplines such as quantum information theory.

2.6 Uncertainty Relation

In classical mechanics, particles move along certain orbits, and any point on the orbits has a determined position and momentum. However, for microscopic particles, the wave property makes microscopic particles have no definite orbit, that is, coordinate and momentum cannot take the definite values at the same time, and there is an uncertainty relation. Let's first take the electron single slit diffraction experiment as an example to illustrate it.

As shown in Figure 2 – 55, an electron beam penetrates perpendicularly to a single slit with slit width a, and a single slit diffraction pattern appears on the screen. Although we don't know at which point each electron passes through the slit, that is, we cannot accurately determine the x coordinate where the electron passes through the single slit. However, we can say that the uncertainty range of the position of the electron in the x direction when it passes through the single slit, that is, the uncertainty is $\Delta x = a$. On the other hand, because of diffraction effect, after the electron passes through the slit, its momentum p_x in the x direction is also uncertain. Δp_x is used to represent the momentum uncertainty in the x direction. Since most of the electrons fall within the central principal maximum, roughly speaking, the uncertainty of p_x is

$$\Delta p_x = p\sin\theta_1 \quad (2-66)$$

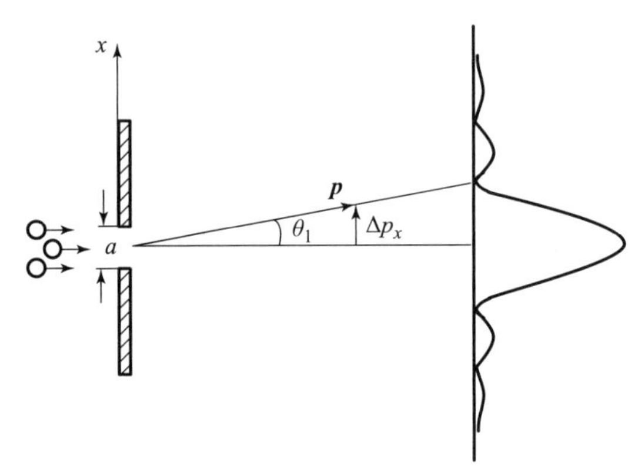

Figure 2 – 55 Using electron diffraction to explain uncertainty relation

where θ_1 is the half-angular width of the central principal maximum, which satisfies $a\sin\theta_1 = \lambda$. From this we can get

$$\Delta x \cdot \Delta p_x = ap\sin\theta_1 = p\lambda \quad (2-67)$$

Using the de Broglie relation $p = h/\lambda$ again, it can be obtained that

$$\Delta x \cdot \Delta p_x = h \quad (2-68)$$

Considering that some electrons may fall to the secondary maximum, therefore, the actual momentum uncertainty will be larger, which can be obtained

$$\Delta x \cdot \Delta p_x \geqslant h \quad (2-69)$$

This is the basic relation that must be satisfied between the uncertainty of the x coordinate and the uncertainty of the corresponding momentum component p_x, which is called uncertainty relation. It applies not only to electron, but also to other microscopic particles. This relation shows that the position and momentum of microscopic particles cannot be measured accurately at the same time. The more accurate the coordinate measurement of microscopic particles in the x direction, that is, the smaller the uncertainty of coordinate Δx, the less accurate the measurement of momentum in this direction, that is, the greater the uncertainty of momentum Δp_x, the more obvious the diffraction phenomenon is when the electron single slit experiment is carried out, and vice versa. Therefore, an attempt to determine the position and momentum of a microscopic particle at the same time is impossible and doesn't make sense. Because of this, the concept of orbit is meaningless for microscopic particles.

In 1927, the German physicist Werner Heisenberg (Figure 2 – 56) gave a more rigorous conclusion by quantum mechanics. In the x direction, the more accurate result about the uncertainty relation between position and momentum should be

$$\Delta x \cdot \Delta p_x \geqslant \frac{\hbar}{2} \qquad (2-70)$$

There is also uncertainty relation in the y and z directions

$$\Delta y \cdot \Delta p_y \geqslant \frac{\hbar}{2} \qquad (2-71)$$

$$\Delta z \cdot \Delta p_z \geqslant \frac{\hbar}{2} \qquad (2-72)$$

Figure 2-56 Heisenberg, 1932 Nobel laureate in physics

In quantum mechanics, there is a similar uncertainty relation between energy and time. According to the relation between energy E and momentum p in the theory of relativity

$$E = (c^2p^2 + m_0^2c^4)^{1/2}$$

Differentiating both sides at the same time, the energy uncertainty is

$$\Delta E = \frac{1}{2}(c^2p^2 + m_0^2c^4)^{-1/2} \cdot 2c^2p\Delta p = \frac{c^2p\Delta p}{E} = \frac{p\Delta p}{m} = v\Delta p$$

Suppose the particle moves in the x direction, then

$$\Delta E = v\Delta p_x \qquad (2-73)$$

And by the uncertainty of the particle's position

$$\Delta x = v\Delta t$$

The time uncertainty is

$$\Delta t = \frac{\Delta x}{v} \qquad (2-74)$$

Multiply Equation (2-73) with Equation (2-74),

$$\Delta E \cdot \Delta t = \Delta x \cdot \Delta p_x$$

Using the uncertainty relation between coordinate and momentum [equation (2-70)], we have

$$\Delta E \cdot \Delta t \geqslant \frac{\hbar}{2} \qquad (2-75)$$

Where ΔE represents the undetermined range, that is, the uncertainty of the energy of the microscopic particle in a certain state, Δt represents the time interval that the particle stays in the energy state.

If the product of two physical quantities has the same dimension (J · s) as Planck's constant h, they are called conjugate quantities, such as position and momentum, energy and time, etc. It can be proved that all two conjugate quantities satisfy the uncertainty relation.

The uncertainty relation is caused by the wave property of matter particle. In microscopic problems, it is often used to estimate the order of magnitude. Suppose the lifetime of an atom in an excited state energy level $\tau = \Delta t$ and the corresponding natural width of the energy level $\Gamma = \Delta E$, according to equation (2-75), we have

$$\tau\Gamma \geqslant \frac{\hbar}{2} \tag{2-76}$$

Theoretically, this relation can be used to estimate the width Γ of the energy level by calculating the average lifetime of the state. Experimentally, this relation can be used to estimate the average lifetime τ of the state according to the width of the energy level.

The uncertainty relation is an inevitable reflection of the wave-particle duality and a manifestation of the inherent properties of microscopic particles. It must be emphasized that the two conjugate quantities cannot be measured accurately at the same time is not caused by the measurement method and measuring instrument, and it is also different from the measurement error, because such error can be reduced by improving the experimental means. The uncertainty relation is the objective law of the motion of microscopic particles. No matter how high the precision of the measuring instrument is, the uncertainties Δx, Δp_x or Δt, ΔE of the two conjugate physical quantities cannot be reduced indefinitely at the same time.

Interestingly, the Austrian physicist W. Pauli (1900—1958) explained popularly in a letter to Heisenberg in October 1926, "One may view the world with the p-eye (Figure 2-57(a)), and one may view it with the q-eye (Figure 2-57(b)), but if one opens both eyes simultaneously then one gets crazy (Figure 2-57(c))." Where the p-eye refers to momentum and the q-eye refers to the position.

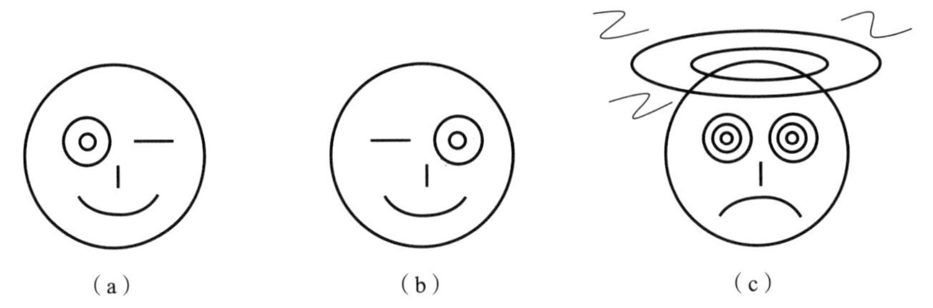

(a) (b) (c)

Figure 2-57 Pauli's popular explanation on the uncertainty relation

The uncertainty relation also indicates the dividing line between macrophysics and microphysics. In a specific problem, if the reduced Planck constant $\hbar$ is a trivial quantity, it can be regarded as $\hbar = 0$, which means that coordinate and momentum can be measured accurately at the same time, and then classical mechanics is applicable. If $\hbar$ cannot be negligible, the uncertainty relation must be used, the wave-particle duality must be considered for microscopic particles, and the method of quantum mechanics must be adopted. Heisenberg was awarded the Nobel Prize in Physics in 1932 for his contributions to quantum theory. It is said that Heisenberg made an epitaph for himself: "he lies somewhere here". Literally, he is here and elsewhere.

Example 2-9: Assuming the mass of a bullet is 0.01 kg and the muzzle diameter is 0.5 cm, find the uncertainty of the bullet's velocity.

Solution: Let the direction of the bullet shooting from the muzzle along the barrel be the y-axis, and the x-axis be perpendicular to this direction, that is, the transverse direction. The muzzle diameter can be regarded as the uncertainty of transverse position when the bullet is ejected out of the muzzle, $\Delta x = 0.5$ cm. By the uncertainty relation (equation (2 – 70)), the uncertainty of the transverse velocity is

$$\Delta v_x \geqslant \frac{\hbar}{2m\Delta x} = \frac{1.05\times10^{-34}}{2\times10^{-2}\times0.5\times10^{-2}}\ \mathrm{m\cdot s^{-1}} = 1.05\times10^{-30}\ \mathrm{m\cdot s^{-1}}$$

This is also the maximum transverse velocity of the bullet, which is far less than the velocity of several hundred meters per second when the bullet is shot out of the muzzle. Here, the reduced Planck constant $\hbar$ is an extremely small quantity, its order of magnitude is about 10^{-34}. Therefore, the uncertainty relation doesn't have any practical effect on shooting and aiming with macroscopic objects such as bullets. The motion of the bullet hardly shows wave-particle duality. So for macroscopic objects, the concept of orbit is meaningful.

Example 2 – 10: The linearity of an atom is estimated by 10^{-10} m, the kinetic energy of the electron in the atom is estimated by 10 eV. Find the uncertainty of the velocity of the electron in the atom.

Solution: The linearity of an atom is the uncertainty of the position of the electron in the atom, i. e. $\Delta x = 10^{-10}$ m. By the uncertainty relation

$$\Delta x\cdot\Delta p_x \geqslant \frac{\hbar}{2}$$

The uncertainty of the velocity of the electron is

$$\Delta v \geqslant \frac{\hbar}{2m\Delta x} = \frac{1.05\times10^{-34}}{2\times9.11\times10^{-31}\times10^{-10}}\ \mathrm{m\cdot s^{-1}} = 0.6\times10^{6}\ \mathrm{m\cdot s^{-1}}$$

According to classical mechanics, the velocity of the electron has the magnitude

$$v = \sqrt{\frac{2E_\mathrm{k}}{m}} = \sqrt{\frac{2\times10\times1.6\times10^{-19}}{9.11\times10^{-31}}}\ \mathrm{m\cdot s^{-1}} = 2\times10^{6}\ \mathrm{m\cdot s^{-1}}$$

It can be seen that since the uncertainty of the velocity Δv and the velocity v itself have the same order of magnitude, we can think that the velocity of the electron is completely uncertain. Therefore, it is meaningless to talk about its velocity, and the position of the particle at the next moment is completely uncertain. So for the movement of electrons in atoms, the concept of orbit has lost its meaning.

In some problems, one of position and momentum has a large uncertainty, and the other can be considered to be accurately determined. For example, observing particles in a cloud chamber takes advantage of this. After the electron enters the cloud chamber, a string of small droplets are formed at the place where it passes to show the orbit of the electron. The orbit width is about 10^{-5} m, and the position uncertainty can be considered to be $\Delta x = 10^{-5}$ m, then

$$\Delta v_x \geqslant \frac{\hbar}{2m_e\Delta x} = \frac{1.05\times10^{-34}}{2\times9.11\times10^{-31}\times10^{-5}}\ \mathrm{m\cdot s^{-1}} = 5.8\ \mathrm{m\cdot s^{-1}}$$

And the speed of electron can be close to the speed of light, Δv_x can be ignored, so its speed can be considered to be accurately determined.

Example 2 – 11: A photon propagates in the x direction with a wavelength of 500 nm. It is known that the uncertainty of the wavelength is $\Delta\lambda = 5\times10^{-8}$ nm, find the uncertainty of the coordinate of the photon in the x direction.

Solution: The uncertainty relation should also be applicable to the photon. Because $p_x = h/\lambda$, differentiate both sides, we have

$$\Delta p_x = \frac{h}{\lambda^2}\cdot\Delta\lambda \tag{2-77}$$

Therefore, Δp_x can be obtained from $\Delta\lambda$. When making estimation, if there is no clear requirement, equation (2 – 69) or equation (2 – 70) of the uncertainty relation can be used to make estimation, and $\Delta x\Delta p_x \geqslant h/2$ or $\Delta x\Delta p_x \geqslant \hbar$ can also be used to make estimation. Here, we use the last one to estimate, then

$$\Delta x \geqslant \frac{\hbar}{\Delta p_x} = \frac{\hbar}{\frac{h}{\lambda^2}\Delta\lambda} = \frac{\lambda^2}{2\pi\ \Delta\lambda} = \frac{(500\times10^{-9})^2}{2\times3.14\times5\times10^{-8}\times10^{-9}}\ \text{m} = 800\ \text{m}$$

The x direction is the propagation direction of the wave train, Δx can represent the extension range of the wave train along the propagation direction, and it can be regarded as the length of the wave train. According to the viewpoint of the particle property that an atom emits a photon in the process of one transition between energy levels, or the viewpoint of the wave property that a wave train is emitted, the conclusion that the position uncertainty of a photon is also the length of the wave train can be understood.

This example tells us that the higher the wavelength accuracy of light, that is, the better the monochromaticity, the worse the accuracy of the photon position. When the wavelength is extremely accurate, the photon coordinate is very uncertain. At this time, the wave property of light is prominent, while the particle property is not significant.

Figure 2 – 58 shows the waveforms of three wave trains with different wave train lengths at a certain time. It can be seen that for these three wave trains the uncertainty of the coordinate in the x-direction Δx is very different. The wave train (a) is in a very uncertain coordinate range x_1 to x'_1, the wave train (b) has a smaller uncertainty coordinate range x_2 to x'_2, while the wave train (c) has fairly certain coordinate value, that is, it is in a small interval near x_3. However, for these three wave trains the uncertainty of the wavelength is completely opposite. The wave train (a) is a longer wave train, which contains more complete waveforms. Within a considerable range, each wave regularly occupies a certain space interval, that is, it has relatively certain wavelength. In other words, the longer the wave train, the more complete waveforms it contains, and the more certain the wavelength is. The wave train (b) contains only a few complete waveforms, and its wavelength is less certain than that of the wave train (a), but its coordinate is more certain than that of the wave train (a). The wave train (c) has a very certain position and a very uncertain wavelength, which also explains the constraint relation between the uncertainty of coordinate and the uncertainty of wavelength. In this example, the length of the wave train is about 800 m, and the wavelength is 5×10^{-7} m. This wave train contains about 10^9 complete waveforms, so the degree of certainty of its wavelength is very high, $\Delta\lambda/\lambda$ reaches the order of 10^{-10}. It can be considered that a wave with a fully certain wavelength is an infinitely long sinusoidal wave train. At this time, the uncertainty of the wavelength is $\Delta\lambda = 0$, and the uncertainty of the momentum is also zero, but the length of the wave train, that is, the coordinate, extends to infinity, $\Delta x \to \infty$, which is completely uncertain.

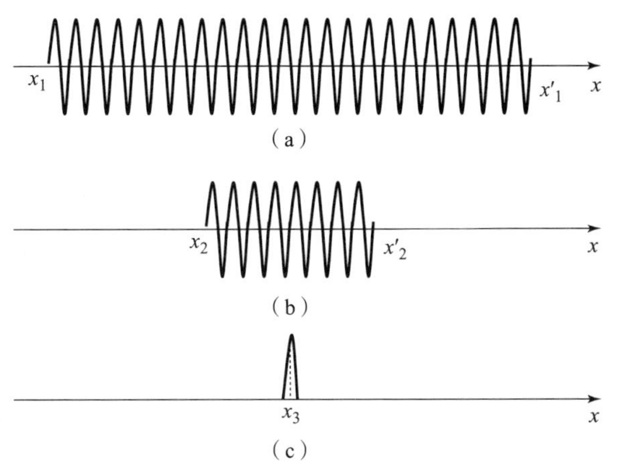

Figure 2 - 58 Three wave trains with different wave train lengths

The above discussion shows that through the de Broglie relation, the uncertainty of momentum Δp_x can be transformed into the uncertainty of wavelength $\Delta\lambda$, and the uncertainty relation between coordinate and momentum can be transformed into the constraint relation between coordinate and wavelength. For a classical wave train, there is no uncertainty relation between coordinate and momentum at all, because momentum is only a quantity related to the motion of particle. Therefore, it can be seen that the uncertainty relation between coordinate and momentum is the reflection of wave-particle duality of microscopic particles.

Example 2 - 12: Find the minimum possible energy, that is, the zero-point energy of a linear harmonic oscillator with a natural frequency ν or a natural angular frequency ω.

Solution: The linear harmonic oscillator vibrates near the equilibrium position along a straight line, and both the coordinate x and momentum p have certain limits. Therefore, the minimum energy can be calculated by the uncertainty relation between coordinate and momentum.

The energy of a linear harmonic oscillator with mass m and moving in the x direction can be expressed as

$$E = \frac{p_x^2}{2m} + \frac{1}{2}m\omega^2 x^2$$

Since the oscillator vibrates near the equilibrium position, we can take $\Delta x \approx x$, $\Delta p_x \approx p_x$, and substitute them into the above formula, then

$$E = \frac{(\Delta p_x)^2}{2m} + \frac{1}{2}m\omega^2(\Delta x)^2$$

According to the uncertainty relation between coordinate and momentum [equation (2 - 70)]

$$E \geqslant \frac{\hbar^2}{8m(\Delta x)^2} + \frac{1}{2}m\omega^2(\Delta x)^2$$

When calculating the minimum possible energy of the linear harmonic oscillator, the above formula takes the equal sign, we can get

$$E = \frac{\hbar^2}{8m(\Delta x)^2} + \frac{1}{2}m\omega^2(\Delta x)^2 \quad ①$$

Let

$$\frac{dE}{d(\Delta x)} = -\frac{\hbar^2}{4m(\Delta x)^3} + m\omega^2(\Delta x) = 0$$

That is,

$$(\Delta x)^2 = \frac{\hbar}{2m\omega} \quad ②$$

Substituting equation ② into equation ①, the minimum possible energy of the linear harmonic oscillator can be obtained as

$$E_{\min} = \frac{1}{2}\hbar\omega = \frac{1}{2}h\nu \quad (2-78)$$

A Short Story

In 1930, at the sixth Solvay Conference on Physics held in Brussels, Einstein proposed an ideal experiment for a photon box, as shown in the left sketch of Figure 2 – 59. The structure of the photon box is very simple. It is a box with an ideal reflecting wall, there is full of radiation inside the box. There is a small hole on the box, and there is a shutter that can control its opening and closing, which is controlled by the clock in the box. The opening and closing time of the shutter can be as short as allowing only one photon to fly out from the box through the small hole at a time. As long as before and after the photon is released the change in the mass of the whole box is measured, the energy change ΔE in the box can be accurately measured according to the mass-energy relation $E = mc^2$ in the theory of relativity. Because the mass measurements are made before and after the shutter being opened and closed, and have nothing to do with Δt, so ΔE and Δt are very certain, the uncertainty relation as given by equation (2 – 75) does not hold, and there is no problem that time and energy can be determined at the same time.

In the next morning after Einstein raised the issue, Bohr also spoke at the Solvay conference. He first drew a sketch on the blackboard similar to the one on the left of Figure 2 – 59, as shown on the right in Figure 2 – 59, which was actually an improvement on Einstein's sketch. Bohr pointed out that a photon ran away and the box became lighter by Δm. We can measure this Δm with a spring scale. Suppose the box is hung under the spring scale, there is a pointer installed on the box, and the position of the pointer can be read from the ruler to measure the mass of the box. After the photon box emits a photon, its mass decreases, even though it can be accurately measured. However, the box and the clock in it have displacement in the gravitational field of the earth, because only in the gravitational field can it be possible to measure the change in mass according to the change of spring scale (change of gravity). According to the gravitational theory of general relativity, a clock moving in a gravitational field will walk slowly. In this way, when we very accurately measure Δm or ΔE, the clock in the box is changed to a great extent, resulting in a great uncertainty Δt, so the uncertainty relation[equation(2 – 75)] still holds. Einstein lost this time. In 1962, when Bohr died, there was still a sketch of Einstein's photon box on the blackboard in his studio.

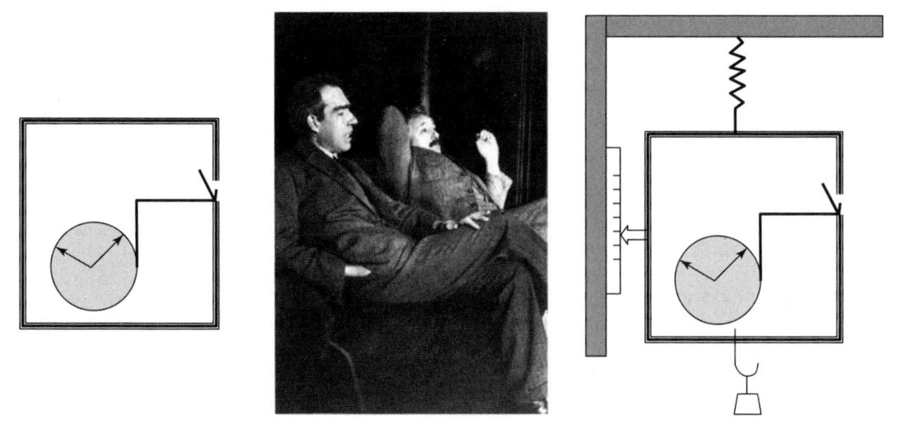

Figure 2 – 59 The debate between Bohr and Einstein

Summary

1. Black Body Radiation

Planck blackbody radiation law

$$M_{\lambda}(T) = \frac{2\pi hc^2}{\lambda^5}\frac{1}{e^{\frac{hc}{\lambda kT}} - 1}$$

Planck's energon hypothesis: The energy of a simple harmonic oscillator is quantized. During emitting radiation or absorbing radiation, the energy can only change in a jumping manner by an integer multiple of $h\nu$. Where $h\nu$ is called quantum of energy, $h = 6.626,070,040(81) \times 10^{-34}$ J · s ≈ 6.626×10^{-34} J · s is called Planck constant.

2. Photoelectric Effect

External photoelectric effect: The phenomenon in which light irradiates the metal surface, causing electrons to escape from the metal surface.

Einstein's photon hypothesis: In a vacuum, a beam of light with frequency ν (wavelength λ) is a stream of particles propagating at speed c. Such particles are called light quanta (later renamed photons), and light quantum has integrity property. Photon has energy $\varepsilon = h\nu$ and momentum $p = mc = h/\lambda$.

Einstein's photoelectric effect equation

$$\frac{1}{2}m_e v_m^2 = h\nu - A$$

Cut-off voltage U_a of photoelectric effect

$$\frac{1}{2}m_e v_m^2 = eU_a$$

Cutoff frequency ν_0 of photoelectric effect

$$\nu_0 = \frac{A}{h}$$

Photoelectric effect is that an electron absorbs a photon at one time, and the interaction process obeys the conservation of energy.

3. Compton Effect

Compton effect: The phenomenon in which X – rays and the like is scattered by substance and its wavelength becomes longer. This is due to the elastic collision between photons and free electrons or weakly bound outer electrons in the scatterer, and the interaction process obeys the conservation of energy and the conservation of momentum.

Relation between the wavelength shift $\Delta\lambda$ and the scattering angle φ:

$$\Delta\lambda = \lambda - \lambda_0 = \lambda_C(1 - \cos\varphi)$$

where $\lambda_C = \frac{h}{m_e c} = 2.426,310,236,7(11) \times 10^{-3}\text{nm} \approx 2.43 \times 10^{-3}\text{nm}$ is called the Compton wavelength of the electron.

4. Hydrogen Atom Spectrum

The generalized Balmer formula (or Rydberg formula)

$$\tilde{\nu} = R_\infty\left(\frac{1}{m^2} - \frac{1}{n^2}\right); m = 1,2,3,\cdots;\ n = m+1,\ m+2,\ m+3,\cdots$$

where $\tilde{\nu}$ is called wave number; $R_\infty = 1.097,373,156,850,8(65) \times 10^7\ \text{m}^{-1} \approx 1.097 \times 10^{-7}\ \text{m}^{-1}$ is called the Rydberg constant.

When a hydrogen atom transitions from a high energy level to a low energy level of $m = 1,2,3,4,5$, the series of the emitted spectral lines is called Lyman series (ultraviolet light), Balmer series (visible light), Paschen series (infrared light), Brackett series (far-infrared light) and Pfund series (far-infrared light) respectively.

Bohr frequency condition

$$h\nu = |E_f - E_i|$$

where, E_i and E_f are the energy values of the initial and the final steady states respectively.

Energy of hydrogen atom

$$E_n = -\frac{m_e e^4}{2(4\pi\varepsilon_0)^2\hbar^2} \cdot \frac{1}{n^2}; n = 1,2,3,\cdots$$

Excited energy: The energy required to transition from the ground state to the excited state.

5. The Wave Property of Particle

De Broglie hypothesis: physical particles also have wave property. The wave associated with physical particle is called de Broglie wave or matter wave. The frequency ν and the wavelength λ of the de Broglie wave associated with a physical particle of mass m and velocity v are respectively

$$\nu = \frac{E}{h} = \frac{mc^2}{h}$$

$$\lambda = \frac{h}{p} = \frac{h}{mv}$$

6. Born's Statistical Interpretation

De Broglie wave is a probability wave.

Wave function $\Psi(\boldsymbol{r},t)$: a mathematical expression describing the motion state of a microscopic particle.

Statistical interpretation of the wave function: The wave function $\Psi(\boldsymbol{r},t)$ of a microscopic particle itself has no direct physical meaning, but the square of its modulus $|\Psi(\boldsymbol{r},t)|^2 = \Psi^*(\boldsymbol{r},t)\Psi(\boldsymbol{r},t)$ represents the (relative) probability density of the particle appearing near the spatial coordinate $\boldsymbol{r}$ at time t.

The normalization condition for the wave function

$$\int_{-\infty}^{\infty} |\Psi(\boldsymbol{r},t)|^2 \mathrm{d}V = 1$$

The standard condition for the wave function: single value, finite and continuous.

The state superposition principle: If $\Psi_1(\boldsymbol{r},t)$, $\Psi_2(\boldsymbol{r},t)$, $\Psi_3(\boldsymbol{r},t)$, $\cdots$ represent a series of different possible states of the system, then their linear combination $\Psi(\boldsymbol{r},t) = C_1\Psi_1(\boldsymbol{r},t) + C_2\Psi_2(\boldsymbol{r},t) + C_3\Psi_3(\boldsymbol{r},t) + \cdots$ is also a possible state of the system, where $C_1, C_2, C_3, \cdots$ are arbitrary complex constants.

7. Uncertainty Relation

If the product of two physical quantities has the same dimension (J · s) as Planck's constant h, they are called conjugate quantities. All conjugate quantities satisfy the uncertainty relation.

The uncertain relation between position and momentum

$$\Delta x \cdot \Delta p_x \geqslant \frac{\hbar}{2},\ \Delta y \cdot \Delta p_y \geqslant \frac{\hbar}{2},\ \Delta z \cdot \Delta p_z \geqslant \frac{\hbar}{2}$$

The uncertain relation between energy and time

$$\Delta E \cdot \Delta t \geqslant \frac{\hbar}{2}$$

Questions

2-1 An incandescent bulb is connected to a dimmer switch. When the bulb works at full power, it appears white, but when it is dimmed, it looks more and more red. Why?

2-2 The dim light of the red bulb in the dark room for making black-and-white film will not damage the film. Why use red bulb instead of white or blue or other colors?

2-3 The light response of the retina in the human eye at low light levels depends on the excitation of photosensitive molecules in rod cells by incident light. These molecules change shape when excited, resulting in other changes in cells that can trigger nerve impulses to the brain. Why do these changes occur even at low light levels? What advantage does the Einstein photon quantum model have over the wave model at this time?

2-4 Someone said: "The greater the intensity of the light, the greater the energy of the photon." Is this right?

2-5 In the photoelectric effect experiment, if the frequency of the incident light remains unchanged but the intensity is doubled, or the intensity remains unchanged but the frequency is doubled, what effect does each have on the experimental results?

2-6 In our daily life, why are we not aware of the wave property of particles?

2-7 What is Compton effect? Can visible light be used to observe Compton effect? Why?

2-8 In both Compton scattering and photoelectric effect, electrons gain energy from incident photons. So what is the essential difference between these two processes?

2-9 How to understand the wave-particle duality of microscopic particles?

2-10 Why should the wave function be normalized? What is the standard condition that a wave function should satisfy? Why does the wave function satisfy the standard condition?

Problems

2-1 The escape work of tungsten is 4.54 eV and that of barium is 2.50 eV. Find the cut-off frequencies of tungsten and barium respectively. Which metal can be used as cathode material for photocell in the visible light range?

2-2 The escape work of sodium is 2.29 eV. What is its cutoff frequency and corresponding wavelength? Now irradiate the sodium surface with light of a wavelength of 500 nm, find the cut-off voltage and the maximum initial velocity of photoelectrons. If the intensity of the incident light is 2.0 W · m^{-2}, how many photons per second hit the metal surface per unit area on average?

2-3 In a photoelectric experiment, the measured data of the wavelength λ of incident light and the cut-off voltage U_a of a metal are as follows:

λ(nm)	253.6	283.0	303.9	330.2	366.3	435.8
U_a (nm)	2.60	2.11	1.81	1.47	1.10	0.57

(1) Draw $U_a - v$ graph on the coordinate paper.

(2) Use the graph to find the red limit frequency and corresponding wavelength of the photoelectric effect of the metal.

(3) Use the graph to find the Planck constant.

2-4 As shown in the figure (Problem 2-4), there is a system in vacuum, M is a metal plate, its red limit wavelength is $\lambda_m = 260$ nm, the uniform electric field with a magnitude of $E = 5 \times 10^3$ V · m^{-1} and the uniform magnetic field with a magnitude of $B = 0.005$ T are perpendicular to each other. If the metal plate M is irradiated with monochromatic ultraviolet light, it is found that photoelectrons are emitted, and the photoelectrons with the

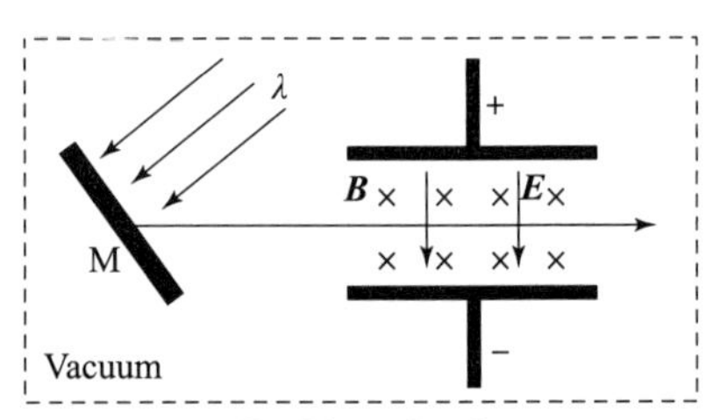

Problem 2-4

maximum velocity can pass through the region of mutually perpendicular uniform electric field and uniform magnetic field in a straight line at a uniform speed. Find: (1) The maximum velocity v_m of photoelectrons; (2) The wavelength λ of monochromatic ultraviolet light.

2-5 Find the energy, momentum, and mass of photons with wavelengths of the following values. (1) Infrared rays with a wavelength of 1 500 nm; (2) Visible light with a wavelength of 500 nm; (3) Ultraviolet rays with a wavelength of 20 nm; (4) X-rays with a wavelength of 0.15 nm.

2-6 The wavelength of a X-ray photon is 6×10^{-3} nm. It collides directly with an electron, and its scattering angle is 180°. (1) Find the change in the wavelength of the X-ray photon. (2) What is the recoil kinetic energy of the impacted electron?

2-7 In the Compton scattering experiment, suppose the wavelength of X-ray incident on paraffin is 0.070 8 nm, what are the wavelengths of the X-rays scattered in the directions of $\pi/2$ and π and the kinetic energy of recoil electrons?

2-8 X-rays with a wavelength of $\lambda_0 = 0.01$ nm is emitted on carbon, resulting in the Compton effect. When observing in the direction at an angle of 90° to the incident direction, find: (1) The scattering wavelength; (2) The kinetic energy and momentum of recoil electron.

*2-9 In 1959, R. V. Pound and Q. A. Rebka conducted a famous "gravitational purple shift" experiment at the Harvard Tower. They put a radioactive source of ^{57}Co that emits γ photon of 14.4 keV on the top of the tower, and measured the frequency ν' of the γ photon at the bottom of the tower. It was found that the frequency ν' was higher than that ν at the top of the tower. The height of the tower is known to be 22.6 m, find $\Delta\nu/\nu$ by using the energy conservation relation of photon in the gravitational field.

2-10 In the Balmer series of hydrogen atom spectrum, there is a spectral line with a wavelength of 434 nm. Try to find: (1) How much eV is the energy of the photon corresponding to this spectral line? (2) The spectral line is generated by the transition from the energy level E_n to the energy level E_k. What are n and k respectively? (3) How many series can a large number of hydrogen atoms with the highest energy level of E_5 emit at most? How many spectral lines are there in total? Show them in the energy level diagram of hydrogen atom, and indicate which spectral line has the shortest wavelength?

2-11 The ground state ionization energy of a hydrogen atom is 13.6 eV. What is the ionization energy when the hydrogen atom is in the first excited state? Which part of the spectral band does the wavelength of a photon with this energy belong to?

2-12 When a hydrogen atom transitions from an initial state to a state where the excitation energy (the energy required from the ground state to the excited state) is $\Delta E = 10.19$ eV, the wavelength of the emitted photon is $\lambda = 486$ nm. Try to find the energy and principal quantum number of the initial state.

2-13 Assuming that the hydrogen atom is originally at rest, what is the approximate recoil speed of the hydrogen atom when it transitions directly from the excited state of $n = 3$ to the ground state by radiation?

2 – 14 When the de Broglie wavelength of an electron is equal to its Compton wavelength, find: (1) The momentum of the electron, (2) the ratio of the speed of the electron to the speed of light.

2 – 15 Suppose that a proton and an electron have the same de Broglie wavelength of 1.00 nm. (1) What are their momentums respectively? (2) What are their relativistic total energies respectively?

2 – 16 If the wavelengths of an electron and a photon are both 0.20 nm, what are their momentums and kinetic energies respectively?

2 – 17 When the kinetic energy of an electron is equal to its rest energy, what is its de Broglie wavelength?

2 – 18 In a uniform magnetic field with the magnitude $B = 0.025$ T, an α particle moves along a circular orbit with a radius of $R = 0.83$ cm. (1) Find its de Broglie wavelength (the mass of α particle $m_\alpha = 6.64 \times 10^{-27}$ kg). (2) If a ball with mass $m = 0.1$ g moves at the same speed as the α particle, what is its wavelength?

2 – 19 It is known that the wave function of a free electron is $\psi(x) = A\cos(5.0 \times 10^{10} x)$, and the unit of x in the expression is m. Find: (1) The de Broglie wavelength of the free electron; (2) The momentum of the free electron; (3) The kinetic energy of the free electron.

2 – 20 The linearity of an uranium nucleus is 7.2×10^{-15} m, find the uncertainty of the velocity of one of the nucleons (proton or neutron).

2 – 21 For a photon with a wavelength of 300 nm, the measurement accuracy of its wavelength is 10^{-5} m, how much can the absolute error of its position measurement not be less than?

2 – 22 The atom in the excited state is very unstable. It will soon return to the low-energy state and emit photon. The general average lifetime is $\tau = 10^{-8}$ s. Try to estimate the width of spectral line frequency according to the uncertainty relation.

2 – 23 The neutral π meson (π^0) is very unstable, and its average lifetime is only 8.4×10^{-17} s. Find the uncertainty of the mass of the π^0 meson.

2 – 24 As part of an "uncertainty relation" experiment, Brent is hitting a golf ball with a club and measuring its speed. Meanwhile, his classmate Rebecca messed up the structure of space-time. To Rebecca's surprise, he opened a hole to another world. Both Brett and the golf ball are sucked into this hole and into another world. In this new world, Planck constant is $h = 0.6$ J · s. Brett measured the mass of the golf ball as 0.30 kg and the speed as 20.0 ± 1.0 m · s^{-1}. (1) In this new world, what is the uncertainty of the position of the golf ball for this sport? (2) What is the de Broglie wavelength of the golf ball? (3) What phenomenon would Brett observe?

Chapter 3

Schrödinger Equation and its Applications

Born's statistical viewpoint explains the relation between the wave property and the particle property of microscopic particles, but it does not explain how the wave function $\Psi(\boldsymbol{r},t)$ changes with time. We also need to know what kind of laws the motion of microscopic particles follows.

In the winter of 1925, the young Schrödinger introduced de Broglie's theory of matter wave at a physics seminar at the Swiss Federal Institute of Technology. After Schrödinger finished speaking, the Dutch physical chemist P. Debey (1884—1966, winner of the 1936 Nobel Prize in chemistry) commented: "To seriously discuss the wave, there must be a wave equation." A few weeks later, Schrödinger gave another report. At the beginning, he said excitedly, "I found the wave equation you want!" This equation is the famous Schrödinger equation. From January to June 1926, he published four papers in a row in the German "Annals of Physics", with the general title of "quantization as an eigenvalue problem". He proposed the theory of using wave equation to describe the law of motion of microscopic particle, and laid the foundation of quantum mechanics.

Quantum mechanics is the basic theory for studying the laws of motion and related phenomena of microscopic systems (such as electrons, atoms, molecules, etc.). It is characterized by Planck constant h, and takes wave-particle duality as basic image. And the description of its state is even with the help of the invisible and intangible probability amplitude. Since its birth, quantum mechanics has made brilliant achievements and is constantly enriched and developed.

This chapter introduces the idea for the establishment of Schrödinger equation, the basic solution and the method for dealing with micro-physical problems. In particular, how to deal with some special phenomena and problems in quantum mechanics, including one-dimensional infinite square-well potential, one-dimensional square-potential, simple harmonic oscillator and hydrogen atom. Study the physical meaning of solutions to these problems and be familiar with their practical applications.

3.1 Schrödinger Equation

In classical mechanics, Newton's laws give the laws that the motion state of an object changes with time. Since classical mechanics never involve the wave-particle duality at all, the law followed by the motion of microscopic particles, that is, the law of the wave function $\Psi(\boldsymbol{r},t)$ changing with time and space certainly cannot be described by Newton's law. The equation satisfied by the wave function $\Psi(\boldsymbol{r},t)$ must be re-established according to the experimental phenomena.

3.1.1 Schrödinger Equation for a Free Particle

Schrödinger first found the equation satisfied by the wave function of the free particle. The idea of its establishment is introduced below.

According to equation(2 – 63), the wave function of a free particle moving along the x axis with momentum p_x and energy E has the following form

$$\Psi(x,t) = Ae^{\frac{i}{\hbar}(p_x x - Et)}$$

Take the partial derivative with respect to time on both sides of the above formula, and multiply by $\mathrm{i}\hbar$, it can be obtained that

$$\mathrm{i}\hbar \frac{\partial}{\partial t}\Psi(x,t) = \mathrm{i}\hbar \frac{\partial}{\partial t}\left[Ae^{\frac{i}{\hbar}(p_x x - Et)}\right] = E\Psi(x,t) \tag{3-1}$$

Take the partial derivative twice with respect to the coordinate x for $\Psi(x,t)$, and multiply by $(-\hbar^2)$, then we have

$$-\hbar^2 \frac{\partial^2 \Psi(x,t)}{\partial x^2} = -\hbar^2 \frac{\partial^2}{\partial x^2}\left[Ae^{\frac{i}{\hbar}(p_x x - Et)}\right] = p_x^2 \Psi(x,t) \tag{3-2}$$

The kinetic energy of a free particle is its total energy E. In the non-relativistic case, there is such relation between energy and momentum

$$E = \frac{p_x^2}{2m} \tag{3-3}$$

where m is the mass of the free particle. It is easy to see from equations(3 – 1), (3 – 2) and (3 – 3) that the wave function $\Psi(x,t)$ satisfies the following equation

$$\mathrm{i}\hbar \frac{\partial}{\partial t}\Psi(x,t) = -\frac{\hbar^2}{2m}\frac{\partial^2}{\partial x^2}\Psi(x,t) \tag{3-4}$$

This is the Schrödinger equation for a free particle in one-dimensional motion.

In fact, if we make the following replacements in the form of the relation (equation(3 – 3)) between energy and momentum of a free particle

$$E \rightarrow \mathrm{i}\hbar \frac{\partial}{\partial t} \tag{3-5}$$

$$p_x \rightarrow -\mathrm{i}\hbar \frac{\partial}{\partial x} \tag{3-6}$$

or

$$p_x^2 \rightarrow -\hbar^2 \frac{\partial^2}{\partial x^2} \tag{3-7}$$

Then the corresponding relation between operators is obtained

$$\mathrm{i}\hbar \frac{\partial}{\partial t} \leftrightarrow -\frac{\hbar^2}{2m}\frac{\partial^2}{\partial x^2} \tag{3-8}$$

Then, apply the operators on both sides of the above formula to the wave function $\Psi(x,t)$ respectively, and let the left and right sides be equal to obtain the Schrödinger equation for a free particle in one-dimensional motion, namely equation(3 – 4).

In quantum mechanics, a symbol that performs some operation or action on the wave function is called an operator, and the operator of the mechanical quantity is represented by adding a" ∧ " sign to the text representing the mechanical quantity. For example, the momentum operators $\hat{p}_x$, $\hat{p}_y$, $\hat{p}_z$ corresponding to the momentums p_x, p_y, p_z can be defined as

$$\hat{p}_x = -i\hbar\frac{\partial}{\partial x} \tag{3-9}$$

$$\hat{p}_y = -i\hbar\frac{\partial}{\partial y} \tag{3-10}$$

$$\hat{p}_z = -i\hbar\frac{\partial}{\partial z} \tag{3-11}$$

In addition, the coordinate operator $\hat{x}$ corresponding to the coordinate x is defined as

$$\hat{x} = x \tag{3-12}$$

3. 1. 2 Schrödinger Equation in General

If a particle moves in a one-dimensional potential field $U(x,t)$, the total energy of the system is the sum of kinetic energy and potential energy, which can be expressed as

$$E = \frac{p_x^2}{2m} + U(x,t) \tag{3-13}$$

Using the replacement equations (3 - 5), (3 - 6) or (3 - 7) and equation (3 - 12), the corresponding relation between operators can be obtained as

$$i\hbar\frac{\partial}{\partial t}\leftrightarrow -\frac{\hbar^2}{2m}\frac{\partial^2}{\partial x^2} + U(x,t) \tag{3-14}$$

By applying the operators on both sides of the above formula to the wave function $\Psi(x,t)$ respectively, and making the left and right sides equal, we get the Schrödinger equation

$$i\hbar\frac{\partial}{\partial t}\Psi(x,t) = \left[-\frac{\hbar^2}{2m}\frac{\partial^2}{\partial x^2} + U(x,t)\right]\Psi(x,t) \tag{3-15}$$

Similarly, if a particle moves in a three-dimensional potential field $U(x,y,z;t)$, the Schrödinger equation satisfied by the wave function $\Psi(x,y,z;t)$ is

$$i\hbar\frac{\partial}{\partial t}\Psi(x,y,z;t) = \left[-\frac{\hbar^2}{2m}\left(\frac{\partial^2}{\partial x^2} + \frac{\partial^2}{\partial y^2} + \frac{\partial^2}{\partial z^2}\right) + U(x,y,z;t)\right]\Psi(x,y,z;t) \tag{3-16}$$

where $\frac{\partial^2}{\partial x^2} + \frac{\partial^2}{\partial y^2} + \frac{\partial^2}{\partial z^2} = \nabla^2$ is called Laplace operator, then the Schrödinger equation for the moving particle in the three-dimensional potential field is

$$i\hbar\frac{\partial}{\partial t}\Psi(x,y,z;t) = \left[-\frac{\hbar^2}{2m}\nabla^2 + U(x,y,z;t)\right]\Psi(x,y,z;t) \tag{3-17}$$

Introduce Hamiltonian operator, also known as the energy operator

$$\hat{H} = -\frac{\hbar^2}{2m}\nabla^2 + U(x,y,z;t) \tag{3-18}$$

The Schrödinger equation (equation (3 - 16)) in general situation can be rewritten as

$$i\hbar\frac{\partial}{\partial t}\Psi(x,y,z;t) = \hat{H}\Psi(x,y,z;t) \tag{3-19}$$

This is the formula on the bust of Schrödinger in the courtyard arcade of the main building at the University of Vienna, Austria, as shown in Figure 3 – 1.

Figure 3 – 1 Bust of Schrödinger in the courtyard arcade of the main building at the University of Vienna, Austria

It can be seen from equation (3 – 19) that the Schrödinger equation is a linear homogeneous differential equation, which ensures the establishment of the state superposition principle. Specifically, if $\Psi_1(x,y,z;t)$ and $\Psi_2(x,y,z;t)$ are the solutions of the Schrödinger equation, and they describe two possible motion states of a particle, respectively, then their linear superposition is also the solution of the Schrödinger equation, and also describes a possible motion state of the particle.

The motion state of a classical particle is described by coordinate and momentum, while other mechanical quantities are functions of coordinate and momentum. Therefore, the classical wave equation is a second-order partial differential equation with respect to time, i. e.

$$\frac{1}{u^2}\frac{\partial^2 \xi(x,y,z;t)}{\partial t^2} = \nabla^2 \xi(x,y,z;t)$$

To find the solution of this equation $\xi(x,y,z;t)$, two initial conditions $\xi(x,y,z;0)$ and $\left.\frac{\partial \xi(x,y,z;t)}{\partial t}\right|_{t=0}$ need to be known. In contrast, the Schrödinger equation is a first-order partial differential equation with respect to time, so only one initial condition $\Psi(x,y,z;0)$ is needed to determine its solution $\Psi(x,y,z;t)$, and it is used to describe the motion state of microscopic particle at any time. At the same time, for the first-order partial differential equation with respect to time, if there is a solution in wave form, there must be an imaginary factor i in the equation, so the solution $\Psi(x,y,z;t)$ of the equation must be in complex form.

The Schrödinger equation was first established by Schrödinger based on the known wave function of a free particle. It was not derived. However, a large number of results obtained by applying this equation to microscopic systems such as molecules and atoms are all consistent with experiments, which indicates that it is the basic equation of quantum mechanics. Since Schrödinger equation reveals the basic law of the motion of microscopic particle, its position in quantum mechanics is equivalent to that of Newton's equation of motion in classical mechanics. It is a powerful tool to deal with all non-relativistic problems in quantum mechanics, and has been widely used in the fields of atom, molecule, solid state physics, nuclear physics, chemistry and so on. For example, in the field of material science today, the Schrödinger equation of electron system determines the electrical conductivity of materials, thermal conductivity of metals, superconductivity, energy band structure, magnetic properties and so on.

In addition, the Schrödinger equation is only applicable to non-relativistic particle with low velocity. In 1928, the British physicist Paul Dirac established the relativistic quantum mechanical equation, which is called Dirac equation. Schrödinger and Dirac shared the 1933 Nobel Prize in Physics.

3.1.3 Steady-state Schrödinger Equation

Next, we discuss the solution of Schrödinger equation. For simplicity, let's take one dimension as an example. If the potential energy $U(x)$ of a particle in the potential field is only a function of coordinate and has nothing to do with time, that is, it does not contain time t, then the Schrödinger equation (equation (3 - 15)) can be solved by the method of separating variables, and one of its special solutions can be expressed as a product of function $\psi(x)$ of spatial coordinate x and function $f(t)$ of time t, that is

$$\Psi(x,t) = \psi(x)f(t) \tag{3-20}$$

By substituting this special solution into equation (3 - 15), we can get

$$\mathrm{i}\hbar\psi(x)\frac{\mathrm{d}f(t)}{\mathrm{d}t} = \left[-\frac{\hbar^2}{2m}\frac{\mathrm{d}^2}{\mathrm{d}x^2} + U(x)\right]\psi(x)f(t) \tag{3-21}$$

Then divide both sides of the above equation by $\psi(x)f(t)$, we can obtain

$$\mathrm{i}\hbar\frac{1}{f(t)}\frac{\mathrm{d}f(t)}{\mathrm{d}t} = \frac{1}{\psi(x)}\left[-\frac{\hbar^2}{2m}\frac{\mathrm{d}^2}{\mathrm{d}x^2} + U(x)\right]\psi(x) \tag{3-22}$$

where the left side is only related to the time t, and the right side is only related to the spatial coordinate x. Since the time t and the spatial coordinate x are independent variables, equation (3 - 22) holds only when both sides are equal to the same constant. Otherwise, when t or x is changed separately, one side of equation (3 - 22) changes while the other side remains unchanged, and the equal sign cannot hold. If the constant is represented by E, the partial differential equation (3 - 22) is transformed into the following two ordinary differential equations

$$\mathrm{i}\hbar\frac{\mathrm{d}f(t)}{f(t)} = E\mathrm{d}t \tag{3-23}$$

$$\left[-\frac{\hbar^2}{2m}\frac{\mathrm{d}^2}{\mathrm{d}x^2} + U(x)\right]\psi(x) = E\psi(x) \tag{3-24}$$

The solution of the first ordinary differential equation, i. e. equation (3 - 23), is

$$f(t) = A\mathrm{e}^{-\frac{\mathrm{i}}{\hbar}Et} \tag{3-25}$$

where A is an undetermined complex constant. Because the exponential part can only be a quantity with dimension of 1, therefore, E must have the dimension of energy, that is, take the unit of energy J as its unit. By analogy with the wave function expression of a free particle, i. e. equation (2 - 63), it can be known that E represents the energy of the particle.

By Substituting equation (3 - 25) into equation (3 - 20), we can obtain the special solution of Schrödinger equation

$$\Psi(x,t) = \psi(x)\mathrm{e}^{-\frac{\mathrm{i}}{\hbar}Et} \tag{3-26}$$

where, the complex constant A is attributed to the function $\psi(x)$ of spatial coordinate x. $\psi(x)$ is the special solution of equation (3 - 24), which is called the steady-state wave function. The equation (3 -

24) it satisfies is called one-dimensional steady-state Schrödinger equation, which can also be expressed as

$$\frac{d^2\psi(x)}{dx^2}+\frac{2m}{\hbar^2}[E-U(x)]\psi(x)=0 \qquad (3-27)$$

Because equation (3 – 24) or equation (3 – 27) contains the potential energy function $U(x)$, its solution is related to the specific system, and the standard condition of the wave function should be taken into account.

The characteristic of the wave function represented by equation (3 – 26) is that its dependence on time has a completely definite form $e^{-\frac{i}{\hbar}Et}$. Therefore, for the one-dimensional steady-state problem in which the potential energy function U is independent of time t, it is only necessary to solve the one-dimensional steady-state Schrödinger equation, that is, equation (3 – 24) or equation (3 – 27), to obtain the steady-state wave function $\psi(x)$. The wave function $\Psi(x,t)$ can be obtained by equation (3 – 26). And because the square of the modulus of the factor $e^{-\frac{i}{\hbar}Et}$, i. e. $|e^{-\frac{i}{\hbar}Et}|^2$, is equal to 1, so there is $|\Psi(x,t)|^2=|\psi(x)|^2$, that is, the probability distribution of microscopic particle does not change with time in the steady-state state. This is the origin of the name steady-state state.

By extending the above one-dimensional steady-state case to the three-dimensional steady-state problem, the specific solution of Schrödinger equation can be obtained as

$$\Psi(x,y,z;t)=\psi(x,y,z)e^{-\frac{i}{\hbar}Et} \qquad (3-28)$$

The steady-state Schrödinger equation in three-dimensional rectangular coordinate system is

$$\left[-\frac{\hbar^2}{2m}\left(\frac{\partial^2}{\partial x^2}+\frac{\partial^2}{\partial y^2}+\frac{\partial^2}{\partial z^2}\right)+U(x,y,z)\right]\psi(x,y,z)=E\psi(x,y,z) \qquad (3-29)$$

The separated variable constant E in equations (3 – 28) and (3 – 29) is the total energy of the particle.

If that an operator acts on a wave function is equal to a value multiplied by the wave function, then the wave function is called the eigenfunction of the operator, and the value is called the eigenvalue of the operator, and the equation is called eigenequation of the operator. By solving the equation, the eigenfunction and eigenvalue corresponding to the operator can be obtained. Therefore, the steady-state Schrödinger equation, ie, equation (3 – 24) or equation (3 – 29), is also called eigenequation of Hamiltonian operator or energy operator.

If the particle is in a bound state, the energy of the system can only take some specific values. For the convenience of discussion, assume that it takes discrete values E_n, where $n=1,2,3,\cdots$, the corresponding steady-state wave functions $\Psi_n(x,y,z;t)$ are physically acceptable, that is, they meet the standard condition of the wave function and can be expressed as

$$\Psi_n(x,y,z;t)=\psi_n(x,y,z)e^{-\frac{i}{\hbar}E_nt};n=1,2,3,\cdots \qquad (3-30)$$

E_n is the energy that the system can have in each steady-state, and the steady-state also refers to the state in which the energy does not change with time and has a certain value. The general solution of the Schrödinger equation can be written as the superposition of a series of steady-state solutions

$$\Psi(x,y,z;t) = \sum_n C_n \Psi_n(x,y,z;t) = \sum_n C_n \psi_n(x,y,z) e^{-\frac{i}{\hbar}E_n t} \tag{3-31}$$

where C_n is called the expansion coefficient.

In quantum mechanics, many practical problems eventually comes down to solving the steady-state Schrödinger equation. As long as the mass m of the particle and the form of its potential energy function U in the potential field are known, the Schrödinger equation can be formulated. Then, the wave function Ψ can be obtained according to the initial condition and boundary conditions. The wave function Ψ itself has no practical physical meaning, and $|\Psi|^2$ gives the probability density of the particle appearing at any position at any time.

Some practical examples are presented below. We mainly applies Schrödinger equation to solve the energy eigenvalue problem of one-dimensional infinite square-well potential, and then briefly introduces the main results of one-dimensional square-barrier potential, simple harmonic oscillator and hydrogen atom. These examples not only allow us to understand the general steps of solving the steady-state Schrödinger equation, but also help us deepen our understanding of the quantization of energy and the significance of the Schrödinger equation.

Erwin Schrödinger (1887—1961, Figure 3-2), Austrian theoretical physicist and one of the important founders of quantum mechanics. In 1926, he proposed the famous non-relativistic Schrödinger equation. In 1933, he shared the Nobel Prize in Physics with British physicist Paul Dirac. In 1935, he proposed the most famous thought experiment—Schrödinger cat. In 1937 he was awarded the Max Planck medal. In 1944, Schrödinger published "what is life", in which he proposed the concept of negentropy. Due to his influence, many physicists have participated in the research work of biology, and the combination of physics and biology has formed one of the most remarkable features of modern molecular biology. In addition, he has made great achievements in the research of specific heat of solid, statistical thermodynamics, atomic spectroscopy, radium radioactivity and so on. He also published many popular science papers, which are still the best guides to enter the world of general relativity and statistical mechanics.

Figure 3-2 Schrödinger, 1926 Nobel laureate in physics

3.2 A Particle in a One-dimensional Square-well

3.2.1 A Particle in a One-dimensional Infinite Square-well

The infinite square-well is a simple theoretical model. The movement of free electrons in metals can be roughly described by this model to explain the physical properties of metals. Due to the free electrons on the surface of the metal material needs to overcome the work function to escape the

metal surface. So for electrons, the potential energy outside the metal is higher than that inside the metal. As an ideal model, it can be considered that the electrons are in an infinite square-well bounded by the metal surface.

As shown in Figure 3 – 3, the potential energy expression of one-dimensional infinite square-well is

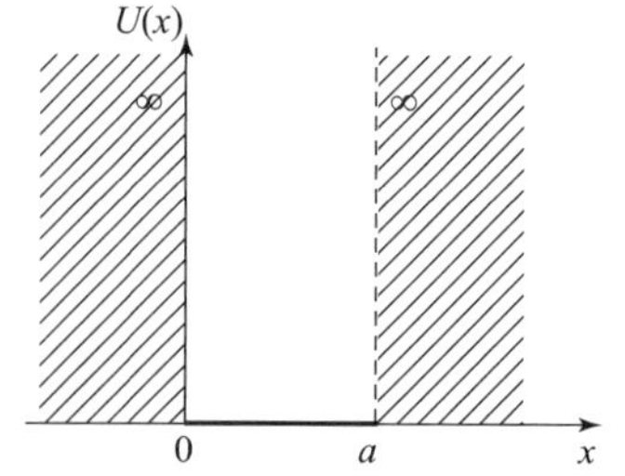

Figure 3 – 3 The potential energy of a one-dimensional infinite square-well

$$U(x) = \begin{cases} 0, & 0 < x < a \\ \infty, & x \leqslant 0, x \geqslant a \end{cases} \tag{3-32}$$

In the regions of $x \leqslant 0$ and $x \geqslant a$, the potential energy $U(x) \to \infty$. The physical meaning of infinite potential energy is that a particle with finite energy cannot appear in this region. Therefore

$$\psi(x) = 0;\ x \leqslant 0, x \geqslant a \tag{3-33}$$

In the region of $0 < x < a$, that is, in the potential well, the potential energy $U(x) = 0$, indicating that the particle moves freely without force. At this time, the one-dimensional steady state Schrödinger equation, i. e. equation (3 – 27) can be written as

$$\frac{\mathrm{d}^2\psi(x)}{\mathrm{d}x^2} + \frac{2mE}{\hbar^2}\psi(x) = 0 \tag{3-34}$$

where m is the mass of the particle and E is the total energy of the particle. Let

$$k = \frac{\sqrt{2mE}}{\hbar} \tag{3-35}$$

Then write equation (3 – 34) as

$$\frac{\mathrm{d}^2\psi(x)}{\mathrm{d}x^2} + k^2\psi(x) = 0 \tag{3-36}$$

The above formula is a second-order differential equation with constant coefficients, and its general solution can be written as

$$\psi(x) = A\sin kx + B\cos kx \tag{3-37}$$

where A and B are undetermined constants, which can be determined by the standard condition of being single-valued, continuous and finite and the normalization condition that a wave function must satisfy.

According to the continuity condition of the wave function, at the well wall of $x = 0$, it requires $\psi(0) = 0$. In equation (3 – 37), only $B = 0$ can make $\psi(0) = 0$, so equation (3 – 37) only remains the first term, that is

$$\psi(x) = A\sin kx \tag{3-38}$$

At the well wall of $x = a$, according to the continuity condition of the wave function, there should also be $\psi(a) = 0$, that is, it requires $\sin ka = 0$. From this we can have

$$k = \frac{n\pi}{a};\ n = 1, 2, 3, \cdots \tag{3-39}$$

Substituting equation (3 – 39) into equation (3 – 38), the corresponding wave function is obtained as

$$\psi_n(x) = A\sin\frac{n\pi}{a}x;\ 0 < x < a \tag{3-40}$$

By the normalization condition

$$\int_{-\infty}^{\infty} |\psi(x)|^2 \mathrm{d}x = \int_0^a |\psi_n(x)|^2 \mathrm{d}x = 1$$

Substituting equation (3 - 40) into the above equation, it is easy to obtain the normalization constant A as

$$A = \sqrt{\frac{2}{a}} \tag{3-41}$$

Substituting equation (3 - 41) into equation (3 - 40), the normalized steady-state wave function when the particle is bound to move in a one-dimensional infinite square-well is obtained as

$$\psi_n(x) = \begin{cases} \sqrt{\frac{2}{a}}\sin\frac{n\pi}{a}x; & n = 1,2,3,\cdots; \quad 0 < x < a \\ 0; & x \leqslant 0, x \geqslant a \end{cases} \tag{3-42}$$

Then substitute equation (3 - 39) into equation (3 - 35), the energy of the particle in the one-dimensional infinite square-well is

$$E_n = \frac{k^2\hbar^2}{2m} = \frac{n^2\pi^2\hbar^2}{2ma^2}; \quad n = 1,2,3,\cdots \tag{3-43}$$

where n can be any positive integer, which is called quantum number.

The state described by a wave function that is zero at infinity is usually called bound state. Equation (3 - 43) shows that when a particle is bound in an infinite square-well, the energy of the particle cannot take continuous values, but can only take a series of discrete values, which are called energy levels. Each value of n corresponds to an energy level. This result is completely different from that of classical particles. In classical mechanics, the energy of a particle can take continuous values. However, in quantum mechanics, the energy of a particle is quantized. And while solving the Schrödinger equation to obtain the steady-state wave function $\psi(x)$, the allowable values of energy is naturally determined, which does not need to be introduced by artificial hypothesis as in the early days of quantum theory. Such a problem is called eigenvalue problem in mathematics, E_n is called energy eigenvalue, and $\psi_n(x)$ is called steady-state eigenwave function corresponding to the energy eigenvalue E_n. It can also be seen from the solving process of this example that the standard condition of wave function plays a role in the process of determining the wave function and energy level.

When $n = 1$, the particle is in the lowest energy state, which is called the ground state. The ground state energy is

$$E_1 = \frac{\pi^2\hbar^2}{2ma^2} \tag{3-44}$$

That is, it is not equal to zero, which is called the zero-point energy, indicating that the particle in the well can never be at rest. This is incomprehensible for classical physics, but according to quantum theory, this is exactly the requirement of the uncertainty relation and the reflection of the wave property of the particle. In fact, if the width a of the well is regarded as the uncertainty range Δx of the particle position, it can be known from the uncertainty relation that the momentum cannot take a definite value. Therefore, the particle cannot come to rest completely.

When n takes other values that are not equal to 1, the energy state is called the excited state, and its energy is n^2 times that of the ground state, that is

$$E_n = n^2 E_1 \tag{3-45}$$

It can be seen from the above formula that the difference between two adjacent energy levels, that is, the energy level interval is

$$\Delta E = E_{n+1} - E_n = (2n+1)E_1 = (2n+1)\frac{\pi^2 \hbar^2}{2ma^2} \tag{3-46}$$

And there is

$$\frac{\Delta E}{E_n} = \frac{(2n+1)}{n^2} \tag{3-47}$$

Although ΔE increases as n increases, but $\Delta E/E_n$ decreases as n increases. When $n \to \infty$, the energy becomes continuous. In addition, it can be seen from equation (3-46) that the larger the particle mass m and the well width a are, the smaller ΔE is. Therefore, for the mass m and the well width a of the object on the macro-scale, the energy can be regarded as continuous change, which corresponds to classical physics. Thus, energy quantization is only a feature of the microscopic world.

Figure 3-4 shows the steady-state wave functions ψ_1, ψ_2 and ψ_3 corresponding to the first three energy levels E_1, E_2 and E_3 of the particle, and the corresponding probability densities $|\psi_1|^2$, $|\psi_2|^2$ and $|\psi_3|^2$, in the one-dimensional infinite square-well. It can be seen from Figure 3-4 or equation (3-42) that each energy eigenstate corresponds to a standing wave of a specific wavelength of the de Broglie wave. In a limited space, the waves of the microscopic particle are reflected back and forth at the edge of the well and superimposed, they exist stably in the form of standing waves. For example, when the particle is at the energy level of $n = 1$, the wave function is

$$\psi_1(x) = \sqrt{\frac{2}{a}}\sin\frac{\pi}{a}x$$

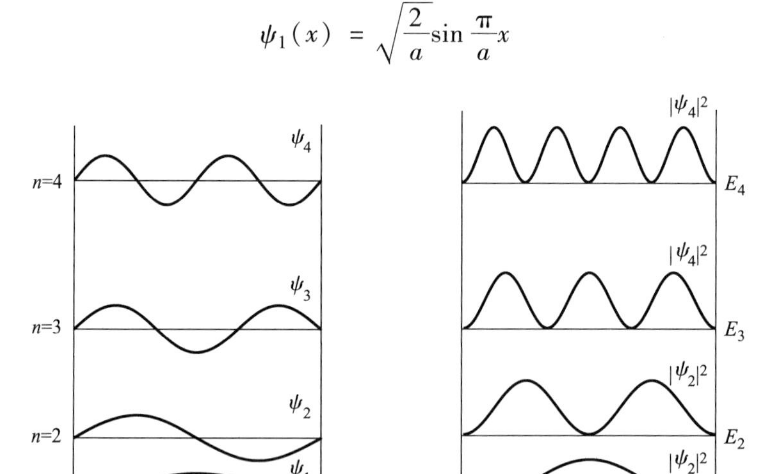

Figure 3-4 Energy levels and wave functions of a particle in a one-dimensional infinite square-well

Its shape is the standing wave containing half the de Broglie wavelength with the nodes at the well edge($x = 0, a$). It is easy to find that the momentum of the particle in the ground state has a magnitude of

$$p_1 = \sqrt{2mE_1} = \frac{\pi\hbar}{a} = \frac{h}{2a} \tag{3-48}$$

From the de Broglie relation $p = h/\lambda$, it can be obtained that $a = \lambda/2$. Therefore, the steady-state matter wave of the particle in the one-dimensional infinite square-well is equivalent to a standing wave in a string with both ends fixed. The wavelength λ_n satisfies

$$a = n\frac{\lambda_n}{2}; n = 1,2,3,\cdots \tag{3-49}$$

That is, the potential well width a must be equal to an integer multiple of the half wavelength of the de Broglie wave.

According to the physical meaning of wave function, $|\psi|^2$ is the probability density of the particle appearing everywhere, and its mathematical expression is

$$|\psi_n(x)|^2 = \begin{cases} \frac{2}{a}\sin^2\frac{n\pi}{a}x; n = 1,2,3,\cdots; & 0 < x < a \\ 0; & x \leqslant 0, x \geqslant a \end{cases} \tag{3-50}$$

It can be seen from Figure 3 – 4 that the probability distribution $|\psi|^2$ of the particle in the well is not uniform. It varies with x and is related to n, and there are several nodes with zero probability. This is very different from classical mechanics. According to classical mechanics, the motion of a particle in the well is unrestricted, and the probability of the particle appearing everywhere in the well should be equal.

Example 3 – 1: A particle moving in one dimension is bound in the range of $0 < x < a$, and its wave function is known to be $\psi(x) = A\sin\frac{\pi x}{a}$. Find, (1) The constant A; (2) The probability of the particle appearing in the region from 0 to $a/2$; (3) Where does the particle appear with the greatest probability?

Solution: (1) By the normalization condition

$$\int_{-\infty}^{\infty} |\psi|^2 \mathrm{d}x = A^2 \int_0^a \sin^2\frac{\pi x}{a}\mathrm{d}x = 1$$

The constant is solved as

$$A = \sqrt{\frac{2}{a}}$$

(2) The probability density of the particle is

$$|\psi|^2 = \frac{2}{a}\sin^2\frac{\pi x}{a}$$

The probability of the particle appearing in the region from 0 to $a/2$ is

$$P = \int_0^{a/2} |\psi|^2 \mathrm{d}x = \frac{2}{a}\int_0^{a/2} \sin^2\frac{\pi x}{a}\mathrm{d}x = \frac{1}{2}$$

(3) The position with the greatest probability should satisfy

$$\frac{\mathrm{d}}{\mathrm{d}x}|\psi|^2 = \frac{4\pi}{a^2}\sin\frac{\pi x}{a}\cos\frac{\pi x}{a} = 0$$

That is, when $\cos\frac{\pi x}{a} = 0$, the probability of the particle appearing is the greatest. Because $0 < x < a$, so $x = a/2$, where the probability of the particle appearing is the greatest.

Example 3 - 2: The steady-state matter wave of a particle bound in a one-dimensional infinite square-well is equivalent to a standing wave in a string with both ends fixed, so the well width a must be equal to an integer multiple of the half wavelength of the de Broglie wave.

(1) Try to find the energy eigenvalue of the particle with mass m from this, where $n = 1, 2, 3, \cdots\cdots$ and a series of positive integer values.

(2) It is relatively common for microscopic particles moving in a potential well. For example, protons and neutrons in the nucleus (with linear degree of 1.0×10^{-14} m) can be regarded as being in an infinite potential well and cannot escape out. Their movement in the nucleus is considered free. According to the estimation of one-dimensional infinite square-well, how much the energy in MeV released by the proton when it transitions from the first excited state to the ground state?

Solution: (1) For a particle bound in a one-dimensional infinitely square-well, the relation between the well width a and the de Broglie wavelength λ should be

$$a = n\lambda_n/2; n = 1, 2, 3, \cdots$$

Therefore, the possible value of the de Broglie wavelength of the particle in the well is

$$\lambda_n = 2a/n$$

The momentum of the particle is related to the de Broglie wavelength of the particle, which magnitude is

$$p_n = h/\lambda_n = hn/(2a) = \pi\hbar n/a$$

The energy of a free particle is the kinetic energy, which is

$$E_n = \frac{p_n^2}{2m} = \frac{\pi^2\hbar^2}{2ma^2}n^2; n = 1, 2, 3, \cdots$$

(2) The energy of the ground state ($n = 1$) of the proton is

$$E_1 = \frac{\pi^2\hbar^2}{2m_\mathrm{p}a^2} = \frac{3.14^2 \times (1.05 \times 10^{-34})^2}{2 \times 1.67 \times 10^{-27} \times (1.0 \times 10^{-14})^2}\ \mathrm{J} = 3.3 \times 10^{-13}\ \mathrm{J}$$

It shows that there is also a minimum energy of non-zero for the proton in the nucleus. In other words, the kinetic energy of a bound particle will not be zero, and the smaller width of the well the bound particle in, the larger ground state energy it has.

The energy of the proton's first excited state ($n = 2$) is

$$E_2 = 4E_1 = 4 \times 3.3 \times 10^{-13}\ \mathrm{J} = 13.2 \times 10^{-13}\ \mathrm{J}$$

The energy released during the transition from the first excited state to the ground state is

$$E_2 - E_1 = (13.2 \times 10^{-13} - 3.3 \times 10^{-13})\mathrm{J} = 9.9 \times 10^{-13}\ \mathrm{J} = 6.2\ \mathrm{MeV}$$

It is observed in the experiment that the energy difference between the two steady-states of the nucleus is generally a few MeV, and the above estimation is roughly consistent with this fact.

3.2.2 A Particle in a One-dimensional Finite Square-well

In fact, free electrons in a metal may escape from the metal surface. Because what they actually encounter is the well with a finite height of the potential energy U_0, as shown in Figure 3 - 5. The potential energy function is

$$U(x) = \begin{cases} 0, & x < 0 \\ U_0, & x > 0 \end{cases} \tag{3-51}$$

Classical mechanics believes that a particle with energy less than U_0 can only move in the well of $x < 0$, and cannot enter the region $x > 0$ where its energy is less than U_0, otherwise its kinetic energy will be negative.

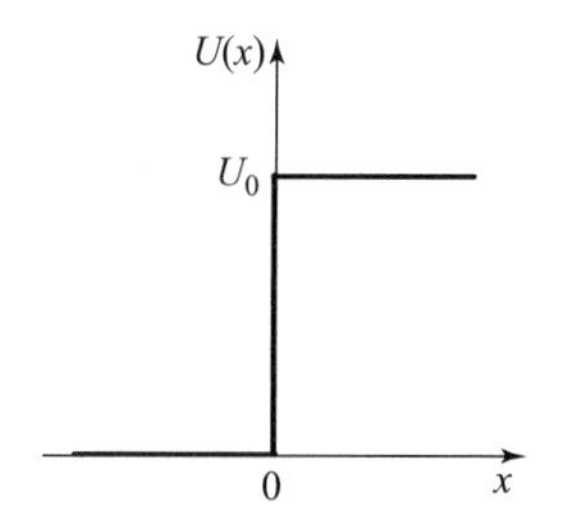

Figure 3 - 5 One-dimensional finite square-well

According to the theory of quantum mechanics, because the potential energy $U(x)$ is independent of time, this is a steady-state problem, which can be solved by one-dimensional steady-state Schrödinger equation (equation (3 - 27)).

In the well where $x < 0$, the steady-state Schrödinger equation is

$$\frac{d^2\psi_1(x)}{dx^2} + \frac{2m}{\hbar^2}E\psi_1(x) = 0 \tag{3-52}$$

Let $k_1^2 = \frac{2m}{\hbar^2}E$, then the special solution of equation (3 - 52) is

$$\psi_1(x) = Ae^{ik_1x} + Be^{-ik_1x} \tag{3-53}$$

where A and B are undetermined constants. From equation (3 - 26), it can be obtained that the wave function of microscopic particle moving freely in the one-dimensional well where $x < 0$ is

$$\Psi_1(x) = \psi_1(x)e^{-i\frac{E}{\hbar}t} = A\,e^{-i\left(\frac{E}{\hbar}t - k_1x\right)} + Be^{-i\left(\frac{E}{\hbar}t + k_1x\right)} \tag{3-54}$$

In the above formula, the first term represents a plane monochromatic wave propagating in the positive direction of x, which can be regarded as an incident wave for the interface at $x = 0$; the second term represents a plane monochromatic wave propagating in the negative direction of x, which can be regarded as a reflected wave for the interface at $x = 0$.

Outside the well where $x > 0$, the steady-state Schrödinger equation is

$$\frac{d^2\psi_2(x)}{dx^2} + \frac{2m}{\hbar^2}(E - U_0)\psi_2(x) = 0 \tag{3-55}$$

Let $k_2^2 = \frac{2m}{\hbar^2}(U_0 - E)$, then the special solution of equation (3 - 55) is

$$\psi_2(x) = Ce^{-k_2x} + De^{k_2x} \tag{3-56}$$

where C and D are undetermined constants. The "finite" condition of the wave function requires $D = 0$, otherwise when $x \to \infty$, the wave function ψ_2 is infinite and becomes meaningless. Therefore, $\psi_2(x)$ has the following decay solution

$$\psi_2(x) = Ce^{-k_2x} \tag{3-57}$$

But it still has a certain value. This shows that the particle may still appear outside the well. Therefore, free electrons in metal always have a certain probability of escaping from the metal surface.

3.2.3 One-dimensional Square Potential Barrier and Barrier Penetration

The so-called one-dimensional square potential barrier refers to the potential energy function as shown in Figure 3 - 6

$$U(x) = \begin{cases} 0, & x < 0, x > a \\ U_0, & 0 \leqslant x \leqslant a \end{cases} \tag{3-58}$$

Where U_0 represents the height of the potential barrier, which is greater than zero, and a is the width of the potential barrier.

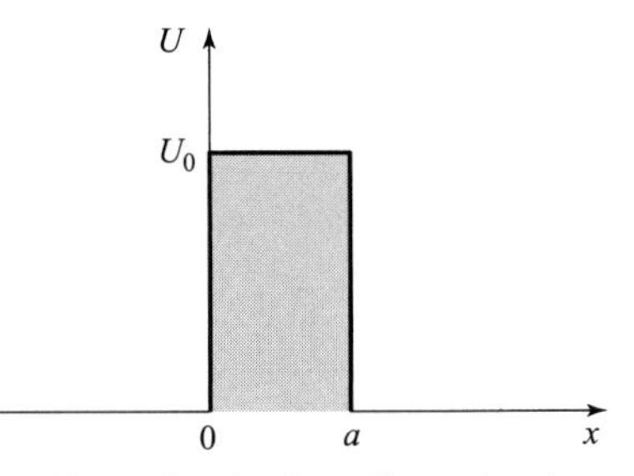

Figure 3 - 6 One-dimensional square potential barrier

Suppose a particle with mass m and energy E is incident towards the barrier from the left along the x - axis, and it is assumed that the energy of the particle is not lost during the interaction between the particle and the barrier. Below we will only discuss the case of $E < U_0$.

According to classical mechanics, since the total energy E of the particle is less than the height of the potential barrier U_0, the classical particle will be completely reflected back in the case of $E < U_0$, so it cannot enter the interior of the barrier. Moreover, it cannot pass through the potential barrier to reach the other side.

According to quantum mechanics, the motion law of the microscopic particle satisfies the Schrödinger equation. For the time independent steady-state problem of the potential energy function shown in equation (3 - 58), there is the following steady-state Schrödinger equation

$$\begin{cases} \dfrac{d^2\psi_1(x)}{dx^2} + \dfrac{2m}{\hbar^2}E\psi_1(x) = 0, & x < 0 \\ \dfrac{d^2\psi_2(x)}{dx^2} + \dfrac{2m}{\hbar^2}(E - U_0)\psi_2(x) = 0, & 0 \leqslant x \leqslant a \\ \dfrac{d^2\psi_3(x)}{dx^2} + \dfrac{2m}{\hbar^2}E\psi_3(x) = 0, & x > a \end{cases} \tag{3-59}$$

Let $k = \dfrac{2m}{\hbar^2}E$ and $k'^2 = \dfrac{2m}{\hbar^2}(U_0 - E)$, then equation (3 - 59) can be written as

$$\begin{cases} \dfrac{d^2\psi_1(x)}{dx^2} + k^2\psi_1(x) = 0, & x < 0 \\ \dfrac{d^2\psi_2(x)}{dx^2} - k'^2\psi_2(x) = 0, & 0 \leqslant x \leqslant a \\ \dfrac{d^2\psi_3(x)}{dx^2} + k^2\psi_3(x) = 0, & x > a \end{cases} \tag{3-60}$$

Although the form of the Schrödinger equation in the region where $x < 0$ is the same as that in the region $x > a$, there are incident wave and reflected wave in the region $x < 0$. While in the region $x > a$, there is only a transmitted wave, that is, a plane monochromatic wave propagating in the positive direction of x, and the probability of the particle appearing near $x = 0$ is greater than that near $x = a$. Then the general solution of equation (3 - 60) is

$$\begin{cases} \psi_1(x) = A_1 e^{ikx} + B_1 e^{-ikx}, & x < 0 \\ \psi_2(x) = A_2 e^{-k'x}, & 0 \leqslant x \leqslant a \\ \psi_3(x) = A_3 e^{ikx}, & x > a \end{cases} \tag{3-61}$$

where A_1, B_1, A_2 and A_3 are undetermined constants. They can be determined by the following boundary condition that the wave function should be smoothly connected at $x = 0$ and $x = a$, that is, $\psi(x)$ and its first derivative $\frac{d\psi}{dx}$ are continuous at $x = 0$ and $x = a$:

$$\begin{cases} \psi_1(0) = \psi_2(0) \\ \psi_2(a) = \psi_3(a) \\ \left.\frac{d\psi_1(x)}{dx}\right|_{x=0} = \left.\frac{d\psi_2(x)}{dx}\right|_{x=0} \\ \left.\frac{d\psi_2(x)}{dx}\right|_{x=a} = \left.\frac{d\psi_3(x)}{dx}\right|_{x=a} \end{cases} \tag{3-62}$$

The wave function solved is shown in Figure 3 – 7.

The results of quantum mechanics show that due to the wave property of microscopic particle, when a microscopic particle is incident and encounters a potential barrier of finite-height with $U_0 > E$, part of the wave is reflected back, another part of the wave enters the potential barrier, and there is some probability of the particle passing the potential barrier and entering the region where $x > a$. This phenomenon that the microscopic particle in front of the potential barrier (the region where $x < 0$) has some probability to penetrate the potential barrier is called barrier penetration or tunneling effect. The tunneling effect originates from the wave-particle duality of microscopic particle. Its existence has been confirmed by a large number of experiments. For example, the emission of α particle from a radioactive nucleus is a kind of tunneling effect, the quantum evaporation of black hole is also the result of tunneling effect.

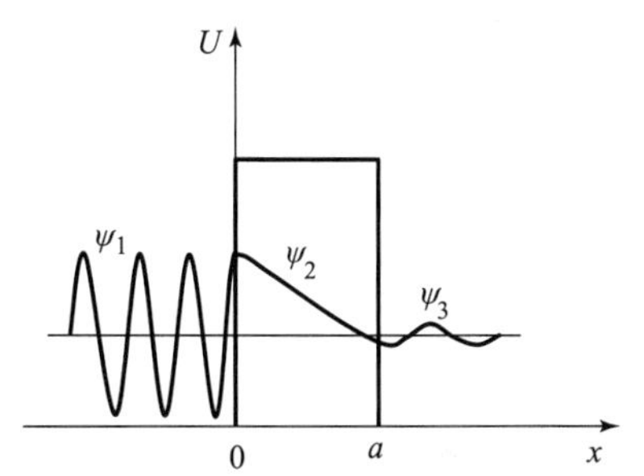

Figure 3 – 7 Quantum tunneling effect

The ratio of the probability density of the transmitted wave to the probability density of the incident wave is defined as the probability T of the particle penetrating the barrier, that is

$$T = \frac{|\psi_3(a)|^2}{|\psi_1(0)|^2} \tag{3-63}$$

Using the boundary condition shown in equation (3 – 62), then

$$T = \frac{|\psi_2(a)|^2}{|\psi_2(0)|^2} = \frac{A_2^2 e^{-2k'a}}{A_2^2 e^{-2k'0}} = e^{-2k'a} = e^{-\frac{2a}{\hbar}\sqrt{2m(U_0-E)}} \tag{3-64}$$

T is also called the penetration coefficient.

It can be seen from equation (3 – 64) that as long as the potential barrier is not infinitely high or infinitely wide, the penetration coefficient will not be equal to zero, and tunneling effect will occur. From equation (3 – 64), it can also be seen that when the barrier width a increases, the penetration coefficient T decreases rapidly and exponentially, so the tunneling effect is hardly observed in the macroscopic field. For example, when the barrier width is above 50 nm, and $U_0 - E = 5$ eV, the penetration coefficient will be less than 10^{-6}, At this time, the tunneling effect is actually meaningless.

The solution of steady-state Schrödinger equation includes two types of problems. One type of problem, such as infinite square potential well, belongs to solving eigenvalue problem, or is called the bound state problem. At this time, $\psi = 0$ at infinity, the particle is bound in a limited space, and its energy is quantized. Another type of problem, such as tunneling effect, belongs to the scattering problem, which calculates the probability of a particle penetrating a potential barrier. At this time, the particle can be incident from infinity and can move all the way to infinity. Correspondingly, the total energy can be given in advance and can be valued continuously.

The Scanning Tunneling Microscope

Scanning tunneling microscopy (STM for short) is an important application of tunneling effect. In 1982, G. Binning (1947 – , German, right in Figure 3 – 8) and H. Rohrer (1933 – , Swiss, left in Figure 3 – 8) from IBM Zurich Laboratory in Switzerland developed the world's first scanning tunneling microscope by using the quantum tunneling effect. As shown in Figure 3 – 9, due to the tunneling effect, electrons can penetrate the potential barrier on the surface of the sample and form an electron cloud near the surface. When a small voltage is applied between the probe and the sample surface, there will be a tunneling current flowing between the probe and the sample. The tunneling current is very sensitive to the distance between the pointed tip and the sample. Through the change of tunneling current, the microstructure information of the sample surface can be recorded, and the arrangement image of atoms and molecules can be obtained.

Figure 3 – 8 Rohrer (left) and Binning (right), 1986 Nobel laureates in physics, and the first generation of STM

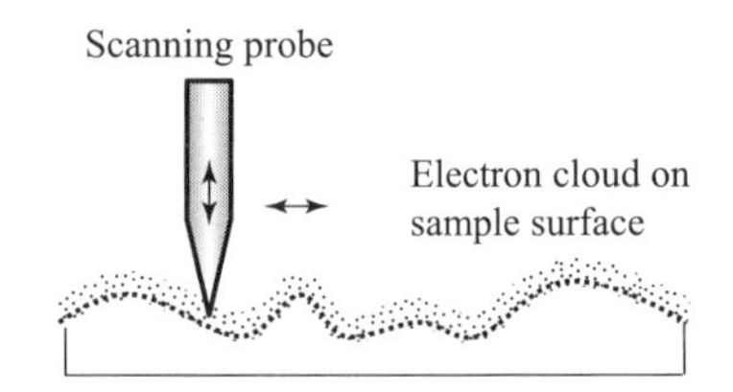

Figure 3 – 9 Scanning tunneling microscope

For example, Figure 3 – 10 shows the STM image of iodine atoms adsorbed on the surface of a platinum single crystal, from which the adsorption positions of iodine atoms and lattice defects on the surface of platinum crystal can be clearly distinguished. STM can not only detect the microstructure of material surface, but also can be used to manipulate atoms. We can use its probe tip to inhale an atom and then put it in another position. For example, Figure 3 – 11 shows the letters IBM, which was formed by moving xenon atoms one by one on

Figure 3 – 10 Adsorption of iodine atoms on a platinum crystal

the surface of a nickel single crystal with STM in 1991, each letter is 5 nm long.

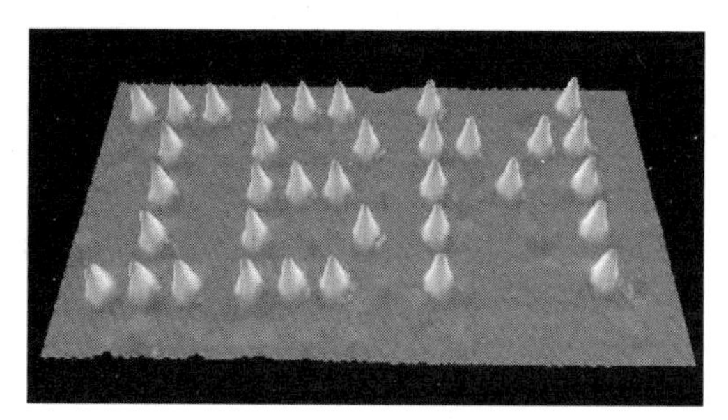

Figure 3 – 11 The letter IBM made by moving atoms with STM

STM has great significance and broad application prospects in the fields of surface science, material science and life science. Binning and Rohrer shared the 1986 Nobel Prize in Physics for their invention of STM with the German physicist E. Ruska (1906—1988), the inventor of the electron microscope in 1932.

3.3 The Simple Harmonic Oscillator

Simple harmonic oscillator is one of the most important models in physics. Many practical problems can be approximated by simple harmonic oscillator. For example, the micro-vibration of atoms on the lattice in a solid can be approximated by simple harmonic motion, the protons and neutrons vibrating in the nucleus can also be described by the simple harmonic oscillator model.

Suppose a particle with mass m is in one-dimensional simple harmonic motion, and the coordinate origin is taken as the equilibrium position, then its potential energy function is

$$U(x) = \frac{1}{2}kx^2 = \frac{1}{2}m\omega^2 x^2 \tag{3-65}$$

where k is the stiffness coefficient and $\omega = \sqrt{k/m}$ is the natural angular frequency. Substituting equation (3 – 65) into the one-dimensional steady-state Schrödinger equation (equation (3 – 27)), we have

$$\frac{d^2\psi(x)}{dx^2} + \frac{2m}{\hbar^2}\left(E - \frac{1}{2}m\omega x^2\right)\psi(x) = 0 \tag{3-66}$$

Equation (3 – 66) is a one-dimensional ordinary differential equation with variable coefficients, and solving it requires a lot of mathematical tools. We shall not go into detail here, but we shall give the main conclusions directly. When $x \to \pm\infty$, there is $U(x) = kx^2/2 \to \infty$, then the wave function is required to satisfy the bound state condition, that is, $\psi(x \to \pm\infty) \to 0$. In order to make the wave function meet the standard condition of being single-valued, continuous and finite, the energy of the simple harmonic oscillator can only take a series of specific values, that is, its energy is quantized, and the energy level formula is

$$E_n = \left(n + \frac{1}{2}\right)\hbar\omega; n = 0, 1, 2, \cdots \tag{3-67}$$

where n can be 0 or any positive integer, which is called quantum number. Because

$$E_{n+1} - E_n = \hbar\omega \tag{3-68}$$

Therefore, the energy levels of the simple harmonic oscillator are evenly spaced, which is the same as the result of Planck's black body radiation theory. However, the concept of energy quantization of simple harmonic oscillator was put forward as a hypothesis by Planck in order to explain the law of black body radiation, and here, it is the inevitable result of solving the Schrödinger equation in quantum mechanics.

It can be seen from the energy level equation (3 - 67) that when $n = 0$, the simple harmonic oscillator is in the state of the lowest energy $E_0 = \hbar\omega/2$, which is called the ground state. $E_0 = \hbar\omega/2$ is called the zero-point energy. The existence of zero-point energy is quite different from that of the classical harmonic oscillator. This implies that the microscopic simple harmonic oscillator can never stand still, which is the manifestation of its wave-particle duality, and it can also be explained by the uncertainty relation.

The simple harmonic oscillator problem can also be regarded as the motion of a particle in a one-dimensional parabolic potential well. Figure 3 - 12 shows the potential energy curve, energy levels and probability density distribution curve of the simple harmonic oscillator. According to classical mechanics, because the particle has the maximum velocity near the equilibrium position of $x = 0$, the probability of its occurrence is the smallest; and the particle cannot appear in the region of $E < U$. But unlike this, the result of quantum mechanics shows that the probability density of simple harmonic oscillator fluctuates. For example, when the simple harmonic oscillator is in the ground state of $n = 0$, the probability of the particle appearing near $x = 0$ is the largest, and there is also a probability of the particle appearing in the region of $E < U$.

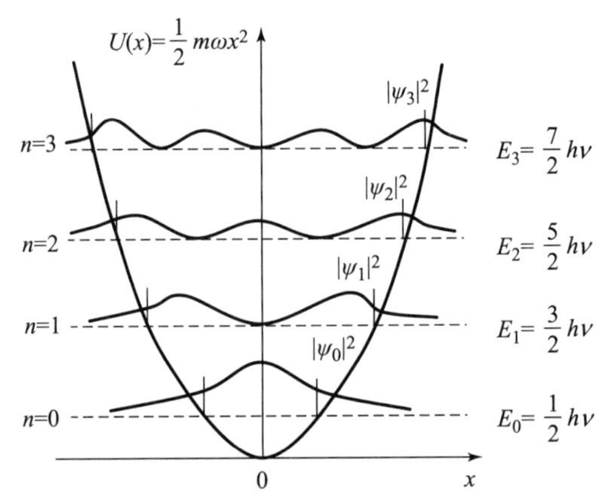

Figure 3 - 12 Potential energy curve, energy levels and probability density distribution curve of simple harmonic oscillator

Example 3 - 3: For a simple harmonic oscillator, the steady-state wave function of its ground state is $\psi_0 = Ae^{-ax^2}$, where A and a are constants. Substitute this formula into the one-dimensional steady-state Schrödinger equation, and try to obtain the zero-point energy $E_0 = h\nu/2$ of the simple harmonic oscillator according to the condition that the obtained formula holds when x takes any value, where ν is the natural frequency of the simple harmonic oscillator.

Solution: The one-dimensional steady-state Schrödinger equation satisfied by the steady-state wave function ψ of the particle is known as

$$-\frac{\hbar^2}{2m}\frac{\mathrm{d}^2\psi}{\mathrm{d}x^2} + U(x)\psi = E\psi$$

where m and E are the mass and energy of the particle, respectively, and $U(x)$ is the potential energy function. For a one-dimensional simple harmonic oscillator

$$U(x) = \frac{1}{2}m\omega^2x^2$$

where ω is the natural angular frequency. Therefore, for this problem, we have

$$\frac{\mathrm{d}^2\psi}{\mathrm{d}x^2} + \frac{2m}{\hbar^2}\left(E - \frac{1}{2}m\omega^2x^2\right)\psi = 0$$

Substitute $\psi_0 = A\mathrm{e}^{-ax^2}$ into the above equation, after sorting, we can obtain

$$\left(4a^2 - \frac{m^2}{\hbar^2}\omega^2\right)x^2 = 2\left(a - \frac{m}{\hbar^2}E_0\right)$$

Since this equation holds true when x takes any value, the coefficient of the x^2 term on the left side of the equation should be zero, and the right side of the equation should also be zero. From this there are

$$4a^2 - \frac{m^2}{\hbar^2}\omega^2 = 0,\ a - \frac{m}{\hbar^2}E_0 = 0$$

Find a from the former formula, and substitute it into the latter formula, it can be obtained that

$$E_0 = \frac{\hbar\omega}{2} = \frac{h\nu}{2}$$

Example 3 - 4: There is a spring oscillator with the mass $m = 1$ g, the stiffness coefficient $k = 0.1\ \mathrm{N \cdot m^{-1}}$, and the amplitude $A = 1$ mm. Find the energy level interval, and estimate the quantum number n corresponding to the energy.

Solution: The natural angular frequency of the spring vibrator is

$$\omega = \sqrt{\frac{k}{m}} = \sqrt{\frac{0.1}{10^{-3}}}\ \mathrm{rad \cdot s^{-1}} = 10\ \mathrm{rad \cdot s^{-1}}$$

The energy level interval is

$$\Delta E = \hbar\omega = 1.05 \times 10^{-34} \times 10\ \mathrm{J} = 1.05 \times 10^{-33}\ \mathrm{J}$$

The total energy of the spring oscillator is

$$E = \frac{1}{2}kA^2 = \frac{1}{2} \times 0.1 \times (10^{-3})^2\ \mathrm{J} = 5 \times 10^{-8}\ \mathrm{J}$$

The quantum number is obtained by equation (3 - 67)

$$n = \frac{E}{\hbar\omega} - \frac{1}{2} = \frac{5 \times 10^{-8}}{1.05 \times 10^{-34} \times 10} - \frac{1}{2} = 4.7 \times 10^{25}$$

It can be seen that the macroscopic simple harmonic oscillator is at a very high energy level, and the interval between adjacent energy levels is so small that it can be ignored, so its energy changes continuously. At this time, the classical pattern and the quantum pattern tend to be consistent.

Therefore, classical physics can be regarded as the limit case in quantum physics when the quantum number n tends to infinity.

Bohr Correspondence Principle

The theory of quantum mechanics successfully describes the motion laws of objects in the microscopic world (such as electrons, atoms, molecules, etc.), while the motion laws of objects in the macroscopic world (such as bullets, spring oscillators, etc.) can be described by classical mechanics. The same physical world needs two different theories to describe just because of the different sizes of objects. In order to complete the physical theory system, Bohr (Figure 3 - 13) put forward the famous "correspondence principle" in 1927, that is, the correspondence between macroscopic and microscopic theories and between similar problems in different fields. The correspondence principle points out that when the system is "large", classical physics can be regarded as an approximation of quantum physics, and according to the direction pointed out by the correspondence principle, new theories can be derived from old theories. This has been fully verified in the later establishment and development of quantum mechanics. Guided by the correspondence principle, Bohr's student Heisenberg sought various specific correspondence and corresponding quantities in quantum mechanics corresponding to classical mechanics, thus matrix mechanics was established. The correspondence theory also played a guiding role in the development of wave dynamics and quantum mechanics by Dirac and Schrodinger.

Figure 3 - 13 Bohr was talking about the correspondence principle

In 1947, when Bohr designed his own coat of arms for the Order of the Elephant awarded to him by the Danish government, he adopted the advice of his co-worker's wife, Hanna Kobylinski who was a Chinese historian, and used the ancient Chinese Yin and Yang symbol as an image of wave-particle duality and the "correspondence principle" of quantum mechanics, as shown in Figure 3 - 14. The difference is that Bohr changed the original white in the Tai Chi diagram to red. Bohr also added the Latin motto contraria sunt complementa, that is, opposites are complementary.

The essence of Eastern and Western cultures, ancient philosophy and modern science were organically combined in this way. The harmony is amazing!

Figure 3 - 14 Bohr's own coat of arms

3.4 Electrons in Atoms

3.4.1 Hydrogen Atom

After the Schrödinger equation was proposed, it was first used to solve the hydrogen atom. and reasonably solved the problems like the linear spectrum of hydrogen atom observed in the experiment. This was the most convincing achievement in the early days of quantum mechanics.

In a hydrogen atom, electron moves in the Coulomb field of the nucleus, and the potential energy function of the atomic system is

$$U(r) = -\frac{e^2}{4\pi\varepsilon_0 r} \tag{3-69}$$

where r is the distance from the electron to the proton. Because the potential energy function $U(r)$ has spherical symmetry, it is more convenient to use the three-dimensional steady-state Schrödinger equation in the spherical coordinate system as shown in Figure 3 – 15, specifically:

$$-\frac{\hbar^2}{2m_e}\left[\frac{\partial^2\psi}{\partial r^2}+\frac{2}{r}\frac{\partial\psi}{\partial r}+\frac{1}{r^2\sin^2\theta}\frac{\partial}{\partial\theta}\left(\sin\theta\frac{\partial\psi}{\partial\theta}\right)\right]-\frac{\hbar^2}{2m_e}\left(\frac{1}{r^2\sin^2\theta}\frac{\partial^2\psi}{\partial\varphi^2}\right)-\frac{e^2}{4\pi\varepsilon_0 r}\psi = E\psi \tag{3-70}$$

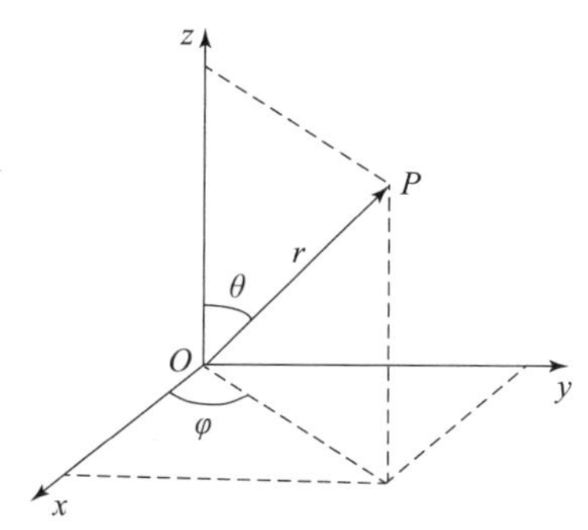

Figure 3 – 15 Coordinate transformation between variables in spherical coordinate system and rectangular coordinate system

where m_e is the rest mass of the electron. The above formula can be solved by the method of separating variables, and the steady-state wave function $\psi(r,\theta,\varphi)$ to be solved can be written as the product of $R(r)$ and $Y(\theta,\varphi)$, that is

$$\psi(r,\theta,\varphi) = R(r)Y(\theta,\varphi) \tag{3-71}$$

where $R(r)$ is the part containing only the radial coordinate r in the steady-state wave function $\psi(r,\theta,\varphi)$, which is called the radial wave function, $Y(\theta,\varphi)$ is the function of the angle part, which is called the angular wave function.

As the solution process is very complex, some important conclusions obtained from the solution are directly given below, and the physical significance will be discussed.

(1) Quantization of Energy.

Since the mass of proton is much larger than that of electron, the proton can be approximately regarded as stationary in hydrogen atom, so the energy of electron represents the energy of the whole hydrogen atom. When the wave function satisfies the standard condition of being single-valued, finite and continuous, in the process of solving the Schrödinger equation, it will come naturally that the energy of the electron (or the whole atom) can only be

$$E_n = -\frac{1}{2}\frac{e^2}{(4\pi\varepsilon_0)a_0}\frac{1}{n^2} = -\frac{m_e e^4}{2(4\pi\varepsilon_0)^2\hbar^2}\frac{1}{n^2}; n=1,2,3,\cdots \quad (3-72)$$

where $a_0 = \frac{4\pi\varepsilon_0\hbar^2}{m_e e^2} = 0.529,177,210,67(12)\times10^{-10}\,\text{m} \approx 0.052,9\ \text{nm}$ is called the Bohr radius; n only can take any positive integer, which is called the principal quantum number. Equation (3-72) is completely consistent with the energy level equation (2-43) of hydrogen atom in Bohr's theory in 1913, but Bohr's conclusion relies on artificial assumptions.

(2) Quantization of Angular Momentum.

Solving the Schrödinger equation for hydrogen atom can also obtain the conclusion that the angular momentum of the electron moving around the nucleus is quantized. The angular momentum of the moving electron is represented by $\boldsymbol{L}$, and its magnitude is expressed by the following formula

$$L = \sqrt{l(l+1)}\hbar; l=0,1,2,\cdots, n-1 \quad (3-73)$$

where n is the principal quantum number, and l is called the orbital quantum number or the angular quantum number. For a certain principal quantum number n, the angular quantum number l has n possible values. When l takes different values, the angular momentum $\boldsymbol{L}$ of the electron has different values, indicating that the state of the electron moving around the nucleus is different, and it also shows different wave functions, that is, the probability distribution of the electron is different everywhere in space. Note that the angular quantum number l can take value of 0, that is, the angular momentum $\boldsymbol{L}$ of the electron can be equal to zero, which is incomprehensible in classical mechanics and is also different from the result of Bohr's semi-classical old quantum theory.

Usually, letters such as s, p, d and f are used to represent the quantum states such as $l=0,1,2$ and 3 respectively. For example, the electrons of $n=4, l=0,1,2$ and 3 are represented by 4s, 4p, 4d, and 4f, respectively.

(3) Spatial Orientation Quantization of Angular Momentum.

Quantum mechanics also points out that the orientation of the electron's angular momentum $\boldsymbol{L}$ in space cannot change continuously, but can only take some specific directions. If the direction of the external magnetic field is taken as the positive direction of the z-axis, the component of the angular momentum $\boldsymbol{L}$ along the direction of the external magnetic field can only take the following discrete values

$$L_z = m_l\hbar;\ m_l = 0, \pm1, \pm2, \cdots, \pm l \quad (3-74)$$

where m_l is called the orbital magnetic quantum number. For a certain angular quantum number l, the orbital magnetic quantum number m_l can take $(2l+1)$ values, which indicates that the orientation of angular momentum $\boldsymbol{L}$ in space can only have $(2l+1)$ possibilities. This conclusion is called the spatial orientation quantization of angular momentum. It is a natural consequence of the Schrödinger equation, rather than the artificial specification as made by the German physicist Arnold Johannes Wilhelm Sommerfeld(1868 – 1951) in 1915.

The possible values of the orbital magnetic quantum number m_l are limited by the angular quantum number l, and the larger l is, the more possible values m_l has. For example, when $l = 1$, the value of m_l can be 0 and ± 1, there are three values, which means that the angular momentum $\boldsymbol{L}$ has three possible orientations in space. When $l = 2$, $m_l = 0, \pm 1, \pm 2$, there are five values, that is, the angular momentum $\boldsymbol{L}$ has five possible orientations in space, as shown in Figure 3 – 16. At this time, the magnitude of the angular momentum $\boldsymbol{L}$ is $\sqrt{6}\hbar$, and the component value of angular momentum $\boldsymbol{L}$ along the direction of the external magnetic field may be $-2\hbar$, $-\hbar$, 0, $\hbar$ and $2\hbar$. Note that the magnitude of the angular momentum $\boldsymbol{L}$ is $L = \sqrt{l(l+1)}\hbar$, and its maximum component along the z-axis is $l\hbar$, it is not equal to L.

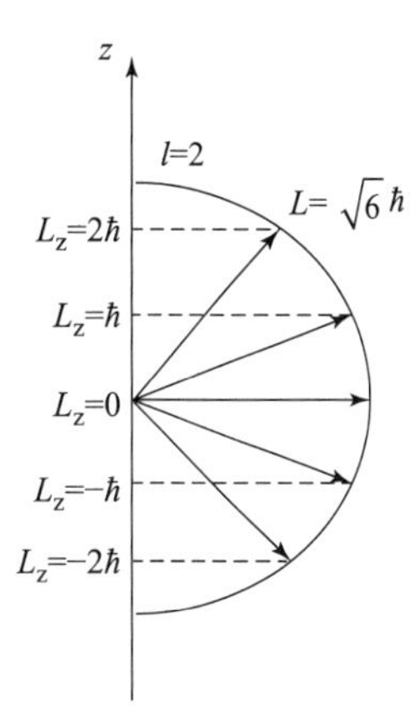

Figure 3 – 16 Vector model for the quantization of spatial orientation of angular momentum

The spatial orientation quantization of angular momentum can satisfactorily explain many physical phenomena, such as the normal Zeeman effect. In 1896, the Dutch experimental physicist P. Zeeman(1865—1943, winner of the 1902 Nobel Prize in Physics) discovered the phenomenon of spectral line splitting in an external magnetic field. In the external magnetic field, there are three quantum states for the energy level of $l = 1$, so when the three quantum states of the energy level of $l = 1$ transition to the energy level of $l = 0$ respectively, three spectral lines are generated. This phenomenon belongs to the normal Zeeman effect. This experiment also confirms that, for a certain value of l, $\boldsymbol{L}$ has $(2l+1)$ orientations and the maximum component along the z-axis is $l\hbar$.

The spatial orientation quantization of angular momentum can also be regarded as the quantization of the component of orbital angular momentum along the direction of magnetic field. This quantization is only possible in the presence of a magnetic field. The theory includes a special direction, that is, the direction of the magnetic field. This cannot be in any direction.

(4) Probability Distribution of Electron.

The energy eigenwave function obtained by solving Schrödinger equation for hydrogen atom is

$$\psi_{nlm_l}(r,\theta,\varphi) = R_{nl}(r)Y_{lm_l}(\theta,\varphi) \tag{3-75}$$

where the principal quantum number $n = 1, 2, 3, \cdots$; the possible value of the angular quantum number l is $0, 1, 2, \cdots$; and the possible value of the orbital magnetic quantum number m_l is $0, \pm 1, \pm 2, \cdots, \pm l$. Corresponding to each set of quantum numbers n, l, m_l, there is a certain wave function describing a certain state. Here $R_{nl}(r)$ represents the radial part of the steady-state wave function, and Table 3 – 1 gives its expression when $n = 1, 2, 3$; $Y_{lm_l}(\theta,\varphi)$ represents the angular part of the steady-state wave function, which is a spherical harmonic function, and Table 3 – 2 gives its expression when $l = 0, 1, 2$.

Table 3 – 1 Some radial wave functions

$n = 1$	$R_{1,0}(r) = \frac{2}{a_0^{3/2}}\exp\left(-\frac{r}{a_0}\right)$
$n = 2$	$\begin{cases} R_{2,0}(r) = \frac{1}{\sqrt{2}a_0^{3/2}}\left(1-\frac{r}{2a_0}\right)\exp\left(-\frac{r}{2a_0}\right) \\ R_{2,1}(r) = \frac{1}{2\sqrt{6}a_0^{3/2}}\frac{r}{a_0}\exp\left(-\frac{r}{2a_0}\right) \end{cases}$
$n = 3$	$\begin{cases} R_{3,0}(r) = \frac{2}{3\sqrt{3}a_0^{3/2}}\left[1-\frac{2r}{3a_0}+\frac{2}{27}\left(\frac{r}{a_0}\right)^2\right]\exp\left(-\frac{r}{3a_0}\right) \\ R_{3,1}(r) = \frac{8}{27\sqrt{6}a_0^{3/2}}\left(1-\frac{r}{6a_0}\right)\exp\left(-\frac{r}{3a_0}\right) \\ R_{3,2}(r) = \frac{4}{81\sqrt{30}a_0^{3/2}}\left(\frac{r}{a_0}\right)^2\exp\left(-\frac{r}{3a_0}\right) \end{cases}$

Table 3 – 2 Some spherical harmonic functions

$l = 0$	$Y_{0,0} = \frac{1}{\sqrt{4\pi}}$
$l = 1$	$\begin{cases} Y_{1,\pm1} = \sqrt{\frac{3}{8\pi}}\sin\theta\exp(\pm i\varphi) \\ Y_{1,0} = \sqrt{\frac{3}{4\pi}}\cos\theta \end{cases}$
$l = 2$	$\begin{cases} Y_{2,\pm2} = \sqrt{\frac{15}{32\pi}}\sin^2\theta\exp(\pm 2i\varphi) \\ Y_{2,\pm1} = \sqrt{\frac{15}{8\pi}}\sin\theta\cos\theta\exp(\pm i\varphi) \\ Y_{2,0} = \sqrt{\frac{5}{16\pi}}(3\cos^2\theta - 1) \end{cases}$

According to the statistical interpretation of the wave function, the probability of an electron appearing in the space volume element $\mathrm{d}V = r^2\mathrm{d}r\sin\theta\mathrm{d}\theta\mathrm{d}\varphi$ is

$$|\psi_{nlm_l}(r,\theta,\varphi)|^2\mathrm{d}V = |R_{nl}(r)|^2\ |Y_{lm_l}(\theta,\varphi)|^2 r^2\mathrm{d}r\sin\theta\mathrm{d}\theta\mathrm{d}\varphi \qquad (3-76)$$

In the above formula, $|R_{nl}(r)|^2 r^2\mathrm{d}r$ represents the probability of the electron appearing in a thin spherical shell with a radius between r and $r + \mathrm{d}r$

$$P_{nl}(r)\mathrm{d}r = \left(\int_0^{4\pi}|Y_{lm_l}(\theta,\varphi)|^2\mathrm{d}\Omega\right)|R_{nl}(r)|^2 r^2\mathrm{d}r = |R_{nl}(r)|^2 r^2\mathrm{d}r \qquad (3-77)$$

where $P_{nl}(r)$ represents the radial distribution of electron probability density when the principal quantum number is n and the angular quantum number is l. Using the steady-state radial wave function in Table 3 – 1, it can be calculated that relation curve between $P_{nl}(r)$ and r when $n = 1, 2, 3$, as shown in Figure 3 – 17. In a hydrogen atom with the ground state, $n = 1$, $l = 0$, the maximum radial probability density P_{10} of electron corresponds exactly to the radius a_0 of the first circular orbit in Bohr's atomic theory. When $n = 2$, there are two states, where for the state of $l = 1$, the maximum position of the radial probability density P_{21} of electron corresponds exactly to the radius $4a_0$ of Bohr's second circular orbit. When $n = 3$, there are three states, where for the state of $l = 2$, the maximum position of the radial probability density P_{32} of electron corresponds exactly to the radius $9a_0$ of Bohr's third circular orbit. However, the electron in Bohr's theory can only be located in circular orbits, and the result of quantum mechanics is that the electron has only the greatest probability of appearing in circular orbits, but it may also appear elsewhere.

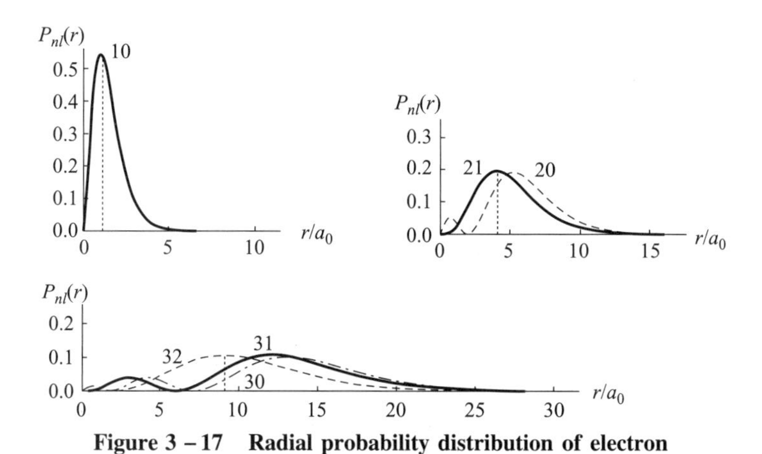

Figure 3 – 17 Radial probability distribution of electron

In equation (3 – 76) $|Y_{lm_l}(\theta,\varphi)|^2\sin\theta d\theta d\varphi$ represents the probability of the electron appearing in the solid angle $d\Omega = \sin\theta d\theta d\varphi$

$$P_{lm_l}(\theta,\varphi)d\Omega = \left(\int_0^{\infty} |R_{nl}(r)|^2 r^2 dr\right)|Y_{lm_l}(\theta,\varphi)|^2 d\Omega = |Y_{lm_l}(\theta,\varphi)|^2 d\Omega \quad (3-78)$$

The angular probability distribution of electron in the states of $l = 0, 1, 2$ can be calculated by using the spherical harmonic functions in Table 3 – 2, as shown in Figure 3 – 18. The results show that in a hydrogen atom with the ground state, $l = 0$, $m_l = 0$, the angular probability distribution of the electron is spherically symmetric. When $l = 1$, $m_l = 0, \pm 1$, the angular probability distribution of the electron in these three states looks like a dumbbell, and they all have axial symmetry with respect to the z-axis.

Because at any time, the place where the electron appears in the space around the nucleus cannot be uniquely determined, it can only be explained by the difference in probability distribution that the chance of the electron appearing somewhere is greater and the chance of the electron appearing somewhere else is smaller. Therefore, the concept of electron cloud is introduced. The so-called electron cloud does not mean that the electron is really like a cloud around the nucleus, but it is just a visual description of the probability distribution of electron, as shown in Figure 3 – 19. If the electron has more opportunities to appear somewhere, the electron cloud there will be denser, while where there are fewer opportunities for the electron to appear, the electron cloud will be sparser.

Example 3 – 5: A hydrogen atom consists of a nucleus and an extranuclear electron. (1) Using

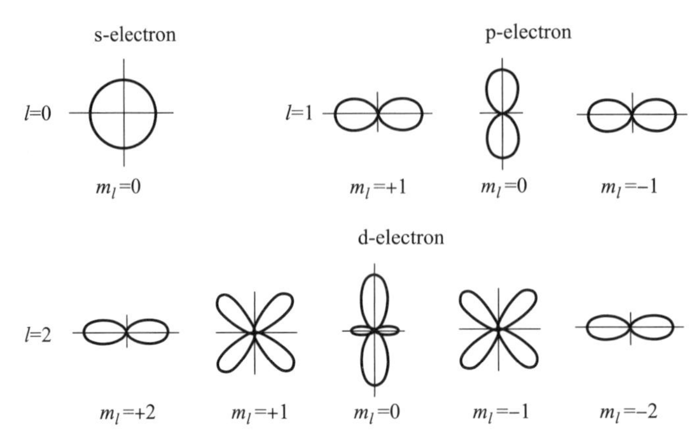

Figure 3 – 18 Angular distribution of probability density of electron

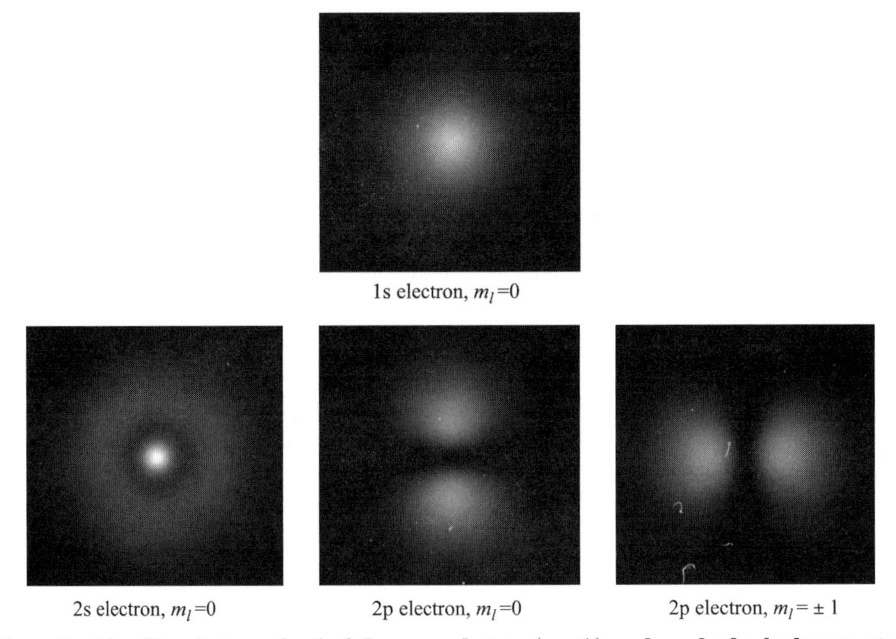

Figure 3 – 19 The electron cloud of the ground state ($n = 1$) and $n = 2$ of a hydrogen atom

the uncertainty relation $\Delta x \Delta p \geqslant \hbar$, estimate the minimum energy of the electron in the hydrogen atom. (2) From the Schrödinger equation, the ground state wave function of hydrogen atom is solved as $\psi_{1,0,0} = \frac{1}{a_0^{3/2}\sqrt{\pi}} e^{-r/a_0}$, where $a_0 = 0.529 \times 10^{-10}$ m is the Bohr radius. Find the probability of the electron appearing within the sphere of radius a_0 when the hydrogen atom is in the ground state.

Solution: (1) When the motion of the nucleus is ignored, the total energy of the hydrogen atom is

$$E = \frac{p^2}{2m_e} - \frac{e^2}{4\pi\varepsilon_0 r} \qquad ①$$

where m_e is the rest mass of the electron and p is the momentum of the electron.

In example 2 – 9, we have calculated that in an atom the velocity v of the electron is of the same order of magnitude as the uncertainty of velocity Δv, so the uncertainty of momentum $\Delta p \approx p$. The electron is bound to move in atom of radius r, so the uncertainty of the coordinate $\Delta x \approx r$. According to the uncertainty relation $\Delta x \Delta p \geqslant \hbar$, the uncertainty of momentum $\Delta p \approx \hbar / r$. Substitute $p \approx \Delta p \approx \hbar / r$ into equation①to get

$$E = \frac{\hbar^2}{2m_e r^2} - \frac{e^2}{4\pi\varepsilon_0 r}$$

In order to find the minimum energy of the electron, let

$$\frac{dE}{dr} = -\frac{\hbar^2}{m_e r^3} + \frac{e^2}{4\pi\varepsilon_0 r^2} = 0$$

From the above formula, the radius of the hydrogen atom corresponding to the minimum energy of the electron is

$$r = \frac{4\pi\varepsilon_0 \hbar^2}{m_e e^2} = \frac{4\pi \times 8.85 \times 10^{-12} \times (1.05 \times 10^{-34})^2}{9.1 \times 10^{-31} \times (1.6 \times 10^{-19})^2} \text{ m} = 0.529 \times 10^{-10} \text{ m}$$

At this time, the energy of hydrogen atom in the ground state is

$$E_{\min} = -\frac{m_e e^4}{2(4\pi\varepsilon_0)^2 \hbar^2} \approx -13.6 \text{ eV}$$

This example also shows that a quantum system has the so-called zero-point energy. Because if the kinetic energy of the bound state is zero, that is, the uncertainty range of the velocity is zero, then the particle tends to infinity in space, that is, it is not bound. This is contrary to the fact.

(2) The probability of the electron appearing within the sphere with a Bohr radius is

$$P = \int_0^{a_0} |\psi|^2 4\pi r^2 dr = \int_0^{a_0} \frac{1}{\pi a_0^3} e^{-2r/a_0} 4\pi r^2 dr = -\frac{2}{a_0^2} \int_0^{a_0} r^2 \, de^{-2r/a_0} = 0.32$$

3.4.2 The Stern-Gerlach Experiment and Electron Spin

(1) The Stern-Gerlach Experiment.

In 1921, German physicists O. Stern (1888—1969, winner of the 1943 Nobel Prize in Physics, Figure 3-20) and W. Gerlach (1889—1979) did experimental observation on the spatial quantization of angular momentum of electron. The device used is shown in Figure 3-20. The atoms emitted from the high-temperature furnace are collimated by the slit to form a beam of atomic rays, which then pass through a non-uniform magnetic field, and finally fall on the screen.

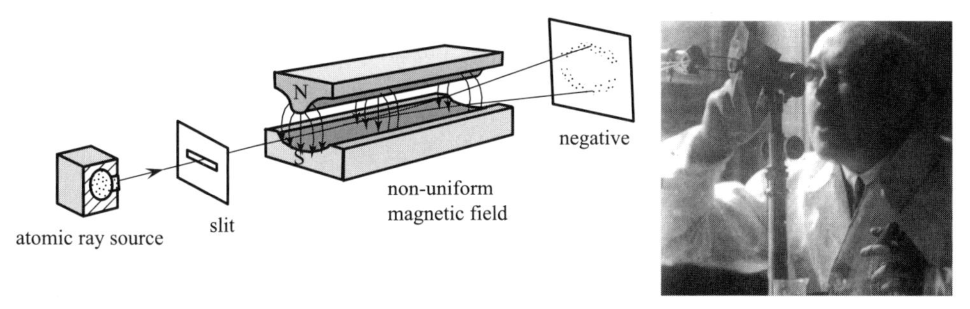

Figure 3-20 Schematic diagram of the Stern-Gerlach experimental device, and the 1943 Nobel Prize winner in Physics Stern was observing

Because the electron is charged, when an electron moves around the nucleus in a circular orbit with a radius of r at a speed v (here we still make use of the concept of "orbit"), it is equivalent to a closed current-carrying coil with a current intensity of $i = \frac{v}{2\pi r}e$. Then the magnitude of the orbital magnetic moment of the electron with rest mass m_e is

$$\mu = \frac{ve}{2\pi r}\pi r^2 = \frac{e}{2m_e}mvr = \frac{e}{2m_e}L \tag{3-79}$$

where e is the absolute value of the electron charge. Equation (3 – 79) shows that the orbital magnetic moment of an electron in a hydrogen atom is proportional to its angular momentum. In fact, in any atom, the orbital magnetic moment of an electron is proportional to its orbital angular momentum. Since the electron is negatively charged, the vector relation between magnetic moment and angular momentum is

$$\boldsymbol{\mu} = -\frac{e}{2m_e}\boldsymbol{L} \tag{3-80}$$

Since the angular momentum of electron is spatially quantized, the orbital magnetic moment is also spatially quantized, and its component along the z-direction in space is also quantized. By equation (3 – 74), there should be

$$\mu_z = -\frac{e}{2m_e}L_z = -m_l\frac{e\hbar}{2m_e};\ m_l = 0, \pm 1, \pm 2, \cdots, \pm l \tag{3-81}$$

The idea behind the design of the Stern-Gerlach experiment is that because electron has magnetic moment, when an atom passes through a non-uniform magnetic field, it will be deflected due to magnetic force. If the direction of the electron magnetic moment can be oriented arbitrarily, a black area will be formed on the screen. The experiment found that several clear black spot lines formed on the screen, indicating that the magnetic moment can only take a few specific directions, thus confirming that the component of electron angular momentum is quantized.

Because the temperature in the high-temperature furnace is not enough to excite most atoms from the ground state to the excited state, the Stern-Gerlach experiment mainly shows the angular momentum and magnetic moment of electrons in the ground state atoms. Experiments on the ground state atomic rays of silver, copper, gold, lithium, sodium, potassium, etc. respectively showed that a beam of the atomic rays was divided into two beams of upper and lower symmetry after passing through the non-uniform magnetic field, leaving two deposition lines separated from each other on the photo negative, as shown in Figure 3 – 21. What was puzzling at the time was that, according to the law of spatial quantization of electron angular momentum, when the quantum number of electron's orbital angular momentum was l, its orientation in space should have $(2l + 1)$ possibilities, and the deflection of atomic rays in the magnetic field should generate $(2l + 1)$, that is, an odd number of deposition lines. For atoms such as silver, copper, gold, lithium, sodium and potassium in the ground state, $l = 0$, the orbital angular momentum and orbital magnetic moment of the electron are both zero, and the atomic deposition lines on the screen should be $2l + 1 = 1$ deposition line. However, the Stern-Gerlach experiment observed two ground state atomic deposition lines, thus it showed that there should be other form of angular momentum as well as orbital angular momentum in an atom.

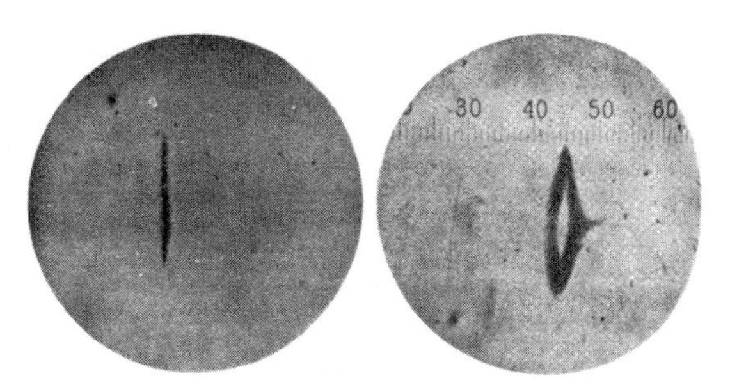

Figure 3 - 21 A beam of silver atoms splits into two beams when passing through a non-uniform magnetic field

(2) Electron Spin.

In 1925, two Ph. D. students from Leyden University in the Netherlands, G. E. Uhlenbeck (1900—1974) and S. A. Goudsmit (1902—1979) put forward a bold view on the basis of the Stern-Gerlach experiment. In addition to orbital motion around the nucleus, electron also has spin motion. Therefore, electron has spin angular momentum in addition to orbital angular momentum. Since the electron is charged, it has both orbital magnetic moment and spin magnetic moment. For atoms such as silver, copper, gold, lithium, sodium and potassium in the ground state, although the orbital magnetic moment of the electron is zero, there is still a spin magnetic moment, which explains the Stern-Gerlach experiment well.

Similar to the magnitude of the electron orbital angular momentum L, the magnitude of the electron spin angular momentum S can be set as

$$S = \sqrt{s(s+1)}\hbar \tag{3-82}$$

where s is called spin quantum number.

The Stern-Gerlach experiment shows that the spin magnetic moment is also spatially quantized in the external magnetic field, and the component of the spin magnetic moment along the direction of the magnetic field has only two values. Therefore, the experiment also shows that the spin angular momentum is also spatially quantized, and the component of the spin angular momentum along the direction of the magnetic field has only two possible values.

Imitating that the component of the electron's orbital angular momentum along the direction of the external magnetic field can only take $2l+1$ values, the component of the electron's spin angular momentum along the direction of the external magnetic field can also take only $2s+1$ values, but the Stern-Gerlach experiment points out that there are only two values. Thus, let

$$2s+1=2$$

The spin quantum number of the electron is obtained as

$$s = \frac{1}{2} \tag{3-83}$$

Therefore, the magnitude of the electron spin angular momentum is

$$S = \sqrt{\frac{1}{2}\left(\frac{1}{2}+1\right)}\hbar = \sqrt{\frac{3}{4}}\hbar \tag{3-84}$$

Since the spin angular momentum of electrons in all atoms is the same, s is no longer presented as a quantum number.

Similar to the component of the electron's orbital angular momentum $\boldsymbol{L}$ along the direction of the external magnetic field, $L_z = m_l\hbar$, the component of the electron's spin angular momentum $\boldsymbol{S}$ along the direction of the external magnetic field can be set as

$$S_z = m_s\hbar \tag{3-85}$$

Imitating the electron's orbital magnetic quantum number $m_l = 0, \pm 1, \pm 2, \cdots, \pm l$;

$$m_s = \pm s = \pm\frac{1}{2} \tag{3-86}$$

where m_s is called the electron's spin magnetic quantum number, which can only take two values.

Therefore, the spin angular momentum of the electron has only two components along the direction of the external magnetic field, and the values are

$$S_z = \pm\frac{1}{2}\hbar \tag{3-87}$$

Then the spin magnetic moment of an electron is also spatially quantized, and it also has two orientations. The two spin motion states of the electron in an external magnetic field are often described visually by Figure 3-22.

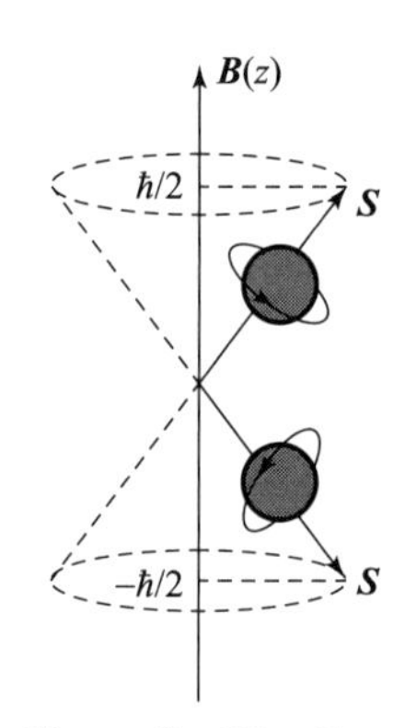

Figure 3-22 Gyro motion image of electron spin

After the introduction of spin, the "even number deposition" in the Stern-Gerlach experiment is naturally explained. However, classical physics cannot understand that an electron has an internal structure. The gyro motion image in Figure 3-22 is just like the orbit motion image, which makes use of the macro image and is very inaccurate.

In 1928, Dirac naturally concluded that an electron has spin by using the relativistic Dirac equation that describes the law of high-speed motion of microscopic particle. The above equations (3-82) - (3-87) obtained by analogy can all be given by quantum mechanics. It should be emphasized that the spin motion of the electron has no classical correspondence. It is a relativistic quantum effect and an intrinsic property of microscopic particle.

Quantum mechanics also shows that the spin magnetic moment $\boldsymbol{\mu}_s$ of an electron has the following relation with the spin angular momentum $\boldsymbol{S}$

$$\boldsymbol{\mu}_s = -\frac{e}{m_e}\boldsymbol{S} \tag{3-88}$$

Its component along the z direction is

$$\mu_{s,z} = \frac{e}{m_e}S_z = \frac{e}{m_e}m_s\hbar$$

where the spin magnetic quantum number m_s can only take the two values $\pm 1/2$, then we have

$$\mu_{s,z} = \pm \frac{e\hbar}{2m_e} = \pm \mu_B \tag{3-89}$$

where μ_B is called Bohr magneton, and its value is $\mu_B = 9.274,009,994(57) \times 10^{-24}\ \mathrm{J \cdot T^{-1}}$.

According to the distance between the two deposition lines of the beam of the silver atoms in the ground state passing through the non-uniform magnetic field on the photo negative, the component value of the spin magnetic moment along the z-direction can be obtained, which is consistent with the above result of theoretical calculation, that is, the value of Bohr magneton. The Stern-Gerlach experiment was the first to demonstrate that the electron has spin.

It should be pointed out that it is incorrect to regard the electron spin as the rotation of a small ball about its own axis. Electron has spin and spin magnetic moment are the basic properties of the electron. Their presence signifies a new degree of freedom for the electron.

A Short Story

In 1925, R. L. Kronig, a Dutch-American physicist who was visiting Copenhagen at the time shortly after the Pauli exclusion principle was proposed, thought of a possibility that an electron rotates around its own axis. He went to discuss it with Pauli, but Pauli strongly opposed it. Pauli said to Kronig, "It is indeed very clever but of course has nothing to do with reality." Because if the electron spin is a kind of spatial motion, that is, it rotates around its own axis, when its angular momentum is $\hbar/2$, the speed of the surface of the electron (regarded as a small ball with a radius of 10^{-16} m) will exceed the speed of light in vacuum many times. This violates the theory of relativity, so Pauli thought it was impossible. Therefore, Kronig abandoned the idea that would led to a major discovery.

Half a year later, without knowing the above circumstance, two students of Dutch physicist P. Ehrenfest (1880—1933), Uhlenbeck and Goudsmit also put forward the same idea. They wrote a short paper and handed it to Ehrenfest. Ehrenfest thought their ideas were very important and advised them to consult the Dutch physicist H. A. Lorentz (1853—1928, 1902 Nobel Prize winner in Physics), who lived not far away. Lorentz received them warmly and made simple estimation. His conclusion was the same as Pauli's. So they went back immediately and asked Ehrenfest to return the paper to them. But Ehrenfest has already sent the paper to the journal Nature, and it would probably be published soon. They were very upset. Ehrenfest comforted them and said, "you are both young enough to allow yourself some follishness!"

It was later recognized that electron spin is an inherent property (intrinsic property) of electron, which cannot be explained by its spatial motion. Two years later, Dirac established relativistic quantum mechanics, and naturally came to the important conclusion that electron has intrinsic angular momentum.

The discovery of electron spin shows the role of chance and status in the development of physics. Uhlenbeck and Goudsmit were very lucky to get their paper on spinning electron published. Although Kronig also believed in the spinning electron before them, but the first person he showed it was Pauli. Pauli ridiculed the whole thing so much that no one else heard anything of it from Kronig. Uhlenbeck once said: they were very lucky and honored to become Ehrenfest's students.

3.5 Four Quantum Numbers and Atomic Shell Structure

3.5.1 Four Quantum Numbers

To sum up, in the description of the structure of hydrogen atom, we introduce four quantum numbers n, l, m_l, m_s to describe the motion state of an electron. In fact, according to quantum theory, the motion state of each electron in a more complex atom can still be described by these four quantum numbers.

(1) The principal quantum number n.

$n = 1, 2, 3, \cdots$. It basically determines the energy E_n of the electron in the atom. The larger the n, the larger the value of E_n.

(2) The angular quantum number l.

$l = 0, 1, 2, \cdots, (n-1)$. It determines the magnitude of the angular momentum $\boldsymbol{L}$ of the electron moving around the nucleus. Generally, the energy of each electron in the state of the same principal quantum number n but different angular quantum number l is slightly different. When n is given, there are n possible values of l.

(3) The magnetic quantum number m_l.

$m_l = 0, \pm 1, \pm 2, \cdots, \pm l$. It determines the orientation of the angular momentum $\boldsymbol{L}$ of the electron moving around the nucleus in space, and affects the energy of the atom in the external magnetic field. When l is given, there are $(2l+1)$ possible values of m_l.

(4) The spin magnetic quantum number m_s.

$m_s = \pm 1/2$. It determines the orientation of spin angular momentum $\boldsymbol{S}$ of the electron in the external magnetic field, and also affects the energy of the atom in the external magnetic field.

3.5.2 The Pauli Exclusion Principle and the Principle of the Lowest Energy

(1) The Pauli Exclusion Principle.

Considering the spin angular momentum, four quantum numbers n, l, m_l, m_s are needed to fully determine the motion state of an electron in an atom. In January 1925, the Austrian physicist W. Pauli (1900—1958, Figure 3-23), after carefully analyzing the atomic spectra and other experimental facts, published a paper entitled "On the connexion between the completion of electron groups in an atom with the complex structure of spectra" in the German journal "Zeitschrift für Physik", proposing the Pauli exclusion principle, that is, it is impossible for two or more electrons in

Figure 3-23 Pauli, 1945 Nobel laureate in physics, was giving a lecture

an atom to be in the same quantum state, or there cannot be two electrons in an atom with exactly the same 4 quantum numbers n, l, m_l, m_s. This principle successfully solved many problems related to atomic structure at that time, for which Pauli was awarded the Nobel Prize in Physics in 1945.

According to the Pauli exclusion principle, the number of electrons with the same principal quantum number n in an atom is at most

$$Z_n = \sum_{l=0}^{n-1} 2(2l+1) = 2n^2 \tag{3-90}$$

(2) The Principle of the Lowest Energy.

When an atom is in its normal state, each electron in it tends to occupy the lowest energy level. This is the principle of the lowest energy. When the energy of electrons in the atom is the lowest, the energy of the whole atom is the lowest, and the atom is in the most stable state, that is, the ground state. According to the principle of the lowest energy, all electrons in an atom are filled in order from low to high according to the energy of each state.

Now it is known that all microscopic particles have intrinsic spin. Particles can be divided into two broad categories based on their spin state.

Particles whose spin quantum numbers are half integers such as 1/2, 3/2, ⋯ are called fermions. For example, the spin quantum number of proton and neutron is 1/2, which is the same as that of electron, so they are all fermions. The neutrino is uncharged, it has an extremely small mass, and its spin quantum number is 1/2. The spin quantum number of anti-sigma negative hyperon($\bar{\Sigma}^-$) discovered by Mr. Wang Ganchang(1907—1998), a famous Chinese nuclear physicist and one of the founders and pioneers of nuclear science in China, is also 1/2. Fermions obey the Pauli exclusion principle.

Particles whose spin quantum numbers are integers such as 0, 1, 2, ⋯ are called bosons. For example, pion has a spin quantum number of 0, photon has a spin quantum number of 1, so they are both bosons. Bosons do not obey the Pauli exclusion principle. A single particle state can hold multiple bosons, known as Bose condensation.

Fermion is named after the Italian physicist Enrico Fermi(1901—1954, Nobel Prize winner in Physics in 1938). Boson is named after the Indian physicist Satyendra Nath Bose(1894—1974). The difference between the characteristics of these two kinds of particles is most obvious at extremely low temperature. When matter cools, the bosons can all gather in the same quantum state, they behave like a big super atom. Fermions, on the contrary, are more like "individualists", each occupies a different quantum state called a "fermion condensate". Numerous experiments have shown that spin is a very important physical quantity that characterizes various particles such as electron, proton, neutron, etc.

3.5.3 Shell Structure of Atoms

Based on the Pauli exclusion principle and the principle of the lowest energy, the German physicist W. Kossel (1888—1956) proposed the shell structure model of multi-electron atoms in

1916. He believed that electrons moving around the nucleus form many shells, that is, in a multi-electron atom, the distribution of electrons is hierarchical, and this layer of electron distribution is called the electron shell. Electrons with the same principal quantum number n in the atom are in the same main shell, and the main shells corresponding to the states of $n = 1, 2, 3, 4, 5, 6, \cdots$, etc. are represented respectively by capital letters K, L, M, N, O, P, ⋯, etc. The maximum number of electrons that the n shell can hold is $2n^2$. In the same main shell, electrons with the same orbital angle quantum number l are in the same sub-shell or branch shell, and the sub-shells corresponding to $l = 0, 1, 2, 3, 4, 5, \cdots$ etc. are represented respectively by lowercase letters s, p, d, f, g, h, ⋯ etc. The maximum number of electrons that the l sub-shell can hold is $2(2l + 1)$.

Since the energy level mainly depends on the principal quantum number n, in general, the smaller the n, the lower the energy level. Therefore, according to the principle of the lowest energy, electrons fill the energy levels generally in the order of n from small to large. The shell closest to the nucleus is usually filled with electrons first. However, since the energy level is also related to the angular quantum number l, in some cases, when the shell with smaller n is not filled, electrons begin to fill the shell with larger n. This situation began to appear in the fourth cycle of the periodic table.

Regarding the energy level of states with different n and l, Chinese scientist Xu Guangxian (1920—2015, winner of China's science and technology top award in 2008) summed up the rule. For the outer electrons of an atom, the energy level depends on $n + 0.7l$, the smaller the value, the lower the energy level. For example, electrons usually fill in the energy level of $n = 4, l = 0$ first, and then fill in the energy level of $n = 3, l = 2$.

Electrons on the sub-shell are usually represented by a lowercase letter representing l after the value of n. For example, the electron of $n = 1, l = 0$ is called the electron of 1s state. The electron of $n = 2, l = 1$ is called the electron of 2p state. The electron of $n = 3, l = 2$ is called electron of 3d state. The combination of electrons with specific values of n and l in an atom is called the electron configuration. For example, the electron configuration of an argon atom in the ground state is $1s^2 2s^2 2p^6 3s^2 3p^6$, where the superscript of the sub-shell symbol indicates how many electrons are in that sub-shell. In the ground state argon atom, there are two electrons in the 1s sub-shell, two electrons in the 2s sub-shell, six electrons in the 2p sub-shell, two electrons in the 3s sub-shell, and six electrons in the 3p sub-shell. This abbreviation is used to express the arrangement of electrons in an atom. Table 3 - 3 shows the extranuclear electron configuration of some atoms in the normal state. In the periodic table of elements, the elements are arranged in ascending order of atomic number Z. It can be seen that it is precisely because of the internal regularity of the electron arrangement in the atomic shell structure that leads to the periodic change of the physical and chemical properties of the elements in the periodic table.

Table 3 – 3 Electron configuration of atoms in their ground states

Element	Z	Electron configuration	Element	Z	Electron configuration
hydrogen	1	$1s$	sodium	11	$1s^2 2s^2 2p^6 3s$
helium	2	$1s^2$	magnesium	12	$1s^2 2s^2 2p^6 3s^2$
lithium	3	$1s^2 2s$	aluminum	13	$1s^2 2s^2 2p^6 3s^2 3p$
beryllium	4	$1s^2 2s^2$	silicon	14	$1s^2 2s^2 2p^6 3s^2 3p^2$
boron	5	$1s^2 2s^2 2p$	phosphorus	15	$1s^2 2s^2 2p^6 3s^2 3p^3$
carbon	6	$1s^2 2s^2 2p^2$	sulfur	16	$1s^2 2s^2 2p^6 3s^2 3p^4$
nitrogen	7	$1s^2 2s^2 2p^3$	chlorine	17	$1s^2 2s^2 2p^6 3s^2 3p^5$
oxygen	8	$1s^2 2s^2 2p^4$	argon	18	$1s^2 2s^2 2p^6 3s^2 3p^6$
fluorine	9	$1s^2 2s^2 2p^5$	potassium	19	$1s^2 2s^2 2p^6 3s^2 3p^6 4s$
neon	10	$1s^2 2s^2 2p^6$	calcium	20	$1s^2 2s^2 2p^6 3s^2 3p^6 4s^2$

Example 3 – 6: Find the magnitude of the orbital angular momentum of each electron outside the phosphorus nucleus in its normal state.

Solution: The atomic number of phosphorus is $Z = 15$, and the arrangement of extranuclear electrons is $1s^2 2s^2 2p^6 3s^2 3p^3$. Where the magnitude of the orbital angular momentum of 1s, 2s and 3s electrons is

$$L_{1s,2s,3s} = \sqrt{l(l+1)}\,\hbar = \sqrt{0(0+1)}\,\hbar = 0$$

The magnitude of the orbital angular momentum of 2p and 3p electrons is

$$L_{2p,3p} = \sqrt{l(l+1)}\,\hbar = \sqrt{1(1+1)}\,\hbar = \sqrt{2}\,\hbar$$

Quantum mechanics was co-founded by a large number of physicists such as Planck, Einstein, de Broglie, Born, Heisenberg, Schrödinger, Bohr, Pauli, and Dirac in the early 20th century. Through quantum mechanics, many phenomena can be truly explained, and new phenomena that cannot be intuitively imagined are predicted. Through quantum mechanics, many phenomena were precisely calculated, and later also obtained very accurate experimental proofs. Through the development of quantum mechanics, people's understanding of nature has achieved a major leap from the macro world to the micro world.

Group photo of the 5th Solvay Conference

Belgian chemist Ernest Solvay (1838—1922) was a person who is very similar to the Swedish chemist Alfred Bernhard Nobel (1833—1896). They were both scientists and wealthy industrialists with a strong family background. Nobel established a science prize fund named after himself, while Solvay provided fund for the high-level academic conferences in the world.

The first Solvay Physics Conference was held in Brussels from October 30 to November 3, 1911. The theme of the conference was radiation and quantum theory. Solvay invited 18 outstanding physicists in the world at that time including Lorentz, Planck, Madam Curie, Sommerfeld, Wien, Rutherford, Langevin, Einstein, Brillouin, Poincaré to attend the conference, as shown in Figure 3 –

24. The 2nd to 5th Solvay Physics Conferences were held in 1913, 1921, 1924 and 1927 respectively, and has continued until now.

Figure 3 – 24 The first Solvay Physics Conference, seated in the front row from left to right: Walther Nernst, Louis Marcel Brillouin, Solvay, Lorentz, Emil Gabriel Warburg, Perrin, Wien, Madam Curie, Jules Henri Poincaré; standing in the back row from left to right: Robert Goldschmidt, Planck, Rubens, Sommerfeld, Frederick Lindemann, de Broglie, Martin Hans Christian Knudsen, Friedrich Hasenöhrl, Georges Hostelet, Edouard Herzen, Kings, Rutherford, Heike Kamerlingh Onnes, Einstein, Langevin

It is worth mentioning that the 5th Solvay Physics Conference in 1927 was star-studded and famous. Among the 29 participants (Figure 3 – 25) in this conference, 17 won the Nobel Prize successively. Among them, those who have won the Nobel Prize in physics were: Lorentz 1902, Madame Curie 1903, Bragg 1915, Planck 1918, Einstein 1921, Bohr 1922, Wilson 1927, Compton 1927, Richardson 1928, De Broglie 1929, Heisenberg 1932, Schrödinger 1933, Dirac 1933, Pauli 1945, Born 1954. The Nobel Prize in chemistry was awarded to Madame Curie in 1911, Langmuir in 1932, and Debye in 1936. This conference was held during the critical period of succession of early quantum theory to quantum mechanics, and the theme was "electrons and photons". Through this meeting, Bohr and Born led the "Copenhagen School" (which main members were Heisenberg, Pauli, Dirac) to have a fierce debate with Einstein, Schrödinger, de Broglie on the physical image and philosophical interpretation of quantum mechanics. After this conference, with the rapid development of physics in the 20th century, this ideological confrontation to promote the development of the theory of physics in depth was brilliant, and the climax occurred one after another, and it still continues to inspire future generations.

Figure 3 – 25 The fifth Solvay Physics Conference, seated in the front row from left to right: I. Langmuir, Planck, Madam Curie, Lorentz, Einstein, Langevin, ch. E. Guye, C. T. R. Wilson, O. W. Richardson; in the middle row from left to right: P. Debye, Knudsen, W. L. Bragg, H. A. Kramers, Dirac, Compton, de Broglie, Born, Bohr; in the back row from left to right: A. Piccard, E. Henriot, Ehrenfest, Herzen, Th. de Donder, Schrödinger, E. Verschaffelt, Pauli, Heisenberg, R. H. Fowler, L. Brillonin

*3.6 Laser

Laser is an emerging technology developed in the early 1960s. Laser is the abbreviation of "light amplification by stimulated emission of radiation". Since the American T. H. Maiman (1927—2007) made the world's first laser in 1960, laser has been widely used. For example, laser surgery, automatic hemostasis, holographic laser photography, optical cable information transmission, and triggering thermonuclear reaction, etc. So, how is laser produced? What are its characteristics? In this section, we will briefly introduce the generation principle and main characteristics of laser through He-Ne laser.

3.6.1 The Generation of Laser

There are mainly three basic processes in the interaction between light and atom: light absorption, spontaneous emission and stimulated emission.

(1) Stimulated absorption.

Let the two energy levels of the atom be E_1 and E_2, and $E_2 > E_1$. If a photon with energy $h\nu = E_2 - E_1$ is irradiated, the atom in E_1 may absorb the energy of this photon and make transition from the low energy level E_1 to the high energy level E_2. This process is called light absorption, also known as stimulated absorption.

The characteristic of stimulated absorption is that this process is not spontaneous, it must be stimulated by external photon, and the external photon must meet the condition of $h\nu = E_2 - E_1$.

(2) Spontaneous emission.

After the atom is excited, the state of high energy level E_2 is unstable. Generally, it can only stay for about 10^{-8} s. Without external influence, it will automatically return to the state of low energy level E_1 and at the same time radiate a photon with energy $h\nu = E_2 - E_1$, as shown in Figure 3 - 26. The process by which an atom spontaneously returns from a high energy level to a low energy level and emits a photon is called spontaneous emission.

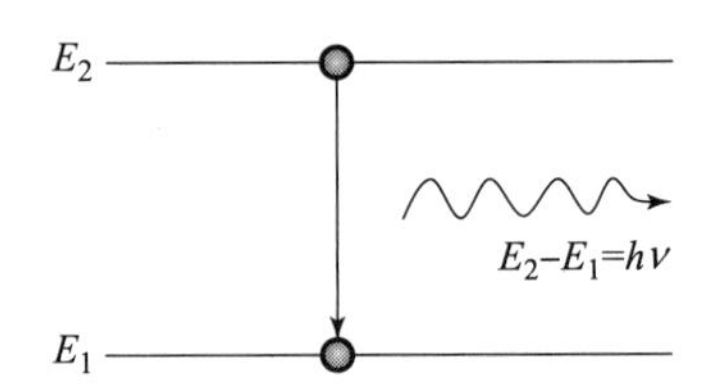

Figure 3 - 26 Spontaneous emission

The characteristic of spontaneous emission is that the transition of each atom is spontaneous and independent, which has nothing to do with external effect. The vibration directions and phases of the light emitted by atoms are not necessarily the same. Therefore, the light emitted by these light sources is not coherent light, such as fluorescent lamps, etc.

(3) Stimulated emission.

In 1916, when Einstein studied the interaction between optical radiation and atom, he predicted that atom would have stimulated emission in addition to stimulated absorption and spontaneous emission. Forty years later, this prediction was convincingly confirmed when the first laser began to operate. If before spontaneous emission, the atom in the state of high energy level E_2 is "stimulated" by the electromagnetic field of incident photon with the energy $h\nu$ which is just equal to the corresponding energy level difference $E_2 - E_1$ of the atom, it may make transition from the state of high energy level E_2 to the state of low energy level E_1, and radiate outward a photon with the same frequency, phase, and polarization direction as the incident photon. This process of making transition from a high-level state to a low-level state and radiating photon due to the stimulation of external photon is called stimulated emission.

The characteristic of stimulated emission is that it is not spontaneous, it must be stimulated by external photon, and the frequency of external photon must meet the condition of $h\nu = E_2 - E_1$. It is particularly important that the photon emitted by stimulated emission is exactly the same as the photon of external stimulation in terms of frequency, emission direction, phase and polarization state.

In this way, if one photon in the material induces a stimulated emission, two identical photons will be produced. If these two photons cause other atoms to produce stimulated emission, thus four identical photons will be produced. By analogy, under the action of one incident photon, a large number of photons with identical characteristics can be obtained. This phenomenon is called light amplification, as shown in Figure 3 – 27. It can be seen that in stimulated emission, the light emitted by each atom is the amplified coherent light, which is called laser. Some novel properties of laser beams stem mainly from the fact that a large number of photons have exactly the same state.

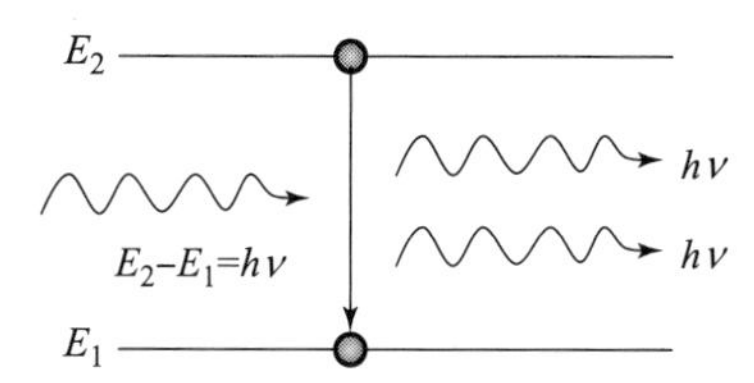

Figure 3 – 27 Stimulated emission

Is it easy to get light amplification if there is an appropriate photon incident into a given material? Actually, it's not. Because under normal circumstances, most of the atoms are in the ground state, while there are very few atoms in the excited state. Einstein pointed out that the probability of atomic stimulated emission and stimulated absorption is the same. Therefore, when photons are incident into the material, they are mainly absorbed, and it is difficult to produce continuous stimulated emission.

(4) Population inversion of Particle Number.

Because under normal circumstances, the number of electrons at the low energy level is more than that at the high energy level, so on the whole, the light absorption process is more dominant than the light stimulated emission process. In order to achieve light amplification after light passes through the material, the number of electrons at the high energy level must be greater than that at the low energy level, that is, the material must be in an "abnormal" state, which is called population inversion of particle number, or particle number inversion for short. Reversing the particle number is a necessary condition for realizing stimulated emission and obtaining optical amplification.

It is difficult to achieve population inversion in ordinary materials, because the average lifetime of the excited states of the atoms in these materials is extremely short. When such atom is excited to the high-energy state, it will spontaneously transition back to the ground state immediately. It is impossible to stay in the high-energy state and accumulate enough atoms so that the population inversion occurs. However, in the atomic energy levels of some materials there is a high energy level with a long average lifetime, which is called metastable energy level. The existence of the metastable energy level makes it possible to realize the population inversion.

The He-Ne laser is through a special way to make electrons excited to a higher energy level have a longer lifetime. In the He-Ne laser, helium is an auxiliary substance, neon is an active substance, and the ratio of He to Ne is 5 : 1 ~ 10 : 1. Figure 3 – 28 shows the relevant energy level diagram of helium and neon atoms. Via electron

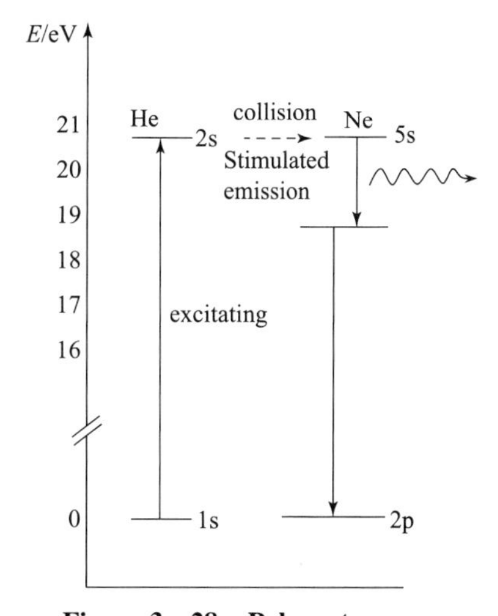

Figure 3 – 28 Relevant energy level diagram of helium and neon atoms

collision, the helium atom is excited to the 2s state, which has a relatively long lifetime and is a metastable energy level. The energy level of the 5s state of the neon atom is very close to that of the 2s state of the helium atom. When such two atoms collide, it is very easy to produce "resonant transfer" of energy, that is, after the collision of helium and neon atoms, the helium atom transfers energy to the neon atom and returns to the ground state, and the neon atom is excited to the 5s energy level. To generate a laser, in addition to increasing the number of particles in the upper energy level, we should also try to reduce the number of particles in the lower energy level. Because the lifetime of 5s state of neon atom is relatively long, and it is a metastable state, while the lifetime of the lower energy level 3p is much shorter than that of the upper energy level 5s. Due to spontaneous emission, the neon atoms in the 3p state will be rapidly reduced. The population inversion of the neon atoms between the 5s state and the 3p state can be realized. Therefore, the working substance (a mixed gas of helium and neon) in the He-Ne laser provides the necessary condition for light amplification. The He-Ne laser was the first successful gas laser.

(5) Optical Resonator.

In order to obtain a laser with a certain lifetime and intensity, there must also be an optical resonator. Figure 3 – 29 is a schematic diagram of a typical structure of the optical resonator, which consists of two mirrors mounted on both sides of the working substance and strictly perpendicular to the axis of the laser tube, one of them is a total reflecting mirror and the other is partially reflecting mirror for laser output. Only the light that is strictly parallel to the axis of the laser tube can be reflected back and forth and strengthened, and the light in other directions will escape out of the resonator after several reflections. When the light reaches the reflector, it is reflected back and passes through the working substance to further obtain light amplification. In this way, the number of photons in the resonator increases continuously, so that a strong laser can be obtained. The main function of the resonator is to maintain photon oscillation and amplification, so that the laser has good directivity and coherence. As an electromagnetic wave, the laser will form a standing wave between the two mirrors, and the distance between the two mirrors controls the wavelength of the standing wave. The light whose wavelength does not meet the standing wave condition will soon weaken and be eliminated. Therefore, the resonator also plays the role of frequency selection, then the frequency width of the output laser is very narrow, so the laser also has a very high monochromaticity.

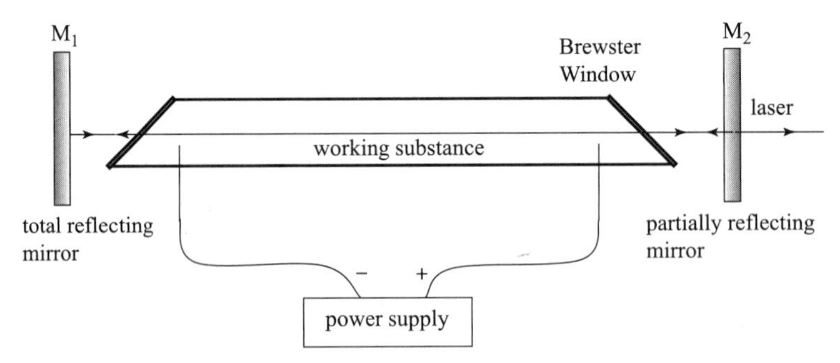

Figure 3 – 29 Optical resonator

3.6.2 Characteristics of Laser

(1) Good directivity.

The laser has good directivity. The divergence angle of the laser beam is very small, about below 0.5×10^{-5} rad. If the laser beam is incident on the moon, which is 3.8×10^5 km away from the earth, the diameter of the light spot is less than 2 km. The good directivity of laser is mainly determined by the optical amplification mechanism of the stimulated emission and the directional restriction effect of the optical resonator. This high directivity of the laser can be used for positioning, guidance, precise distance measurement, etc.

(2) Good monochromaticity.

The laser has good monochromaticity. The monochromaticity of laser ($\Delta\nu/\nu$) is about 10^{10} times higher than that of ordinary light. By the good monochromaticity of laser, the wavelength of laser can be used as the length standard for precise measurement. It can also be used for laser communication, plasma testing, etc.

(3) Energy concentration.

The energy of the laser is highly concentrated in space and time. The light emitted by the ordinary light source (such as incandescent lamp) is scattered in all directions, and the energy is dispersed. Even if it passes through a lens, only part of its light can be concentrated. While the light emitted by the laser device is almost a beam of parallel light due to its good directivity, so the energy of the laser is highly concentrated in space. The energy of the laser is also highly concentrated in time. For example, the energy of a pulsed laser can be concentrated in a very short time and emitted in the form of pulses. A laser emitted by a CO_2 laser-device with a power of about 1 kW, after being concentrated, can burn through a 5 cm thick steel plate in a few seconds. The characteristic of laser energy concentration can be used for drilling, cutting, welding and other precision machining for materials. In medicine, laser can be used as surgical scalpel. In military, laser can be used as a weapon. Laser can also be used to initiate nuclear fusion.

(4) Good coherence.

The luminescence of ordinary light sources is a process of spontaneous emission, and the light emitted is not coherent light. While the luminescence mechanism of laser is stimulated emission, and the light emitted is coherent light, so the laser has good coherence. Laser can be used directly as coherent light source. It is widely used in laser interference, laser holography and so on.

3.6.3 Application of Laser: Laser Cooling

Obtaining low temperature has been a technology deliberately pursued by scientists for a long time. It not only brings tangible benefits to mankind, such as the discovery and research of superconductivity, but also creates unique conditions for studying the structure and properties of matter. For example, at low temperature, the influence of the thermal motion of molecules and atoms can be greatly weakened, and atoms are more likely to expose their "nature". In the past, low temperature was mostly achieved in solid or liquid systems, which include a large number of strongly interacting particles.

In 1975, T. W. Hänsch (1941 – , winner of the 2005 Nobel Prize in Physics) and A. L. Schawlow (1921—1999, winner of the 1981 Nobel Prize in Physics) proposed a method of cooling neutral atoms by laser. In the 1980s, people used laser technology to obtain the extremely low temperature (such as 10^{-10} K) state of neutral gas molecules. This method of obtaining low temperature is called laser cooling.

The basic idea of the laser cooling is that after the moving atom resonantly absorbs an incoming photon, it makes transition from the ground state to the excited state, and the atomic momentum will decrease. The atom in the excited state will spontaneously emit a photon, return to its initial state, and gain momentum due to recoil. Because the photons absorbed by the atom are from a laser beam of the same direction, the momentum of the atom should be reduced. But the direction of the photons emitted spontaneously is random, and the average effect of multiple spontaneous emission does not increase the momentum of the atom. In this way, after the atom absorbs and spontaneously emits photon for many times, its speed will be reduced significantly and the temperature will also drop. In fact, an atom can absorb and emit tens of millions of photons per second on the average, so it can be effectively slowed down. For resonant light with a wavelength of 589 nm that cools sodium atoms, this deceleration effect is equivalent to 100,000 times the gravitational acceleration.

In fact, the motion of an atom is three-dimensional. As shown in Figure 3 – 30, in 1985, in the experiment by the team of Steven Chu (1948—) at Bell Lab, three pairs of laser beams in opposite directions were used to irradiate sodium atoms, and the sodium atom cluster at the intersection of the six lasers was cooled to a temperature of 240 μK.

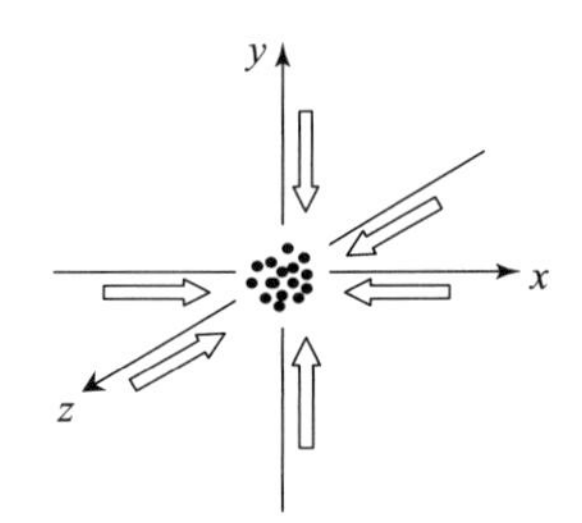

Figure 3 – 30 Schematic diagram of laser cooling

Because the atom continuously absorbs and randomly emits photon, atom and photon exchange momentum with each other, and low-speed atom is bound in every way and cannot escape. Atoms are glued to each other, which is called "optical molasses". This is a method of trapping atoms and focusing them. In 1997, Steven Chu, C. C. Tannoudji (1933—) and W. D. Phillips (1948—) were awarded the Nobel Prize in Physics for their contributions to the study of laser cooling and trapping atoms. Steven Chu was also the fifth Chinese scientist to win the Nobel Prize.

Summary

1. Schrödinger Equation

Schrödinger equation

$$i\hbar \frac{\partial}{\partial t}\Psi(x,y,z;t) = \hat{H}\Psi(x,y,z;t)$$

where Hamiltonian operator $\hat{H} = -\frac{\hbar^2}{2m}\nabla^2 + U(x,y,z;t)$

One-dimensional steady-state Schrödinger equation

$$\left[-\frac{\hbar^2}{2m}\frac{d^2}{dx^2}+U(x)\right]\psi(x) = E\psi(x)$$

This differential equation is linear and satisfies the superposition principle.

2. Application of the Steady-state Schrödinger Equation

Steady state condition: $U(x,y,z;t)$ does not change with time.

(1) A particle in a one-dimensional infinite square-well.

The potential energy function: $U(x) = \begin{cases} 0, 0 < x < a \\ \infty, x \leqslant 0, x \geqslant a \end{cases}$

Conclusion: the energy is quantized, the energy eigenvalue is

$$E_n = \frac{k^2\hbar^2}{2m} = \frac{n^2\pi^2\hbar^2}{2ma^2}; n = 1,2,3,\cdots$$

The energy eigenwave function is

$$\psi_n(x) = \begin{cases} \sqrt{\frac{2}{a}}\sin\frac{n\pi}{a}x; n = 1,2,3,\cdots; & 0 < x < a \\ 0; & x \leqslant 0, x \geqslant a \end{cases}$$

The steady-state matter wave of the particle in the one-dimensional infinite square-well is equivalent to a standing wave in a string with both ends fixed. The wavelength λ_n satisfies

$$a = n\frac{\lambda_n}{2}; n = 1,2,3,\cdots$$

(2) Barrier penetration.

The potential energy function: $U(x) = \begin{cases} 0, & x < 0, x > a \\ U_0, & 0 \leqslant x \leqslant a \end{cases}$

Conclusion: a microscopic particle has some probability to penetrate the potential barrier, and this phenomenon is called barrier penetration or tunneling effect.

(3) One dimensional simple harmonic oscillator.

The potential energy function: $U(x) = kx^2/2 = m\omega^2x^2/2$

Conclusion: the energy is quantized, the energy eigenvalue is

$$E_n = \left(n+\frac{1}{2}\right)\hbar\omega; n = 0,1,2,\cdots$$

The zero-point energy $E_0 = \frac{1}{2}\hbar\omega$

(4) Hydrogen atom.

The potential energy function: $U(r) = -\frac{e^2}{4\pi\varepsilon_0 r}$

Conclusion: the energy eigenwave function is $\psi_{nlm_l}(r,\theta,\varphi) = R_{nl}(r)Y_{lm_l}(\theta,\varphi)$

Where the principal quantum number $n = 1,2,3,\cdots$. It basically determines the energy of an electron in atom.

$$E_n = -\frac{m_e e^4}{2(4\pi\varepsilon_0)^2\hbar^2}\frac{1}{n^2} \approx -13.6 \times \frac{1}{n^2}(\text{eV}); \ n = 1,2,3,\cdots$$

When n is given, the electron in an atom has $2n^2$ possible states.

The angular quantum number $l = 0, 1, 2, \cdots, (n-1)$. It determines the magnitude of the angular momentum of the electron moving around the nucleus.

$$L = \sqrt{l(l+1)}\hbar$$

When the principal quantum number n is the same, L can have n different values of angular momentum. When n and l are given, the electron in an atom has $2(2l+1)$ possible states.

The orbital magnetic quantum number $m_l = 0, \pm 1, \pm 2, \cdots, \pm l$. It determines the $(2l+1)$ spatial orientations of the angular momentum of the electron moving around the nucleus in the external magnetic field.

$$L_z = m_l \hbar$$

When n, l and m_l are given, the electron in an atom has 2 possible states.

3. Electron Spin and Pauli Exclusion Principle

(1) Electron spin.

The magnitude of spin angular momentum of electron is $S = \sqrt{\frac{3}{4}}\hbar$.

The spin magnetic quantum number of electron $m_s = \pm \frac{1}{2}$. It determines the two orientations of the spin angular momentum of the electron in the external magnetic field, and the component of the spin angular momentum in the external magnetic field is $S_z = \pm \frac{1}{2}\hbar$.

The state of an electron is determined by the four quantum numbers n, l, m_l and m_s.

(2) The Pauli exclusion principle.

There cannot be two or more electrons in an atom with exactly the same 4 quantum numbers n, l, m_l, m_s.

4. The Arrangement of Electrons in a Multi-electron Atom

The arrangement of electrons in an atom of ground state is determined by the principle of the lowest energy and the Pauli exclusion principle.

Atomic shell: the electronic shells of $n = 1, 2, 3, 4, 5, 6, \cdots$ are called K, L, M, N, O, P, $\cdots$ shells respectively. The sub-shells of $l = 0, 1, 2, 3, 4, \cdots$ are represented by s, p, d, f, g, $\cdots$, respectively.

5. Laser

The laser is generated by stimulated emission, which requires population inversion in the luminescent material and light amplification in an optical resonator.

The laser is characterized by coherent light, which has the characteristics of concentrated energy, good monochromaticity and strong directivity.

Questions

3-1 What is steady-state? How can the solution of the Schrödinger equation be expressed in a form of steady-state expansion?

3-2 What is tunneling effect? Under what circumstances is the tunneling effect not obvious?

3 – 3 The figure (Question 3 – 3) is from Gamov's "Adventures in the Physical World", in which a car in the garage next door suddenly broke into the living room. What physical phenomenon does this imply?

Question 3 – 3

3 – 4 In 1995, an antihydrogen atom was "made" with an accelerator, which is composed of an antiproton and a positron moving around it. Do you think its spectrum will be exactly the same as that of hydrogen atom? Why?

3 – 5 Briefly describe the Pauli exclusion principle and the principle of the lowest energy.

3 – 6 Among the four quantum numbers describing the state of hydrogen atom, if $n = 3$ is known, what values can be taken by l, m_l and m_s respectively?

3 – 7 What basic conditions should be met to generate a laser? Give examples to illustrate the application of laser in modern science, technology and production.

Problems

3 – 1 There is an electron in a one-dimensional infinite square potential well with a width of 0.20 nm. Find: (1) the energy of the electron at the lowest energy level; (2) when the electron is in the first excited state ($n = 2$), where in the well the probability of the electron appearing is the smallest, and what is its value?

3 – 2 There are many biological particles with a mass of $m = 1.0 \times 10^{-17}$ kg in a cell with a linearity of 1.0×10^{-5} m. if the biological particle is regarded as microscopic particle moving in a one-dimensional infinite square potential well, try to estimate the energy levels of $n_1 = 100$ and $n_2 = 101$ of the particle and difference between the two energy levels?

3 – 3 An electron with mass m is in a one-dimensional infinite square potential well with width a, and its energy value and wave function are expressed as follows

$$E_n = \frac{n^2\pi^2\hbar^2}{2m_e a^2};\ \psi_n(x) = \begin{cases} \sqrt{\dfrac{2}{a}}\sin\dfrac{n\pi}{a}x, & 0 < x < a \\ 0, & x \leqslant 0, x \geqslant a \end{cases};n = 1,2,3,\cdots$$

After absorbing the energy $\Delta E = \dfrac{3\pi^2\hbar^2}{2m_e a^2}$, the electron makes a transition from a low energy level to a high energy level. Find the probability of finding the electron in the region $0 < x < a/4$ before and after the transition, respectively.

3 – 4 The steady-state wave function of an electron in a one-dimensional infinite square-well with a

width of $a=0.1$ nm is $\psi_n(x)=\sqrt{\frac{2}{a}}\sin\frac{n\pi x}{a}$. Find: (1) the energy the electron required to make transition from the ground state to the first excited state; (2) in the ground state, the probability finding the electron between $x=0$ and $x=a/3$.

3 - 5 The wave function of a particle in a one-dimensional infinite square-well is zero at the boundary, and its steady-state is a standing wave. Try to prove that the steady-state kinetic energy of the particle is quantized according to the de Broglie relation and the standing wave condition; find the formulas of quantized energy level and minimum kinetic energy (without considering the relativistic effect).

3 - 6 As shown in the figure (Problem 3 - 6), a particle is confined between two impenetrable walls with a distance of L. The wave function describing the state of the particle is $\psi=cx(L-x)$, where c is the undetermined constant. Find the probability of finding the particle in the range of $0\sim L/3$.

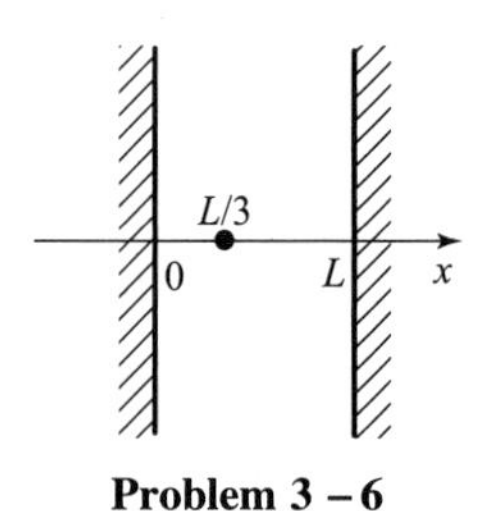

Problem 3 - 6

3 - 7 The wave function of a particle in one-dimensional motion is known to be

$$\psi(x)=\begin{cases}Axe^{-\lambda x}, & x\geqslant 0\\ 0, & x<0\end{cases}$$

where the constant $\lambda>0$. Try to find: (1) the normalization constant A; (2) the probability density of the particle appearing; (3) where is the probability of the particle appearing the greatest? (Hint: Integral formula $\int_0^{\infty}x^2e^{-ax}dx=2/a^3$)

3 - 8 Assume that an atom in the ground state has its outer electrons just filling the M shell. (1) What kind of element is this atom? (2) Write down its electron configuration.

3 - 9 (1) When the principal quantum number $n=6$, how many possible values are there for the angular quantum number l? (2) When $l=4$, what are the possible values of the orbital magnetic quantum number m_l? (3) What is the minimum value of l such that the z component of angular momentum is $3\hbar$?

3 - 10 Write down the electron configurations of lithium ($Z=3$), boron ($Z=5$) and argon ($Z=18$) atoms in the ground state, respetively.

Chapter 4
Electrons in Solids

In the previous chapter, we realized that quantum mechanics can give a good explanation for microscopic problems such as a single atom. The initial impression that quantum mechanics gives us seems to be only used to reveal the laws of the microscopic world, and it seems that quantum mechanics is not necessary when explaining macroscopic physical phenomena. Actually it is not the case. Although it has achieved many successes in solving macroscopic problems, classical physics still has limitations, and its predictions for certain macroscopic physical phenomena are still inconsistent with the experimental results. In fact, when quantum mechanics theory is introduced, the experimental results are well explained, and the microscopic mechanisms hidden behind the macroscopic phenomenon are revealed more fully and profoundly. For example, applying the ideal gas kinetic theory of classical physics to metal free electrons can obtain Ohm's law, but the theory cannot explain the experimental fact that the heat capacity of different metals is basically the same, while quantum mechanics can fundamentally explain Ohm's law and the heat capacity of metals.

A solid is a state of existence of matter, generally referring to a crystal in which atoms are arranged periodically. Solids have many properties, such as strength, hardness, ductility, color, luster, transparency, electrical conductivity, thermal conductivity, magnetism, etc. These properties have important applications in modern science and technology. The physics branch that studies these properties as well as the corresponding solid microstructure, interaction and motion laws of constituent particles (atoms, ions, electrons, etc.) is called solid state physics. Among the properties mentioned above, this chapter will mainly focus on the electric conductive properties of solids. Since the electrons in solids play an important role in the electric conductivity of solids, a brief introduction will be given in this regard. Firstly, the free electron gas model of metals is introduced, then the energy band theory of metals, insulators and semiconductors is briefly introduced, and finally the properties and applications of semiconductor conduction are presented.

4.1 Free Electrons in Metals

The conductivity of metal is strong, because there are many electric charges that can flow freely inside the metal, and these charges are prone to directional movement to form an electric current when a voltage is applied. Positive ions in metals are confined to the crystal lattice and difficult to flow, so they cannot be the charge that forms the current. When atoms are combined to form a metal, the outermost electrons of the metal atoms are weakly bound around the nucleus. The electrons can be easily separated from the atoms and flow freely in the metal, so they are the current carriers. We call electrons that can flow freely in metals as free electrons. The interaction between free electrons is very weak, and they are distributed inside the metal as if gas molecules dispersed in a container. The system constituted with free electrons is called free electron gas.

4. 1. 1 The Free-election-gas Model

Free electrons are not completely free, and they cannot escape from metal at ease, otherwise there is not the problem of work function for metal. The free electrons are attracted by the Coulomb force of the positive ions, which are bound on the periodically arranged lattice in the metal, so that a spatially periodic potential field is provided to the free electrons. The interaction between free electrons and the interaction between free electrons and positive ions are complex. In order to obtain the relevant essential law, we introduce a simple model: the free-electron-gas model.

For the sake of clarity, we consider a metal chain formed by metal atoms connected at equal intervals, ie, the case of one-dimensional metal crystals. If an electron is around a positive ion, it is subject to the Coulomb potential shown in Figure 4 – 1(a), with the potential energy being a function of the distance r of the electron from the ion. If an electron is in a one-dimensional crystal, it will be subjected to the periodic potential field shown by the solid line in Figure 4 – 1 (b), and the corresponding potential energy function is the superposition of the Coulomb potential of each ion represented by the dash lines. It is impossible to accurately solve the Schrödinger equation with such a complex potential energy function. In order to grasp the main characteristics of the problem, we should reasonally simplified the potential energy function.

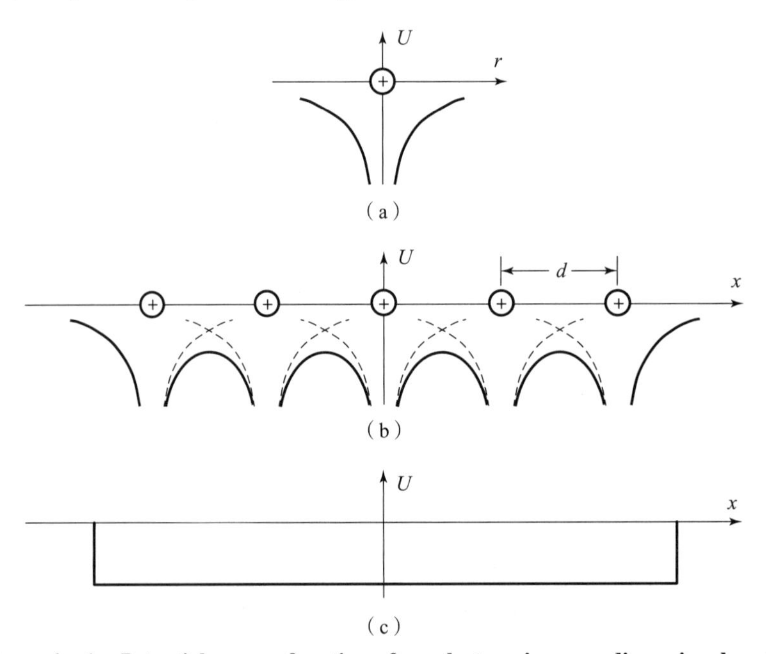

Figure 4 – 1 Potential energy function of an electron in one – dimensional metal

From the viewpoint of wave-particle duality, electrons often behave as waves. A wave presents itself differently when it meets different obstacle. When the obstacle size is much larger than the wavelength, the wave leaves a shadow behind the obstacle, for example, sunlight leaves a shadow behind a person. When the obstacle size is much smaller than the wavelength, the wave is not affected by the obstacle, for example, water wave travels a reed rod in the water without any effect. When the obstacle size is roughly equivalent to the wavelength, the wave has obvious diffraction phenomenon, which is the experimental basis for proving the wave nature of some physical objects. In metals, the electrons exhibit the wave properties; the obstacles they encounter are the periodically arranged ions. In general, the de Broglie wavelength of electrons in metals is much larger than the ion spacing, therefore the behavior of electrons is basically not affected by the complex periodic potential field of the ions in the metal, and the potential field felt by free electrons is just the average of the periodic potential field. Figure 4 – 1 (c) shows the corresponding potential energy function, a one-dimensional square potential well. For simplicity, we further assume that this well is a one-dimensional infinitely-deep square potential well. This means that the electrons are not affected by any potential field inside the one-dimensional crystal, and only feel the infinitely strong binding effect at the crystal boundary.

For sodium metal, the density is $0.97 \times 10^3\ \mathrm{kg/m^3}$ and the molar mass is $23 \times 10^{-3}\ \mathrm{kg/mol}$, then the ionic distance can be estimated as

$$d = \left[1\Big/\left(\frac{0.97 \times 10^3\ \mathrm{kg/m^3}}{23 \times 10^{-3}\mathrm{kg/mol}} \times 6.02 \times 10^{23}/\mathrm{mol}\right)\right]^{1/3} = 3.4 \times 10^{-10}\mathrm{m}$$

At room temperature ($T = 300$ K), the root-mean-square speed of electrons is $\nu = \sqrt{3kT/m}$, then the de Broglie wavelength is

$$\lambda = \frac{h}{m\nu} = \frac{h}{\sqrt{3mkT}} = \frac{6.63 \times 10^{-34}\mathrm{J \cdot s}}{\sqrt{3 \times 9.1 \times 10^{-31}\mathrm{kg} \times 1.38 \times 10^{-23}\mathrm{J/K} \times 300\mathrm{K}}} = 6.2 \times 10^{-9}\mathrm{m}$$

It can be seen that the de Broglie wavelength is much larger than the ion spacing (the latter is about 20 times larger than the former), so the above simplified model, the square potential well, is reasonable.

For three-dimensional metal crystals, free electrons can move freely in the three directions of x, y, and z, so the electrons can be considered to be in a three-dimensional infinitely-deep square potential well. This simplified model, in which electrons in metals are approximated as free electron gas in a three-dimensional infinitely-deep square potential well, is called free-electron-gas model. Many properties of metals is studied based on this model.

4.1.2 Fermi Energy of Free Electron Gas

1. Energy of electrons in free electron gas

In the free-electron-gas model, electrons are in a three-dimensional infinitely-deep square potential well. Inside the metal, the potential energy of the electron $U = 0$, and the stationary Schrödinger equation is

$$-\frac{\hbar^2}{2m}\left(\frac{\partial^2\psi}{\partial x^2}+\frac{\partial^2\psi}{\partial y^2}+\frac{\partial^2\psi}{\partial z^2}\right)=E\psi \tag{4-1}$$

The wave function can be expressed in the form of separated variables $\psi(x, y, z) = X(x)Y(y)Z(z)$, where $X(x)$, $Y(y)$, $Z(z)$ are respectively the solutions of the stationary Schrödinger equations for the one-dimensional infinitely-deep square potential well in the three directions of x, y, and z, and they represent the standing waves in the three directions. Therefore, we can discuss the energy distribution of the electrons in three-dimensional metals by the characteristics of standing waves, as in Section 3.2.1. This can be regarded as an extension of the one-dimensional infinitely-deep square-potential-well problem towards the three-dimensional infinitely-deep square-potential-well problem.

As shown in Figure 4 - 2, let the metal be a cube with side length a, and its three edges that are perpendicular to each other and intersect at a point are taken as the x, y and z axes. The wavelengths λ_x, λ_y, λ_z of the standing waves along the three directions satisfy the following conditions

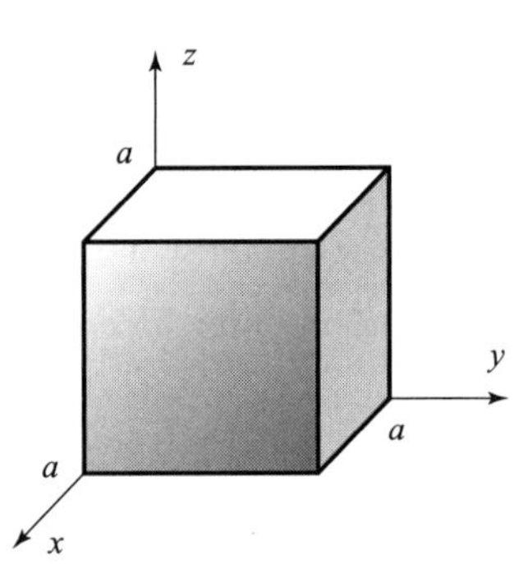

Figure 4 - 2 Cube of metal with side length *a*

$$n_x\frac{\lambda_x}{2}=a, \quad n_y\frac{\lambda_y}{2}=a, \quad n_z\frac{\lambda_z}{2}=a \tag{4-2}$$

In the equations, n_x, n_y, n_z are independent quantum numbers, which can be arbitrarily selected as positive integers 1, 2, 3, $\cdots$. Using the de Broglie formula $p = \frac{h}{\lambda}$ in the three directions respectively, and the equations (4 - 2), we can obtain the components of the electronic momentum in each direction

$$p_x=\frac{h}{2a}n_x, \quad p_y=\frac{h}{2a}n_y, \quad p_z=\frac{h}{2a}n_z \tag{4-3}$$

So the energy of electron is

$$E=\frac{p^2}{2m}=\frac{p_x^2+p_y^2+p_z^2}{2m}=\frac{h^2}{8ma^2}(n_x^2+n_y^2+n_z^2)=\frac{\pi^2\hbar^2}{2ma^2}(n_x^2+n_y^2+n_z^2) \tag{4-4}$$

It is easy to see that the above equation is the sum of the energy eigenvalues of the stationary Schrödinger equations of a one-dimensional infinitely-deep square potential well in three directions.

2. The states of electrons in free electron gas

In Chapter 3, four quantum numbers (n, l, m_l, m_s) are used to describe the state (quantum state) of the electrons in an atom, where (n, l, m_l) represents the space state (also known as orbital state) of the electrons, and m_s represents the spin state of the electrons. Similarly, we use the combination of three quantum numbers (n_x, n_y, n_z) in different forms to represent the space state of free electrons in a metal, and still use m_s to represent the spin state of the electrons, so (n_x, n_y, n_z, m_s) can represent the quantum state of free electrons in a metal, where the space quantum numbers n_x, n_y, and n_z may take any positive integers 1, 2, 3, $\cdots$, respectively, and the spin magnetic quantum number m_s may take two fractions $\pm 1/2$. For example, (3, 3, 4, 1/2), (3, 3, 4, - 1/2) and (3, 6, 5, - 1/2) represent different quantum states. It should be emphasized that the electrons in an atom and in a metal have different forms of space quantum numbers(n,l,m_l) and(n_x,n_y,n_z). The reason is that both are in different potential fields: the former is in a spherically-symmetric Coulomb potential field, the latter is in a square potential field. This makes the behavior and state of electrons in both situations different. But the electrons have a common feature, that is, they are all in three-dimensional potential fields, so they all have three space quantum numbers.

Equation(4 − 4) shows that the energy of the electrons in a metal is proportional to the sum of the squares of n_x, n_y, and n_z, and has nothing to do with m_s. The quantum states represented by different combination patterns of the quantum numbers with the same value $(n_x^2 + n_y^2 + n_z^2)$ have the same energy (energy level), but they still represent different quantum states. The energy level corresponding to multiple quantum states is called a degenerate energy level. For example, the energy level corresponding to quantum states(13,3,4,1/2) and(12,5,5,1/2) is degenerate(because $13^2 + 3^2 + 4^2 = 12^2 + 5^2 + 5^2$). The number of the quantum states corresponding to a degenerate energy level is called the degeneracy of the energy level. For example, all the quantum states corresponding to an energy level are(2,1,1,1/2), (2,1,1, −1/2), (1,2,1,1/2), (1,2,1, −1/2), (1,1,2,1/2) and(1,1,2, −1/2), so the degeneracy of this energy level is 6.

The Pauli exclusion principle is not only applicable to electrons in an atom, but also to electrons in a solid, and is a universal principle for the systems of the fermions including electrons. According to this principle, a quantum state(n_x,n_y,n_z,m_s) can hold one electron at most, namely, it is either empty or filled with one electron. In this way, finitely many free electrons in a metal is at first filled into the lower-energy states, then the higher-energy states, until the filling is completed for all the electrons.

3. Fermi energy

Consider the case of the quantum states filled with electrons. Take the rectangular coordinate system whose three coordinate axes are the quantum numbers n_x, n_y and n_z, as shown in Figure 4 − 3. This space is called the quantum number space. It can be seen from equation (4 − 3) that the three components of the electronic momentum are proportional to the three quantum numbers n_x, n_y and n_z, respectively, so the quantum number space is also called the momentum space.

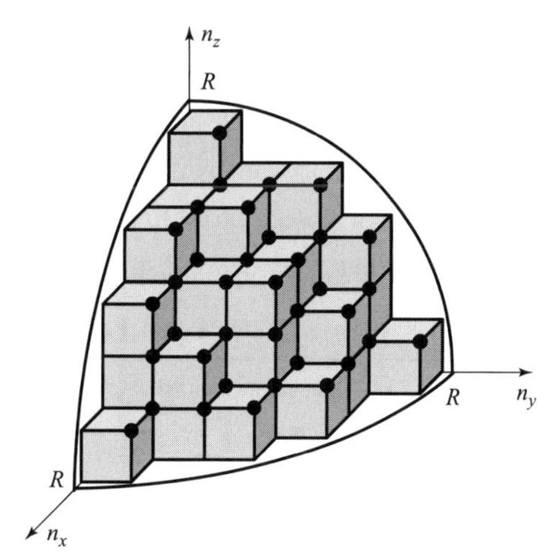

Figure 4 − 3 Quantum states in quantum number space

In the quantum number space, the points on the spherical surface centered at the origin and with radius R have the same value $(n_x^2 + n_y^2 + n_z^2)$, so the states corresponding to these points have the same energy, according to equation (4 − 4). The spherical radius corresponding to energy E is

$$R = \sqrt{n_x^2 + n_y^2 + n_z^2} = \sqrt{\frac{2ma^2}{\pi^2 \hbar^2} E}$$

In the first quadrant of this three-dimensional rectangular coordinate system, any point with positive integer coordinate values corresponds to two quantum states $(n_x, n_y, n_z, 1/2)$ and $(n_x, n_y, n_z, -1/2)$, or in other words, a cube with one unit volume corresponds to two quantum states. Therefore, when R is large enough (namely, when the number of space quantum states is large enough), the number of quantum states with the energy less than E is equal to twice the volume of the 1/8 sphere in the first quadrant

$$N_s = 2 \times \frac{1}{8} \times \frac{4}{3}\pi R^3 = \frac{1}{3\pi^2}\left(\frac{2m}{\hbar^2}\right)^{3/2} a^3 E^{3/2} \tag{4-5}$$

Because the metal volume $V = a^3$, the number of quantum states with the energy less than E in unit volume of metal is

$$n_s = \frac{N_s}{V} = \frac{1}{3\pi^2}\left(\frac{2m}{\hbar^2}\right)^{3/2} E^{3/2} \tag{4-6}$$

According to the lowest energy principle and Pauli exclusion principle, electrons are filled successively into the states from with the lowest energy to with higher energy. The number of quantum states is infinite in principle, but the number of electrons is limited. The highest energy level occupied by electrons is called the Fermi level, and the corresponding energy is called the Fermi energy. Assume that there are n free electrons per unit volume metal (or the number density of free electrons is n), and fill these electrons into the above quantum states. When $n_s = n$ is satisfied, the Fermi energy can be obtained by equation (4 – 6), denoted by

$$E_F = (3\pi^2)^{2/3} \frac{\hbar^2}{2m} n^{2/3} \tag{4-7}$$

This equation shows that in the free-electron-gas model, the Fermi energy is related to the number density of free electrons in a metal. Table 4 – 1 lists the free electron number densities and Fermi energies of some representative metals, where all the Fermi energies are in the order of several eVs. In the narrow energy range from 0 to E_F, a large number of energy levels are densely arranged, and this means that the energy distribution of the free electrons is quasi-continuous.

Table 4 – 1 Fermi parameters of some metals at $T = 0$ K

Metal	Electron number density n/m^{-3}	Fermi energy E_F/eV	Fermi speed $v_F/(\mathrm{m/s})$	Fermi temperature T_F/K
Li	4.70×10^{28}	4.76	1.29×10^6	5.52×10^4
Na	2.65×10^{28}	3.24	1.07×10^6	3.76×10^4
K	1.40×10^{28}	2.12	0.86×10^6	2.46×10^4
Mg	8.56×10^{28}	7.08	1.58×10^6	8.24×10^4
Al	18.1×10^{28}	11.7	2.02×10^6	13.6×10^4
Fe	17.1×10^{28}	11.2	1.98×10^6	13.0×10^4

Continued

Metal	Electron number density n/m^{-3}	Fermi energy E_F/eV	Fermi speed $v_F/(m/s)$	Fermi temperature T_F/K
Cu	8.49×10^{28}	7.05	1.57×10^{6}	8.18×10^{4}
Ag	5.85×10^{28}	5.50	1.39×10^{6}	6.38×10^{4}
Au	5.90×10^{28}	5.53	1.39×10^{6}	6.41×10^{4}

Fermi energy is an important concept in solid state physics. It allows us to obtain a completely different picture of physics against classical physics. In classical theory, any particle at $T=0$ K stops moving, and both the kinetic energy and the speed should be zero. Both according to quantum theory, even at $T=0$ K the energy of free electron in a metal is not zero, but has a range of $0\sim E_F$.

4. Fermi speed and Fermi temperature

Similar to the Fermi energy, the motion speed of free electrons also has a maximum value, which is called the Fermi speed, and this value is

$$\nu_F = \sqrt{2E_F/m} \tag{4-8}$$

Substituting the Fermi energy of several eVs order of magnitude, we can obtain the Fermi speed of as high as 10^6 m/s, which is the same order of magnitude as the electron speed given by the Bohr model of hydrogen atom. This means that even at absolute zero, electrons still move vigorously. In addition, from the Fermi energy we can introduce the Fermi temperature

$$T_F = E_F/k \tag{4-9}$$

where k is the Boltzmann constant. For the E_F of several eVs, T_F is up to 10^4 K. This means that the vigorousness of the electron motion in a metal at $T=0$ K is equivalent to the level of thermal motion of the particles at temperature 10^4 K. The Fermi speeds and Fermi temperatures of some typical metals are also listed in Table 13 – 1.

In the universe exists a class of celestial bodies called neutron stars, which are the final destination of the evolution of massive stars. The electrons in their atoms are pressed into the nucleus and combined with protons to become neutrons, so the matter in the neutron star exists in the form of neutrons. Because the original empty space of atoms is compacted, the density of neutron stars is extremely high, reaching $10^{16}\sim10^{18}$ kg/m^3. Like electrons, neutrons are also fermions with the spin angular momentum $\hbar/2$, so a neutron star can be regarded as Fermi-neutron-gas system. By replacing m in equations(4 – 7) and(4 – 8) with the neutron mass, both the equations can be used to discuss the Fermi energy and Fermi speed of neutron stars.

4.1.3 Density of States and Fermi-Dirac Distribution

The distribution of the quantum states with energy is not uniform. In order to express the distribution, in solid state physics the concept of density of states is introduced, which is defined as the number of quantum states per unit energy interval near energy E per unit volume of solid. By equation(4 – 6), the density of states can be expressed as

$$g(E) = \frac{\mathrm{d}n_s}{\mathrm{d}E} = \frac{(2m)^{3/2}}{2\pi^2\hbar^3}E^{1/2} \tag{4-10}$$

This function shows that the $g(E)$ curve is a parabola, as shown in Figure 4 - 4. This indicates that as E increases, the energy levels corresponding to the quantum states are arranged more and more densely. At $T = 0$ K, the dense levels below the Fermi level E_F are filled with electrons, while the levels above E_F are completely empty. Therefore, the curve OAB is the energy distribution curve of the electrons at $T = 0$ K, i. e., the $\frac{\mathrm{d}n}{\mathrm{d}E} - E$ curve, and it represents the electron number density in the unit energy range near E.

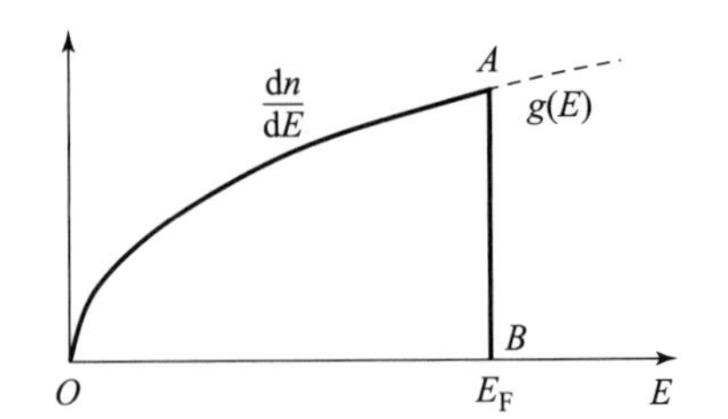

Figure 4 - 4 The density of states of free electrons in metal and the distribution of electron energy at $T = 0$ K

The density of states of real metals can be measured experimentally or calculated using the advanced quantum mechanics theory. Figure 4 - 5 shows the calculation results of the density of states of several metals, where the Fermi energy E_F was set as zero of the abscissa E. It can be seen from the figure that the curves of most main-group-element metals have roughly parabolic shape in the energy range below the Fermi level, while the curves of transition metals is more complicated. This shows that although the free-electron-gas model may be too ideal, it can reflect the distribution of free electrons with energy in metals to some extent.

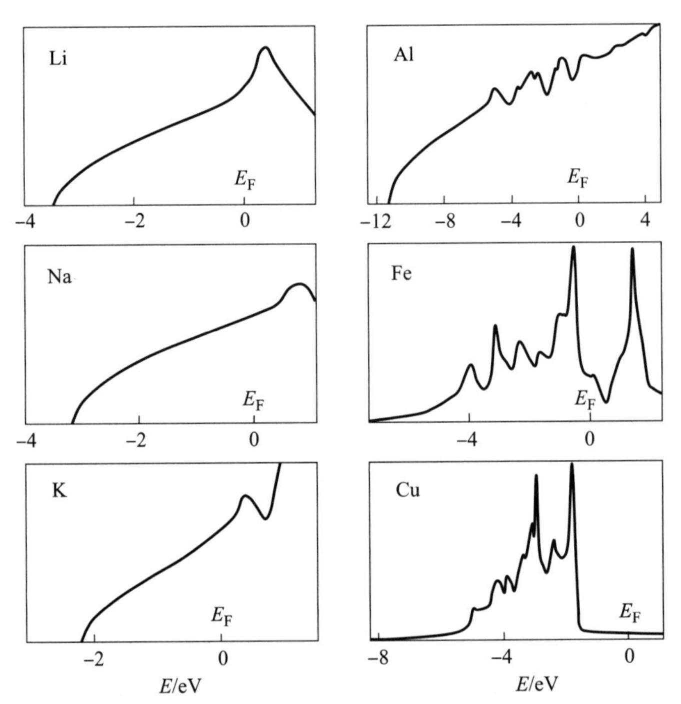

Figure 4 - 5 Curves of density of states of several metals calculated using quantum mechanics theory

Next we consider the energy distribution of electrons when the temperature increases. Quantum statistics theory states that in an electronic system in an equilibrium state, the probability of a quantum state of energy E occupied by an electron obeys the Fermi-Dirac (P. A. M. Dirac) distribution

$$f(E) = \frac{1}{1 + e^{(E-E_F)/kT}} \tag{4-11}$$

In the formula, T is the thermodynamic temperature, and E_F is the Fermi energy (E_F has, with temperature, a complex and small change, which is generally not considered). Figure 4-6 shows the Fermi-Dirac distribution curve.

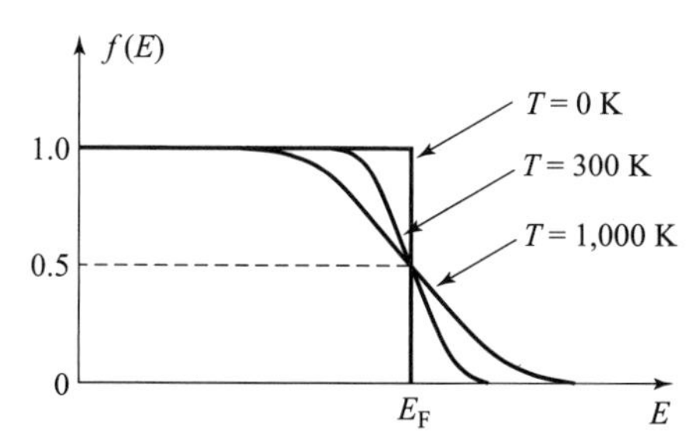

Figure 4-6 Fermi-Dirac distribution curves at different temperatures

If $T \to 0$ K, then, when $E > E_F$, $e^{(E-E_F)/kT} \to +\infty$ and $f(E) \to 0$; when $E < E_F$, $e^{(E-E_F)/kT} = 0$ and $f(E) = 1$. That is, the quantum states above E_F have no electrons, while the quantum states below E_F are filled completely with electrons. The distribution has been shown before.

If $T > 0$ K, then, when $E - E_F \gg kT$, $f(E) = 0$; when $E - E_F \ll kT$, $f(E) = 1$; when $E = E_F$, $f(E) = 1/2$. There is not much difference for $f(E)$ between at room temperature and at 0 K. The higher the temperature is, the greater the difference is.

In this way, when $T > 0$ K, the electron number density in the unit energy interval near E is expressed by the product of the density of states and the Fermi-Dirac distribution function, namely

$$\frac{dn}{dE} = g(E)f(E) = \frac{g(E)}{1 + e^{(E-E_F)/kT}} \tag{4-12}$$

The function at room temperature is shown as the solid curve in Figure 4-7. Obviously it is not very different from the curve at 0 K (Figure 4-4 or the dash line in Figure 4-7), but it is some different around $E = E_F$. The curve deviates from the parabola at E slightly lower than E_F, while is no longer zero at E just above E_F. We can discuss qualitatively why this phenomenon occurs.

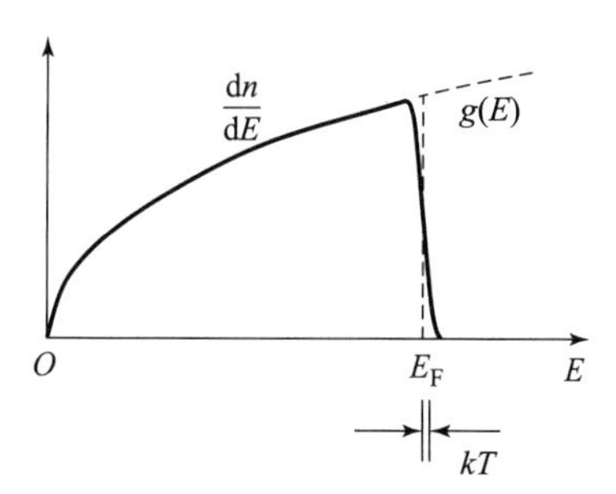

Figure 4-7 Electron energy distribution curve at room temperature

With the increase of temperature, the vibration of metal lattice ions becomes more and more intense. The vibration energy kT of each ion at $T = 300$K is on the order of 0.027 eV on average, which is roughly the energy that electrons can obtain from ions at room temperature when they collide with ions. However, this energy is not always available for electrons by collision, because most quantum states below the Fermi level are already filled with electrons. According to the Pauli exclusion principle, most electrons at lower energy levels cannot transition into the occupied energy levels kT higher than their own energy levels, so they cannot absorb this energy. Only the electrons in the quantum states near the Fermi level may absorb this energy to transition into a higher empty energy level. That is to say, the electrons at lower energy levels at room temperature cannot transition to higher energy levels by colliding with lattices to obtain energy, and the distribution of electrons at lower energy levels cannot change; Only the electrons whose energy is below the Fermi energy roughly within the energy range of 0.027eV may absorb energy to transition, thereby reducing the electron distribution in this energy range and increasing the electron distribution in the quantum states slightly higher than the Fermi energy. Because $kT \ll E_F$ at room temperature, the change of electron distribution only occurs in a small energy range around the Fermi energy, while the electron distribution is indistinguishable from at 0 K in most of the energy range below the Fermi level. To use an analogy, the free electron gas is like an ocean made up of electrons. The ocean is undulating and turbulent on the surface, but calm and tranquil in the depth.

The density of states acts in the scanning tunneling microscope (STM, described in Section 3.2.3), which works with the tunnel effect. In the STM, the surface of the sample is imaged by the tunneling current between the scanning probe and the sample (as shown in Figure 3-9). The density of states of the probe tip and the sample surface plays an important role in the magnitude of the tunneling current. The working probe tip is very close to the sample surface (on the order of nm), and the electron clouds of both sides overlap slightly, forming a structure similar to chemical bonds. Tunneling current is formed by the transition of electrons from one side of the "bonding" to the other side through the tunnel effect. This transition generally occurs between energy levels with the same energy on both sides, and the paired energy levels are like the channels of electron tunneling. The denser the energy levels, the more the channels and the stronger the current. In terms of density of states, the phenomenon means that the more the overlap of the density of states curves of the tip and the surface, the stronger the current. The probe needs to undulate up and down when scanning along the surface, in order to keep the current constant. By this way the shape of the sample surface is described (the undulation precision can reach 0.1nm). Therefore, the sample surface measured by the scanning tunneling microscope is actually a curved surface with a constant density of states. When the atomic composition of the sample surface is single, the iso-density-of-states surface is roughly the atomic-scale fluctuation surface of the sample surface, but when the sample surface composition is complex (such as surface oxidation, adsorption of impurity atoms, etc.), the situation is different. In addition, if the probe tip has only one or a few atoms, only a few discrete energy levels are formed, and the continuous density of states does not need to describe the energy level distribution. When a certain voltage V is applied between the probe and the sample, the discrete energy levels of the tip will be shifted by energy eV. Select a discrete energy level and let the voltage V increase slowly, the energy level will sweep across the density of states curve of the sample surface, so that the density of state can be measured by tunneling current.

Example 4 – 1: Calculate the average energy and average speed of electrons in the free-electron-gas model when $T = 0$ K.

Solution: In thermotics, Maxwell's speed distribution (the distribution function of ideal gas molecules with molecular speed) is introduced. Similarly, in the free-electron-gas model in metal, the distribution of quantum states with energy is described by the density of states $g(E)$, as shown in equation (4 – 10). At $T = 0$ K, each quantum state below the Fermi energy E_F is filled with one electron, and all quantum states above E_F are completely empty, so the distribution function of free electrons with energy is

$$g_E(E) = \frac{dN_E}{dE} = \begin{cases} CE^{1/2}, E \leqslant E_F \\ 0, \quad E > E_F \end{cases}$$

It means the number of electrons distributed in the unit energy interval near E, and the corresponding curve is the same as Figure 4 – 4, where the constant $C = \frac{(2m)^{3/2}V}{2\pi^2\hbar^3}$, V the metal volume. Since the total number of electrons is $\int dN_E$, and the sum of the energies of all electrons is $\int E dN_E$, the average energy of the free electrons is

$$\bar{E} = \frac{\int E dN_E}{\int dN_E} = \frac{\int_0^{E_F} E g_E(E) dE}{\int_0^{E_F} g_E(E) dE} = \frac{\int_0^{E_F} CE^{3/2} dE}{\int_0^{E_F} CE^{1/2} dE} = \frac{3}{5}E_F$$

Using $E = mv^2/2$ and the relationship $E_F = mv_F^2/2$ of Fermi energy and Fermi speed, the average speed of free electrons can be obtained

$$\bar{v} = \frac{\int v dN_E}{\int dN_E} = \frac{\int_0^{E_F} v g_E(E) dE}{\int_0^{E_F} g_E(E) dE} = \frac{\int_0^{E_F} vCE^{1/2} dE}{\int_0^{E_F} CE^{1/2} dE} = \frac{\int_0^{E_F} \sqrt{2/m} E dE}{\int_0^{E_F} E^{1/2} dE} = \sqrt{\frac{9}{8m}} E_F^{1/2} = \frac{3}{4} v_F$$

4.2 Band Theory of Solids

The free electrons in metals are in the periodic potential field of the ion lattice, and the free-electron-gas model simplifies this potential field into a three-dimensional infinitely-deep square potential well. Here the effect of ions on electrons is ignored, and thus the density of states for specific metals cannot be obtained by applying this model alone. If this simplification is removed and periodic potential fields are taken into account, then the energy levels of the electrons will expand into band-like energy structures called energy bands. The degree to which the energy bands are occupied by electrons determines the electrical properties of the solids. For the different periodic structure and potential field of each solid lattice, a unique energy band structure belonging to the solid can be obtained.

4.2.1 Energy Bands of Solids

Two or several atoms are combined into molecules through the chemical bonds between atoms, and a large number of atoms are also combined into solids through the chemical bonds. The chemical bonds in molecules are not different from those in solids in essence, so at first let's discuss the effects of the electrons in diatomic molecules in the formation of the chemical bonds.

Before introducing the probabilistic interpretation of wave function, we carefully analyzed the electron double-slit interference experiment (see Section 2. 5. 3), and mentioned the superposition of wave functions

$$\psi = \psi_1 + \psi_2 \qquad (4-13)$$

where ψ_1 and y_2 are the wave functions of electrons on the screen when the slits 1 or 2 is opened separately, and ψ is the wave function of electrons on the screen when the two slits are opened simultaneously. The corresponding probability distribution is

$$|\psi|^2 = |\psi_1 + \psi_2|^2 = |\psi_1|^2 + |\psi_2|^2 + \psi_1^* \psi_2 + \psi_1 \psi_2^* \qquad (4-14)$$

Obviously, $|\psi|^2$, the probability of electrons appearing somewhere on the screen when the two slits are opened simultaneously is not equal to $|\psi_1|^2 + |\psi_2|^2$, the sum of the probabilities when each slit is opened separately, for an extra exchange term $\psi_1^* \psi_2 + \psi_1 \psi_2^*$ left. This exchange term appears just because of the wave nature of electrons in quantum mechanics. In classical physics, electrons are just particles, and the exchange effect is impossible, because the sum of the probabilities when each slit is opened separately is equal to the probability when both slits are opened simultaneously. Therefore, the concept of the exchange is not available in classical physics, but the peculiar result of quantum mechanics.

The above characteristics are also reflected in the chemical bonds of diatomic molecules. For simplicity, take the example of two Na atoms forming a Na_2 molecule. Let ψ_1 and ψ_2 be respectively the wave functions of the valence electrons (3s electrons) of the two Na atoms, and ψ be the wave function of the shared electrons of the Na_2 molecule (the valence electrons originally belonging to each atom are shared by both atoms when the molecule is formed). When two atoms are isolated, that is, they are infinitely far apart, the two electron clouds exist independently without overlapping, which is expressed as the exchange term $\psi_1^* \psi_2 + \psi_1 \psi_2^* = 0$; when the two atoms are close, their electron clouds overlap, and the exchange term $\psi_1^* \psi_2 + \psi_1 \psi_2^* \neq 0$. When $\psi_1^* \psi_2 + \psi_1 \psi_2^* > 0$, $|\psi|^2 > |\psi_1|^2 + |\psi_2|^2$, the density of electron cloud between the atoms increases, a chemical bonds are formed, the two atoms combine into a molecule, and the energy decrease, as shown in Figure 4 – 8 (a); on the contrary, when $y_1^* y_2 + y_1 y_2^* < 0$, $|y|^2 < |y_1|^2 + |y_2|^2$, the density of electron cloud between the atoms decreases, no chemical bond is formed, the two atoms do not combine into a molecule, and the energy increases, as shown in Figure 4 – 8(b).

With the Schrödinger equation of the multi-electron system, we can calculate the curve of the valence electron energy of the system composed of two Na atoms as a function of the interatomic distance, as shown in Figure 4 – 9 (a). Curve 1 indicates that the valence electron energy of two atoms within a certain distance range is lower than when they exist in isolation (the 3s energy level in the figure represents the energy of valence electrons when the atoms exist in isolation), and a molecule can be formed, corresponding to Figure 4 – 8 (a). Curve 2 indicates that the valence

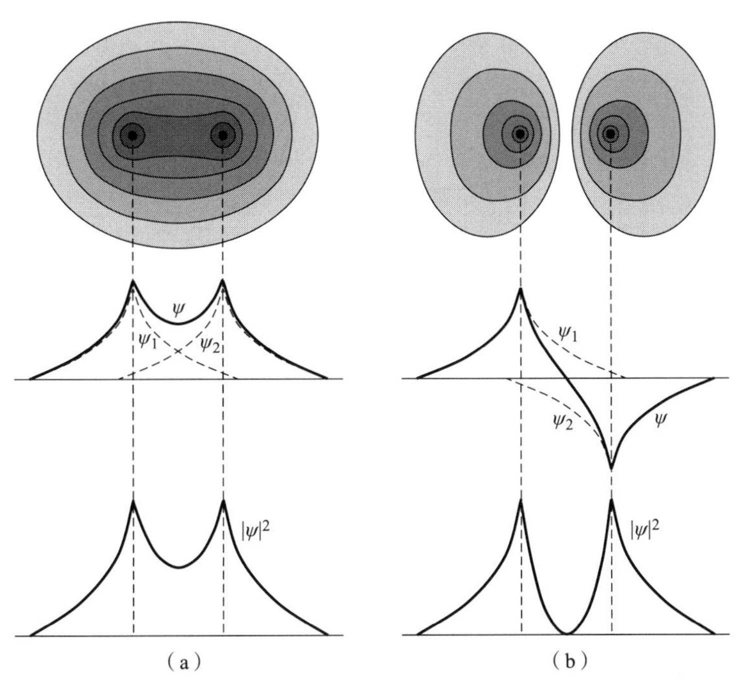

Figure 4 - 8 Electron cloud distribution and wave function when two Na atoms are close to each other

electron energy of two atoms increases at any distance and a molecule cannot be formed, corresponding to Figure 4 - 8(b). The bond length of stable diatomic molecular takes the equilibrium value r_0 (the lowest point of curve 1), and two energy levels E_1 and E_2 are formed in this case. This indicates that the 3s energy level of Na atom is split into two energy levels. The spin directions of the two shared electrons of the molecule are opposite. According to the Pauli exclusion principle, they can occupy the lower energy level E_1 at the same time, while the higher energy level E_2 is unoccupied, in the other words, one energy level is full and the other is empty.

It can be proved that for a system composed of N Na atoms, there are N curves of valence electron energy versus interatomic distance, some of which have a lower energy range and some not, as shown in Figure 4 - 9(b). When N Na atoms form a stable atomic cluster, the Na atoms take the equilibrium distance r_1. The 3s energy level of Na atom are split into N energy levels. The N shared electrons formed by in the way that every Na atom devotes a valence electron occupy the $N/2$ lower energy levels, and the other $N/2$ higher energy levels are free. Still half of the energy levels are full, and the other half are empty.

If N is extremely large (to the order of 10^{23}, for example), a solid is formed, and the curves of the valence electron energy varing with the interatomic distance are as many as $\sim 10^{23}$, as shown in Figure 4 - 9(c). The spacing of Na atoms in solids is r_2, and the 3s energy level of Na atom is split into $\sim 10^{23}$ energy levels. So many energy levels are densely packed in the range of $\Delta E \sim$ a few eVs, inducing the spacing between the adjacent energy levels only $\sim 10^{-23}$ eV on average, this indicates that the energies take values almost or quasi continuously. We call the band-shaped energy range, in which the energy levels are so densely arranged, as the energy band. Here the energy band coming

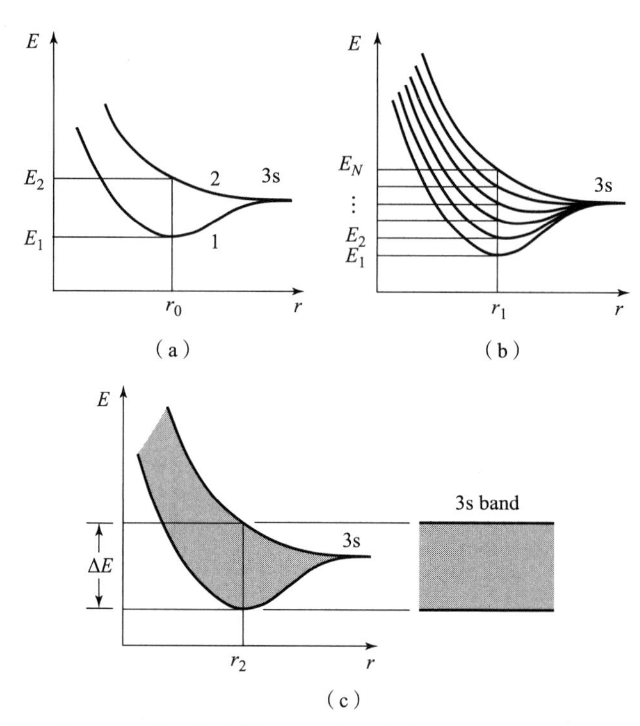

Figure 4 –9 The 3s energy level of Na atom is split into the 3s energy band of Na crystal

from the 3s energy level is called the 3s energy band. As for the distribution of the energy levels in the energy band with energy(i. e. ,the density of states),it is a very complicated problem in solid physics,omitted here. Obviously,when the shared electrons are filled into the 3s energy band of Na metal,half of the energy levels are completely occupied,and the other half are completely empty,in other words,the energy band is half full.

When atoms are combined into a solid,in addition to the splitting of the s energy levels,such as 1s,2s,and 3s,into the s energy bands,the p energy levels,such as 2p and 3p,are also split into the p energy bands,and the d and f energy levels are also split. A s energy level can hold 2 electrons, and a s energy band of a solid composed of N atoms can hold up to $2N$ electrons;A p energy level can hold 6 electrons, and a p energy band can hold up to $6N$ electrons. Generally, for a solid composed of N atoms,an energy band with an orbital angle quantum number of l can accommodate $2(2l+1)N$ electrons at maximum.

We can find, from the above analysis, that the formation of energy bands origins from the interaction between the atoms in solids, or further, from the exchange of wave functions of the electrons in the atoms, as shown in equation(4 – 14). The 3s electron of Na atom is its outermost electrons,which is relatively free and has a relative wide range of motion. When a solid is formed,the corresponding electron clouds overlap greatly,and the wave function exchange effect is strong,so the 3s energy band is wider. The 2p electrons are the inner electrons,which are strongly bound by the atoms and can only move in a local range. The corresponding electron clouds overlap less,and the wave function exchange effect is weak,so the 2p energy band is narrow. Similarly,the 2s,1s energy bands are narrower. Figure 4 – 10 show the difference in the width of the energy bands. When the atomic spacing takes different values, the overlap degree of the electron clouds will also vary, resulting in the changes in the widths of the energy bands.

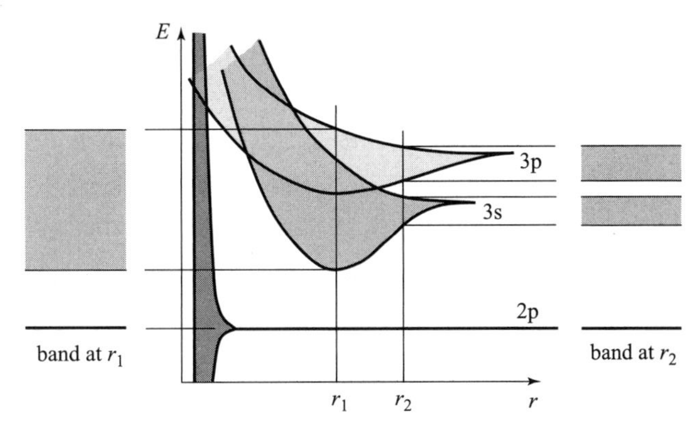

Figure 4 – 10 Energy bands and band overlap of Na crystal

Accurate quantitative calculations with quantum mechanics show that the energy bands of solids often overlap, that is, different energy bands appear in the same energy range. The overlapping energy bands will be filled with electrons as a whole, according to the Pauli exclusion principle and the lowest energy principle. We can see from Figure 4 – 10 that the overlap of the 3s energy band and the 3p energy band of Na metal. Obviously, the overlapping energy band cannot be exactly half filled with electrons. Sometimes the band overlap has a significant effect on the physical properties of solids. For example, a Mg atom has two valence electrons, and its 3s energy level is full; after forming a solid, the 3s energy band will be fully filled with electrons. According to the conclusion in Section 4. 2. 3, it will lead to the wrong conclusion that Mg is an insulator. However, due to the partially filling of the overlapping energy bands, the correct conclusion that Mg is a metal will be drawn.

It should be noted that although the above energy band theory is established by analyzing metals, it is still applicable to all the solids including insulators and semiconductors.

4. 2. 2 Valence Band, Conduction Band and Forbidden Band

A single atom has discrete electron energy levels. When atoms combine into a solid, the energy levels are split into the energy bands, which are arranged in order of energy. Filled into these energy bands, solid electrons follow the lowest energy principle and the Pauli exclusion principle. The electrons are filled into the low energy bands first, and then the high energy bands. The lower energy levels in the energy bands are filled first, and then the high energy levels are filled. Each energy level can be filled with two electrons with opposite spins. If every level of an energy band is filled with electrons, then this energy band is called a full band. If all energy levels of an energy band are free of electrons, then this energy band is called an empty band.

The energy bands are filled successively with electrons in this way, and the energy band in which the filling is just finished is the highest energy band in which electrons exist. This energy band is generally occupied with valence electrons, so it is called the valence band. The valence band of a metal is generally not occupied completely with electrons. When an electric field is applied, the valence band electrons can absorb energy from the electric field and transition to higher empty energy levels in the valence band. This means that the valence band electrons are accelerated by the electric field, forming a current and being able to conduct electricity. So valence bands of metals are also conduction bands. Unlike metals, valence bands of insulators and semiconductors are full bands, and each energy level of them is occupied by electrons. Due to the limitation of the Pauli exclusion principle, other electrons are not allowed to enter valence bands. That is, their valence band electrons cannot transition inside the valence band, so no currents induced. However, if the valence band electrons can transition to higher empty bands for some reason (such as thermal motion, illumination), then under an applied electric field, the electrons can transition into higher energy levels in the empty band and participate in conduction, so this empty band of semiconductors and insulators is called the conduction band.

Combining the above two cases of conduction band, it can be defined as the energy band in which empty quantum states exist at $T = 0\mathrm{K}$. In addition to the overlapping case of energy bands, between the energy bands there is an energy region without electron energy levels, called the forbidden band. The electrons in solids are not allowed to possess energy values within the forbidden band range. Figure 4 – 11 is a schematic diagram of valence band, conduction band and forbidden band of different kinds of solids. Their different characteristics play an important role in their electrical properties.

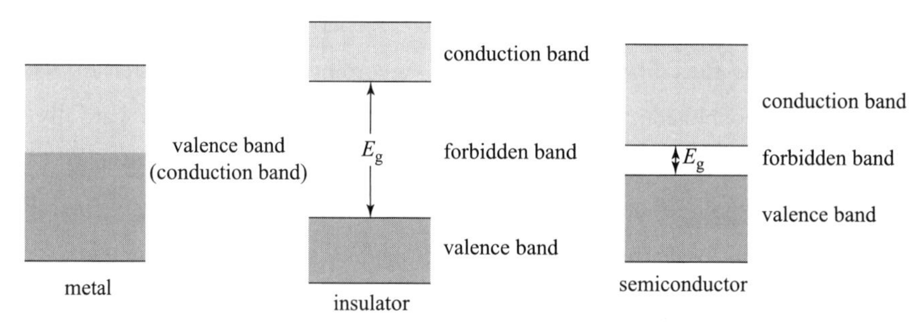

Figure 4 – 11 Comparison of the energy band structures of metals, insulators and semiconductors

4.2.3 Conductor, Insulator and Semiconductor

Solids are divided into conductors, insulators and semiconductors according to their electrical conductivity. The solids with resistivity below $10^{-8}\ \Omega \cdot \mathrm{m}$ are conductors, those with resistivity above $10^{8}\ \Omega \cdot \mathrm{m}$ are insulators, and the resistivity of semiconductors is between both. Typical conductors include various metals, typical insulators include diamond, sodium chloride crystal, etc., and typical semiconductors include silicon, germanium, selenium, gallium arsenide, etc. The different energy band structures of conductors, insulators and semiconductors determine their different conductive properties.

The conduction band of a metal (also its valence band) is partially filled, and the electrons can easily transition inside the conduction band under the an electric field to generate current, so a metal is a conductor. For some divalent metals, such as Mg, Be, Zn, etc. , their valence bands are formed by the overlap of full and empty bands, so they are still partially filled with electrons, thus exhibiting good electrical conductivity.

The free-electron-gas model of metals sets the electrons in a three-dimensional infinitely-deep square potential well, and the electron energy levels are arranged from the bottom up toward the top of the potential well. If we modify this model to adopt a periodic finitely-deep potential well formed by a metal lattice like Figure 4 – 1 (b), then the band structure of the metal can be obtained by solving the multi-electron Schrödinger equation. Because the valence electrons in metals are weakly bound by the periodically arranged ions, this model is called the near-free electron model. Under this model, the Fermi level can also be defined like in the free-electron-gas model, namely, the highest energy level occupied by the electrons in the valence band at $T = 0$ K. Moreover, both models give similar characteristics of the quantum states distribution with energy. In fact, the density of states curves in Figure 4 – 5 were calculated with the near-free electron model. They show the quantum states distribution of metals with energy from the bottom of the conduction band to higher energy.

The energy band structures of insulators and semiconductors are similar. As shown in Figure 4 – 11, at $T = 0$ K, they all have a valence band as a full band, a conduction band as an empty band, and a forbidden band separating the valence and conduction bands. The transition of electrons inside an energy band is also limited by the Pauli exclusion principle. Since there are no empty quantum states, in the valence band, that can accept other electrons, the electrons in the valence band can only transition across the forbidden band toward the conduction band. The forbidden band width E_g of insulators is very wide (about 3 ~ 6 eV). For example, the forbidden band width of diamond is 5.5eV. This is much higher than the energy absorbed in the collision of valence band electrons and lattice ions at room temperature ($kT = 0.027$ eV at $T = 300$ K). Such low energy cannot excite the valence band electrons to the conduction band, so insulators cannot conduct electricity. Under an applied strong electric field, some insulators can absorb enough energy from the electric field to make the valence band electrons transition to the conduction band, so as to conduct electricity. This phenomenon is called the electrical breakdown of the insulator (Reference Example 4 – 4).

The forbidden band width of semiconductors is relatively narrow (about 0.1 ~ 1.5 eV), such as, 1.14 eV for silicon, 0.67 eV for germanium, and 1.43 eV for gallium arsenide. At room temperature, a small amount of valence band electrons are thermally excited into the conduction band so that the semiconductor exhibits some conductivity. When the temperature increases, the number of electrons excited into the conduction band increases approximately exponentially, which greatly enhances the conductivity of the semiconductor. In addition, when the semiconductor is illuminated by the light with appropriate frequency, the number of electrons excited from the valence band to the conduction band can also be increased. Therefore, semiconductors exhibit good thermal sensitivity and photosensitivity, thus can be fabricated into thermistors and photoresistors.

Although both the free-electron-gas theory and the energy band theory are new theories, the basic concepts and basic principles of quantum mechanics used to establish them are those we have learned in Chapters 2 and 3. They mainly include the following aspects:

(1) Electrons are confined in infinitely-deep square potential wells, and their motion behavior can be represented by de Broglie standing waves. It is a one-dimensional standing wave in a one-dimensional infinitely-deep square potential well, and a three-dimensional standing wave in a three-dimensional infinitely-deep potential well of free electron gas.

(2) The quantum states of electrons include space states and spin states. The quantum states of atoms are represented by 4 quantum numbers (n, l, m_l, m_s) and those of free electron gas by 4 quantum numbers(n_x, n_y, n_z, m_s). The first 3 quantum numbers(n, l, m_l) and (n_x, n_y, n_z) represent the space states in the three-dimensional case, and the 4th quantum number m_s represents the spin state.

(3) The Pauli exclusion principle and the lowest energy principle are the basic laws satisfied in the arrangement of electrons. In atoms, the electrons are filled in discrete energy levels, and filling results determine the type of atoms. In solids, the electrons are filled in the energy bands and in the quasi-continuous energy levels of the energy bands, and filling results make solids be conductors, semiconductors, or insulators. Therefore, the Pauli exclusion principle is a deep determinant of the chemical properties of atoms and the electrical conductivity of solids.

(4) The probabilistic interpretation and the superposition principle of wave functions are generally valid in the microscopic field. In the phenomenon of the double slit interference of electrons, an electron can pass through the two slits at the same time, and the interference fringes on the screen are the probability performance of electron interference. For a covalent bond formed between atoms, one electron is shared by two or more atoms. The covalent bond is the area with a high probability of electrons occurrence in the electron interference (the electron clouds overlap). Therefore, a covalent bond is the direct result of the electron wave-particle duality, so it cannot be explained by classical physics.

Example 4 - 2: Quantum mechanics often takes the energies of free electrons as positives, and the energies of bound electrons as negatives, as was done in Chapter 3 where a hydrogen atom was solved. Similarly, the energies of the electrons bound in metal are taken as negatives, and the energies of the stationary electrons which can just escape the metal are taken as zero (this energy level is called the vacuum level), as shown in Figure 4 - 12. In the situation, calculate the Fermi energy E_F and the energy of conduction band bottom E_b of sodium metal by the following experimental data:

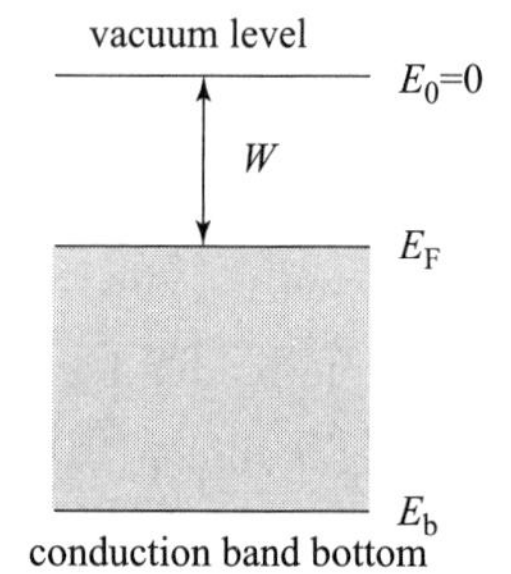

Figure 4 - 12 Diagram for Example 4 - 2

(1) The maximum initial kinetic energy of a photoelectron is 1.84 eV, when sodium metal is irradiated with the monochromatic light with a wavelength of 300 nm;

(2) Density 971 kg/m^3, molar mass 23.0 g/mol.

Solution: The photoelectric effect occurs when sodium metal is irradiated with monochromatic light. With the data, the work function can be calculated

$$W = h\nu - E_{\mathrm{km}} = \frac{hc}{\lambda} - E_{\mathrm{km}}$$

$$= \frac{6.63 \times 10^{-34}\mathrm{J \cdot s} \times 3 \times 10^{8}\mathrm{m/s}}{300 \times 10^{-9}\mathrm{m} \times 1.6 \times 10^{-19}\mathrm{J/eV}} - 1.84\mathrm{eV} = 2.30\mathrm{eV}$$

The active electrons in a metal are the electrons near the Fermi level, and the work function is the energy absorbed by the electrons transitioning from the Fermi level to the vacuum level. So the Fermi energy is

$$E_{\mathrm{F}} = E_0 - W = 0 - 2.30\ \mathrm{eV} = -2.30\ \mathrm{eV}$$

From Figure 4 – 5 we can see that the free-electron-gas model can well describe the energy band structure of sodium metal. With this model, we have $E_{\mathrm{F}} - E_{\mathrm{b}} = (3\pi^2)^{2/3}\frac{\hbar^2}{2m}n^{2/3}$, so the energy of conduction band bottom is

$$E_{\mathrm{b}} = E_{\mathrm{F}} - \frac{\hbar^2}{2m}\left(\frac{3\pi^2 N_{\mathrm{A}}\rho}{M}\right)^{2/3}$$

$$= -2.30\mathrm{eV} - \frac{(1.05 \times 10^{-34}\mathrm{J \cdot s})^2}{2 \times 9.11 \times 10^{-31}\mathrm{kg}}\left(\frac{3 \times \pi^2 \times 6.02 \times 10^{23}/\mathrm{mol} \times 971\mathrm{kg/m^3}}{0.023\mathrm{kg/mol}}\right)^{2/3}\frac{1}{1.60 \times 10^{-19}\mathrm{J/eV}}$$

$$= -5.43\mathrm{eV}$$

Example 4 – 3: What is the maximum wavelength of the light emitted by a GaAsP semiconductor with a band gap $E_{\mathrm{g}} = 1.9$ eV?

Solution: A semiconductor emits light when the electrons of conduction band transition to the valence band across the forbidden band. The maximum wavelength corresponds to the minimum energy difference between the transition energy levels, i. e., the forbidden band width $E_{\mathrm{g}} = h\nu_{\min} = hc/\lambda_{\max}$. We can get

$$\lambda_{\max} = \frac{hc}{E_{\mathrm{g}}} = \frac{6.63 \times 10^{-34}\mathrm{J \cdot s} \times 3.00 \times 10^{8}\mathrm{m/s}}{1.9\mathrm{eV} \times 1.6 \times 10^{-19}\mathrm{J/eV}} = 6.54 \times 10^{-7}\mathrm{m} = 654\mathrm{nm}$$

Example 4 – 4: Estimate the electrical breakdown field strength of diamond. The forbidden band width of diamond is $E_{\mathrm{g}} = 5.5$eV, and the mean free path of electrons is $\lambda = 0.2$ mm.

Solution: Diamond will be electrically broken down, if electrons in diamond obtain enough energy to jump from the valence band to the conduction band, when it is accelerated by an applied electric field in the movement process of a mean free path. Take E_{b} to represent the breakdown field strength, then $E_{\mathrm{g}} = eE_{\mathrm{b}}\lambda$, thus we get

$$E_{\mathrm{b}} = \frac{E_{\mathrm{g}}}{e\lambda} = \frac{5.5\mathrm{eV} \times 1.6 \times 10^{-19}\mathrm{J/eV}}{1.6 \times 10^{-19}\mathrm{C} \times 0.2 \times 10^{-6}\mathrm{m}} = 2.8 \times 10^{7}\mathrm{V/m} = 28\mathrm{kV/mm}$$

This value is nearly 10 times the breakdown field strength of air.

4.3 Semiconductor Conduction

Semiconductors are very important solid materials and play an important role in the electronic information industry. Since the establishment of quantum mechanics, the properties of semiconductors have been deeply studied. A semiconductor physics branch has been developed in solid state physics.

4. 3. 1 Classification of Semiconductors

Semiconductors are similar to insulators in the energy band characteristics, but they have some conductivity due to their small forbidden band width. But how different is their conductive characteristics, compared with metal conductors?

The conductivity of semiconductors is about 11 orders of magnitude lower than that of metals, and this difference is caused by their different energy band structures. In semiconductors, only after electrons transition from the valence band to the conduction band, across the forbidden band, can they absorb the energy of the applied electric field to transition inside the conduction band to act as electrical carriers. Although the band gap is not big, it is dozens of times larger than kT at room temperature, so the proportion of the electrons thermally excited into the conduction band is very small. For silicon as an example, the number of electrons in the conduction band is 12 orders of magnitude lower than those in the valence band (the same order of magnitude as the number of atoms). That is to say, on average, 10^{12} silicon atoms contribute only one conduction electron. Therefore, the number density of the carriers in pure silicon is relatively small. On the other hand, the conductive electrons in metals are the electrons in the unfull valence band (also the conduction band), and their number is of the same order as the number of atoms. For copper, on average, each atom contributes one conduction electron, showing that the carrier number density in metals is very high. Therefore, the difference in the number density of carriers between semiconductors and metals is an important reason for the huge difference in their electrical conductivity.

A small number of valence band electrons in a semiconductor transition to the conduction band, leaving empty quantum states near the top of the valence band. The empty quantum states are called holes. Like conduction band electrons, valence band holes are also carriers. After the holes are formed, the valence band is not strictly full (this is somewhat similar to the valence band of metals). The electrons in the valence band can transition to a higher empty state in the same energy band under the applied electric field, which is equivalent to the holes at higher energy levels transitioning to lower energy levels. Figure 4 – 13 is a schematic diagram of this process. Metaphorically, some cars are parked one by one to form a row. In front of the first car is a vacancy. The car drives out to fill the vacancy, leaving a vacancy between the first and second. The second car then drives out to fill the new vacancy, leaving a vacancy between the second and third. The third car drives out, and so on. Processing in turn, the cars keep moving forward, and the vacancy keeps moving backward. It can be seen that the conduction of holes in the valence band is essentially the conduction of electrons there. The negatively charged electrons in the valence band move against the direction of the applied electric field, which is equivalent to the holes moving along the direction of the field. Therefore, a hole can be regarded as a particle with a positive charge $+e$, and also with the mass of an electron. It should be clear that the valence band holes are not actual particles, but empty quantum states, or a reflection of some valence band electrons. They contribute to the conduction of semiconductors together with the conduction band electrons as two different kinds of carriers. If the conduction band electrons re-transition back to the valence band, the electrons neutralize the holes there.

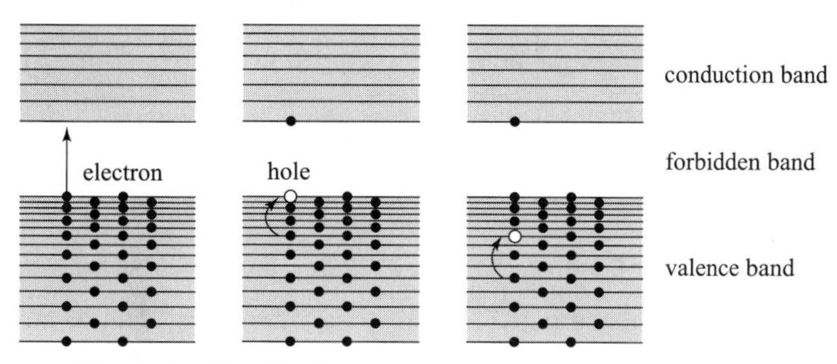

Figure 4 – 13 The formation process and the conduction mechanism of the holes in the valence band of semiconductor

The proportion of the number densities of conduction band electrons and valence band holes has an important influence on the conductive properties of semiconductors. Doping small amounts of impurities into semiconductors can drastically change the number of conducting electrons or holes. Thus in composition and function, semiconductors are divided into pure semiconductors and impurity semiconductors. Pure semiconductors are also called intrinsic semiconductors, in which some electrons transition to the conduction band, leaving the same quantity of holes in the valence band, indicating that the number of conduction band electrons is the same as that of valence band holes (forming electron-hole pairs), as shown in Figure 4 – 14(a). Both electrons and holes contribute exactly the same to intrinsic semiconductor conduction.

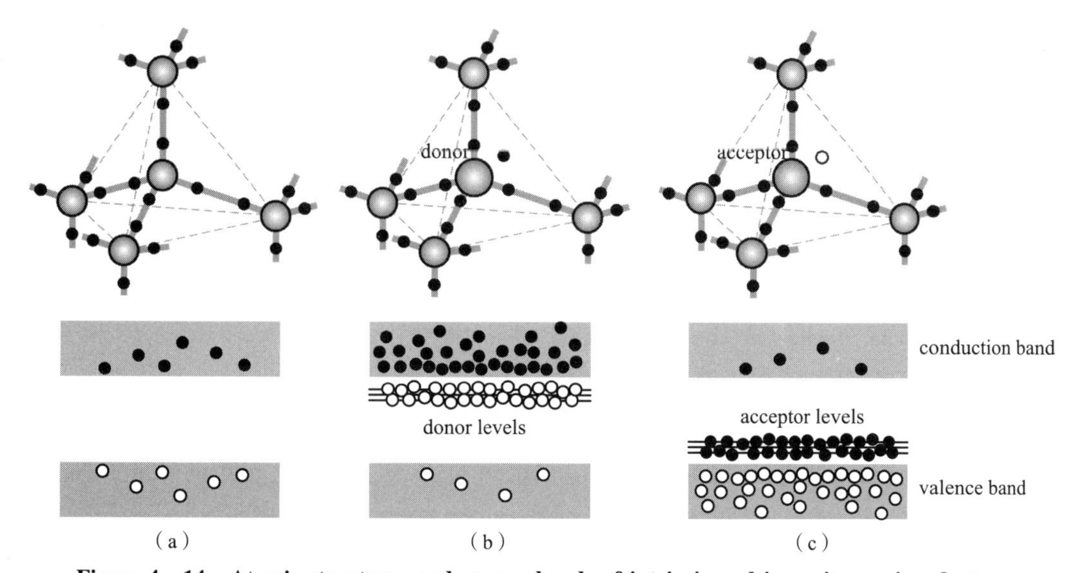

Figure 4 – 14 Atomic structures and energy bands of intrinsic and impurity semiconductors

(a) intrinsic semiconductor; (b) N – type semiconductor; (c) P – type semiconductor

A small amount of impurity atoms are doped into an intrinsic semiconductor to replace some original atoms, and the semiconductor formed in this way is called an impurity semiconductor. In impurity semiconductor silicon, the proportion of impurity atoms to silicon atoms is only about 10^{-7}. According to the types of impurity atoms, impurity semiconductors are divided into N – type and P – type semiconductors. Both silicon and germanium are tetravalent elements, and each atom is bonded to 4 other atoms to form a regular tetrahedral structure. After such an intrinsic semiconductor is doped with 5 – valent impurity elements such as phosphorus and arsenic, each impurity atom replaces a silicon atom, contributing 4 of the 5 valence electrons to be bonded to the surrounding 4 silicon atoms. The remaining one electron is weakly bound to the impurity atom and become free electrons. Expressed in the energy band diagram, new energy levels called the impurity levels appear inside the forbidden band and near the bottom of the conduction band. The electrons on the impurity levels (the remaining electrons from the impurity atoms) are easily excited to the conduction band. Because each impurity atom contributes such an electron, a trace amount of impurities can cause a dramatic increase in conduction band electrons, with little change in the number of valence band holes. This kind of semiconductor mainly conducts by electrons, majority carriers, not by holes, minority carriers, and is called electron-type or N – type semiconductor, the impurities of which are called donors (meaning devoting free electrons). Figure 4 – 14(b) shows the characteristics of a N – type semiconductor.

If a trivalent impurity element such as boron and aluminum is doped into an intrinsic semiconductor, each impurity atom has to be bonded to the surrounding 4 intrinsic atoms, thus lack of one valence electron. If an impurity atom captures an electron, it will leave a free hole around. In the energy band diagram, empty impurity energy levels appear inside the forbidden band and near the top of the valence band, and the electrons in the valence band are easily excited to these new energy levels, leaving a large number of holes in the valence band. This kind of semiconductor mainly conducts by holes, majority carriers, not by electrons, minority carriers, and is called hole-type or P – type semiconductor, the impurities of which are called acceptors, as shown in Figure 4 – 14(c).

In impurity semiconductors, the free electrons or free holes introduced by the donors or the acceptors are much more than the carriers when undoped, so the conductivity of impurity semiconductors is much stronger than that of intrinsic semiconductors. By combining N – type and P – type semiconductors in different ways, a variety of semiconductor devices can be made. Basically, all the semiconductor devices are based on doped materials.

Example 4 – 5: The number density of the conduction electrons n_0 (the electrons entering the conduction band from the valence band) in pure silicon at room temperature is about 10^{16} m^{-3}. How many silicon atoms contribute a conduction electron on average? If a trace amount of phosphorus impurity is doped, and on average one of per 5×10^{6} silicon atoms is replaced by a phosphorus atom, by how many times will the number density of conduction electrons increase? Suppose that each phosphorus atom contributes an "extra" electron to the conduction band. The density and the molar mass of silicon are respectively 2,330 kg/m^3 and 28.1 g/mol.

Solution: According to the given data, the atomic number density of pure silicon can be obtained

$$n_{\mathrm{Si}} = \frac{\rho N_{\mathrm{A}}}{M_{\mathrm{Si}}} = \frac{2,330\mathrm{kg/m^3} \times 6.02 \times 10^{23}/\mathrm{mol}}{0.028,1\mathrm{kg/mol}} = 5 \times 10^{28}\mathrm{m}^{-3}$$

So $n_{\mathrm{Si}}/n_0 = 5 \times 10^{28}/10^{16} = 5 \times 10^{12}$ silicon atoms contribute one conduction electron. Comparing with metals in which each atom contributes at least one conduction electron, we know that the conductivity of semiconductors is much weaker than that of metals.

Using the given data, the number density of phosphorus impurity atoms can be obtained as $n_{\mathrm{P}} = n_{\mathrm{Si}}/5 \times 10^6 = 10^{22}\ \mathrm{m}^{-3}$. It is also the increased number density of conduction electrons due to the doping of phosphorus impurities, which can be seen from the contribution of one conduction electron of each phosphorus atom. So the number density of conduction electrons increases by a factor

$$\frac{n_{\mathrm{P}}}{n_0} = 10^6$$

Such a trace amount of impurities increase conduction electrons by a million times! It can be deduced that the conductivity of impurity semiconductors is significantly enhanced, compared with that of intrinsic semiconductors. But even so, they are much weaker than metals in conductivity.

4.3.2 PN Junction

The most basic and core structure in various semiconductor applications is the so-called PN junction, which is manufactured by doping 5 – and 3 – valent elements respectively in the two adjacent parts of an intrinsic semiconductor. The formed junction structure of the N – type and P – type semiconductors is the PN junction.

A partition is inserted in the middle of a box, and both sides are filled with oxygen gas and nitrogen gas. When the partition is removed, the two gases will move towards each other. This phenomenon is called diffusion. This also occurs at a PN junction. As shown in Figure 4 – 15, the majority carrier holes on the P – type side diffuse toward the N – type side, and the majority carrier electrons on the N – type side also diffuse toward the P – type side. Two kinds of carriers meet to neutralize near the interface. This will result in lack of holes and being negatively charged on the P – type side, while lack of electrons and being positively charged on the N – type side. The space charge distribution generates an electric field E directed from the N region to the P region, which prevents holes and electrons from continuing to diffuse toward each other. With the diffusion of carriers, the electric field becomes stronger and stronger, and the diffusion becomes weaker and weaker. Finally the electric field reaches a stable value, and the diffusion reaches an equilibrium. Due to the neutralization, a thin layer lacking carriers is formed on the interface, called the depletion layer, which has larger resistance. In the equilibrium, the thickness of the depletion layer and the electric field are certain, with the typical values 1 mm and $10^6 - 10^8$ V/m.

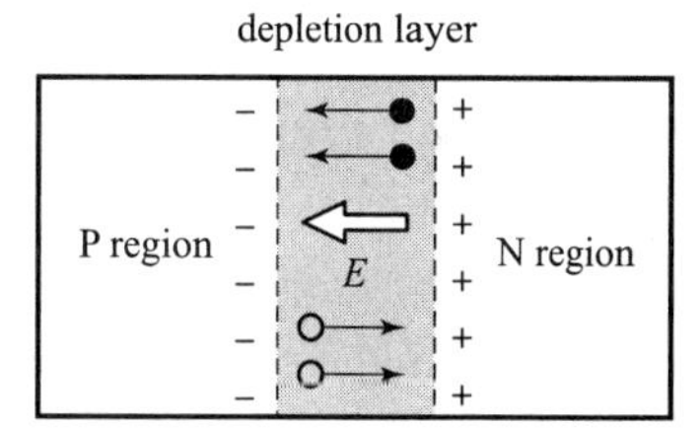

Figure 4 – 15 The formation mechanism of PN junction

An important property of PN junction is its unidirectional conductivity. As shown in Figure 4 –

16(a),a forward voltage is applied to the PN junction, namely, the P terminal is connected to the positive pole of electric source, and the N terminal to the negative pole. The electric field applied by the electric source to the PN junction is opposite to the original electric field in the junction. The balance mentioned above is broken, and the depletion layer becomes thinner. The holes in the P region and the electrons in the N region can continuously diffuse to each other through the depletion layer to form a forward current. The current increases rapidly as the forward voltage increases, corresponding to the segment of positive voltage of the current-voltage curve in Figure 4 – 16(c). If a reverse voltage is applied to the PN junction as shown in 4 – 16(b), the depletion layer will become thicker, so that the holes in the P region and the electrons in the N region are hindered, too difficult to diffuse to the opposite side. However, there are minority carriers with opposite charges in the two regions, and they will generate a weak reverse current along the direction of the electric field. The current tends to saturate quickly with the increase of the reverse voltage, corresponding to the segment with negative voltage as shown in Figure 4 – 16(c). The irregular dependence of current on voltage shown in Figure 4 – 16(c) is a nonlinear relationship. The nonlinearity determines the rich electronic properties of PN junctions, unlike the simple linear dependence of current on voltage (Ohm's law) in metals.

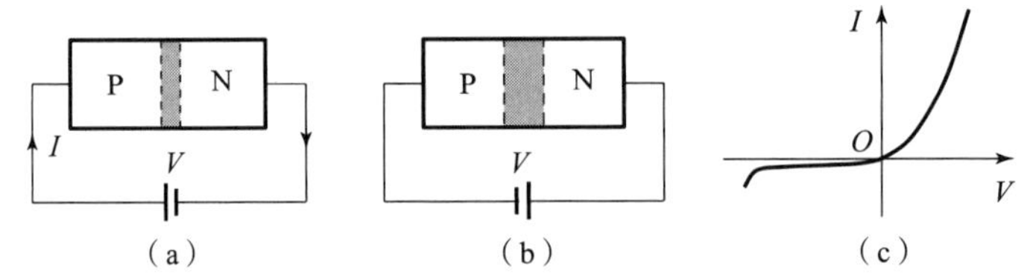

Figure 4 – 16 The current-voltage curve(c) when a forward voltage(a) and a reverse voltage(b) are applied to the PN junction

A significant current flows through a PN junction only when a forward voltage is applied, which is called the unidirectional conductivity of PN junction. Therefore only an unidirectional current can pass a PN junction when an AC voltage is applied, and this is the rectification effect of a PN junction.

4. 3. 3 Semiconductor Devices

Semiconductor devices made of PN junctions are varied, rich, and widely used. Several devices are briefly introduced below.

1. Light emitting diode(LED)

For ordinary semiconductors, when electrons transition from the bottom of the conduction band to the top of the valence band to neutralize the holes there, the energy is mostly released in the form of thermal energy. But for semiconductors such as gallium arsenide, gallium phosphide, and gallium nitride, the energy may be released in the form of light. In order to emit sufficiently strong light, a large number of "electron-hole pairs" like this must be annihilated. Intrinsic semiconductors only have fewer electrons and holes; N – type semiconductors have more conduction band electrons and fewer valence band holes; P – type semiconductors have more valence band holes and fewer conduction band electrons. Therefore, a single type of semiconductor, whether intrinsic or doped, cannot provide enough electron-hole pairs, thus cannot meet the requirements of practically emitting light.

It can meet the requirements to add a forward voltage between both sides of the appropriate PN junction. At this time, the current injects electrons into the N region and holes into the P region. If the dopants is enough many and the current flow is enough strong, there will be enough electron-hole pairs, and the depletion layer will become narrow enough. In this way, a large number of electrons and holes will be neutralized at the PN junction, thereby emitting evident light. Commonly used commercial LEDs are generally gallium-based, and properly doped with arsenic, phosphorus, nitrogen, indium, aluminum to form a PN junction, so that various forbidden band widths can be obtained, and can emit the light with big range of frequency, from the infrared to the ultraviolet (including the visible light of various colors). Figure 4 – 17(a) is a picture and structural diagram of a LED.

Now LEDs are very mature in the manufacturing technology, and widely applied. The bright colored characters displayed on some meters, household electric appliances, elevators, etc. , and some traffic lights are all composed of LEDs. An infrared LED can be installed in a remote control to control the work of electronic instrument. LEDs have the advantages of small size, high brightness, low heat generation, high efficiency, long life, easy control, etc. LEDs have become a new type of lighting lamps, which have basically replaced traditional lamps such as incandescent lamps, fluorescent lamps and iodine tungsten lamps. The super large display screens composed of neatly arranged LEDs have the advantages of high brightness and bright colors, and is widely used in public places such as gymnasiums, shopping malls, and city squares, as shown in Figure 4 – 17(b).

Whether used for lighting or displaying, LEDs are required to emit three primary colors of light to form white or various colors of light. In fact, the LEDs that emit red, yellow, and green light were invented more than 30 years before the blue-emitting LEDs. The reason is that the blue-emitting semiconductor crystals are difficultly prepared technically, due to the high frequency of blue light and the large forbidden band gap E_g (at least 2.6eV, close to the forbidden band gap of an insulator), which electrons cross in transition. After thousands of perseverant experiments, three scientists Isamu Akasaki, Hiroshi Amano and Shuji Nakamura, from Japan and the United States of America, finally succeeded in producing gallium nitride and indium gallium nitride crystals that can emit blue light efficiently. They made blue-emitting LEDs in 1989, for which they won the 2014 Nobel Prize in Physics. "Incandescent lights illuminated the 20th century, and LED lights will illuminate the 21st century", said the judging panel, evaluating their great contributions. In August 2008 Hiroshi Amano gave a lecture on blue-emitting LEDs at Beijing University, and a few days later the opening ceremony of the Beijing Olympics began. The Olympic rings composed of tens of thousands of LEDs rose gently in the night sky of Beijing, and a huge LED display screen was slowly unfolded, radiantly and dazzlingly, on the land of China, becoming the best display of LEDs since the advent. Since then the wide application of LEDs began in the world, as shown in Figure 4 – 17(b).

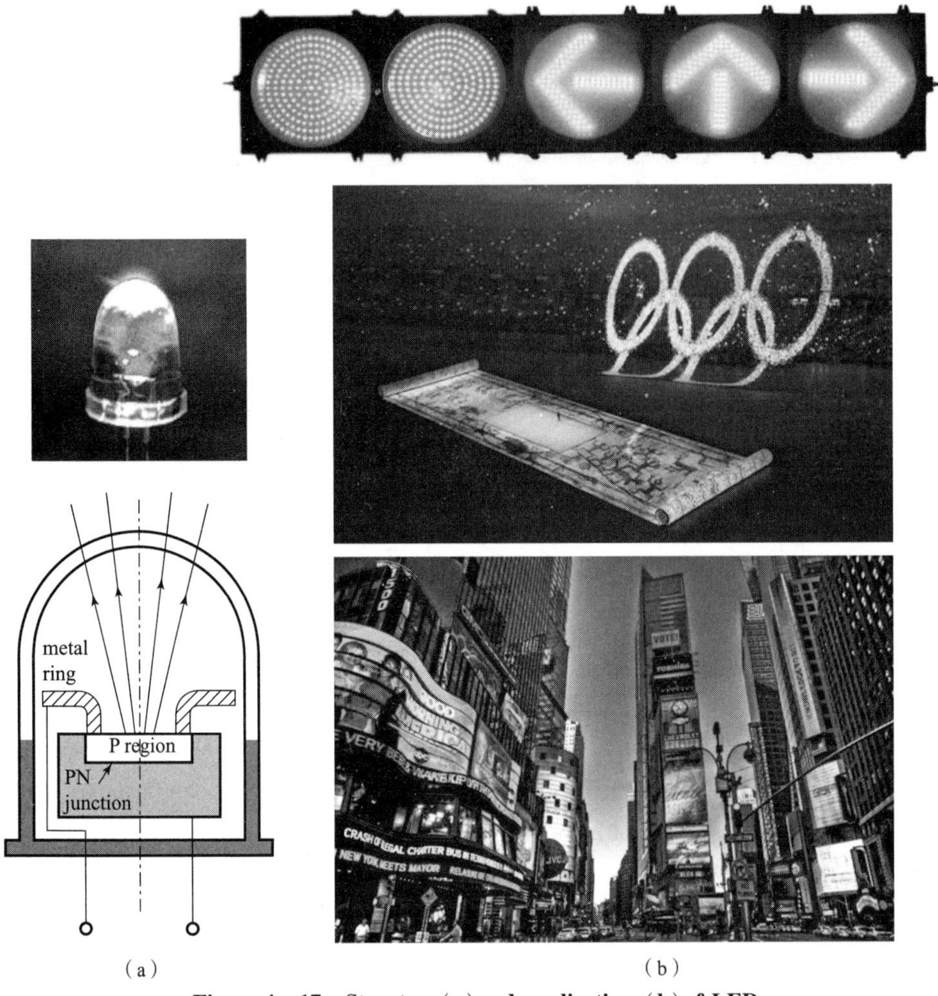

Figure 4 – 17　Structure (a) and applications (b) of LEDs

2. Photocell

If a LED works reversely, it becomes a photocell. When light is irradiated on an appropriate PN junction, the valence band electrons will transition to the conduction band, forming a large amount of electron-hole pairs at the PN junction. Electrons gather in the N region and holes in the P region. As a result, the electric potential of the P region is higher than that of the N region, and the PN junction becomes a power source. This is actually a photoelectric effect. In order to distinguish it from the photoelectric effect of metals, it is called the internal photoelectric effect, or the photovoltaic effect. The device on the TV that receives the infrared signal of the remote control is a small photocell, which converts the received light signal into an electrical signal to control the TV to perform some actions, such as, changing channels and adjusting volume. Solar photovoltaic cells have low power, low efficiency (only 12% – 22%, recently reported efficiency up to 41%), high cost, but less pollution, no need for transmission lines, and are suitable for home, transportation, communication, aerospace, etc. Figure 4 – 18 (a) shows the street lights using solar photovoltaic cells as energy sources.

By a huge number of photovoltaic cells in series and parallel, large current, voltage and power can be obtained, which is photovoltaic power generation if used for large-scale power generation. Now, photovoltaic power generation has become an industry and attracts much attention in China and many other countries in the world, and the goverments are competing to support photovoltaic power integrated into the national grids. Photovoltaic power generation has been developed rapidly in China. China has become the world's largest photovoltaic market. By 2021, China's cumulative installed photovoltaic capacity has reached 306 GW. Figure 4 – 18 (b) is a photovoltaic power station in Qinghai province, China.

(a) (b)

Figure 4 – 18

(a) Street light using solar photovoltaic cells as energy sources;

(b) Photovoltaic power station in Qinghai province, China

3. Junction laser

In a PN junction, there are many electrons in the conduction band of the N region, and many holes and few electrons in the valence band of the P region. This may cause the inversion of the electron population at the PN junctions of semiconductor such as gallium arsenide. This means that the number of electrons on a high energy level is more than that on a low energy level, which is one of the necessary conditions for the generation of Light Amplification by Stimulated Emission of Radiation (Laser). In addition, both ends of the PN junction must be strictly parallel to form a resonant cavity. In this way, the PN junction constitutes a junction laser (as shown in Figure 4 – 19), also known as a laser diode. Junction lasers have the advantages of compact structure and low power consumption, and can emit highly coherent and single-wavelength laser different from LED light. Optical disc drives are equipped with infrared junction lasers. The laser is irradiated at the dense optical track of the disc, is collected after reflected, converted into a digital signal, and then processed by a computer to obtain the information stored on the optical disc. Junction lasers are also widely used in barcode readers, laser pointers, and laser printers. Now, the heterojunction (interface structure of two different semiconductor materials) semiconductor lasers has become a key component of optical cable communication. Zhores I. Alferov from Russia and Herbert Kroemer from USA were awarded the 2000 Nobel Prize in Physics for their pioneering work on the theory and experiment of heterojunction semiconductor lasers.

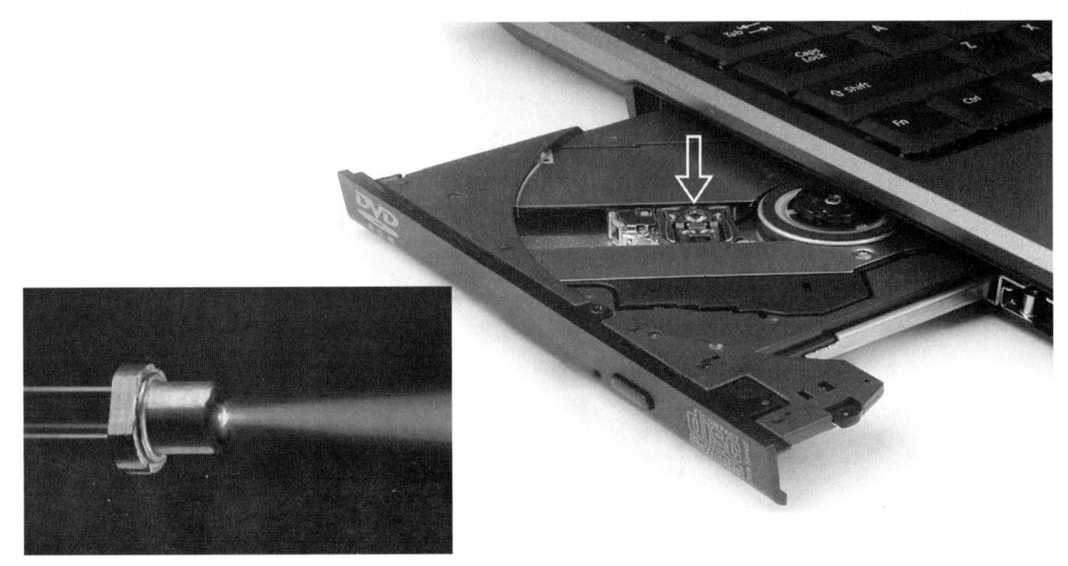

Figure 4 – 19 Junction laser and optical disc drive equipped with a junction laser

4. Metal-Oxide-Semiconductor Field Effect Transistor (MOSFET)

This is a three-terminal semiconductor transistor widely used in digital logic circuits, and has the advantages of fast response, simple technology, low noise and low power consumption. As shown in Figure 4 – 20, on the P – type silicon semiconductor substrate, diffused with N – type impurities, two highly doped N regions are generated, each connected to a metal electrode, called source (S) and drain (D) respectively. At first a layer of silicon dioxide insulating layer is prepared by oxidating the surface, and then a layer of metal is deposited on it to serve as the gate (G). Note that the gate is not electrically contact with the semiconductor, but is separated by a thin layer of insulating oxide. A large number of immobile positive ions are doped into the thin oxide layer, and attracted by the ions, the electrons are enriched on the semiconductor surface. The semiconductor there is changed from P – type to N – type, that is, an N – type channel is formed to connect the source and drain. When a voltage V_{DS} is applied between the source and drain, a current I_{DS} is formed between them.

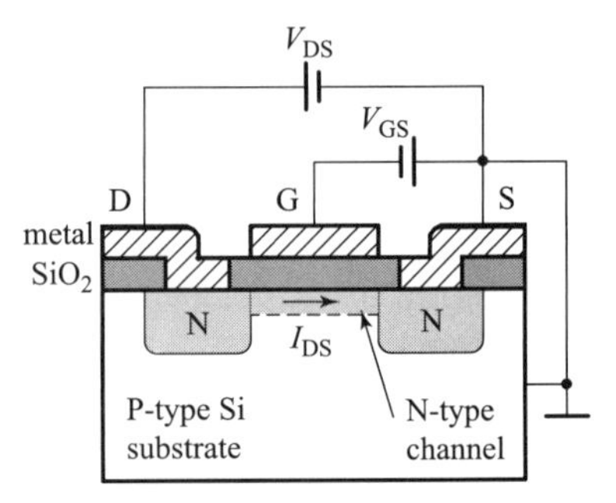

Figure 4 – 20 MOSFET structure diagram

If a negative voltage V_{GS} is applied between the gate and the substrate, namely, the gate potential is lower than the substrate one, an electric field (this is the origin of the name of field effect) will be established, repelling electrons from the channel to the substrate. This makes the PN junction between the channel and the substrate become wider, the channel narrower, the resistance larger, and the I_{DS} smaller. If the V_{GS} is appropriate, the I_{DS} can be reduced to zero. Therefore, by controlling V_{GS}, I_{DS} can be turned on and off, that is to say, the MOSFET can be switched between on and off, which can be used to represent binary 1 and 0.

5. Integrated circuit(IC)

The computers and electronic equipments we commonly use today are composed of thousands, millions, even several billions of electronic components such as transistors, capacitors, and resistors. These components are no longer complicatedly connected exposely, but extremely delicately fabricated and connected together on a small piece of semiconductor to form an integrated circuit. On tiny semiconductor wafers, the field effect transistors are made by means of the precise diffusion of impurities and the proper oxidation, other electronic components are simulated with properly designed microcircuits, and the connecting lines are also interposed between them on the wafer.

The idea of combining multiple transistors on a semiconductor wafer was popular in the 1950s. In the early 1960s, Jack S. Kilby of Texas Corporation and Noyce of Intel Corporation obtained the patents and became the inventors of integrated circuits. Kilby shared the 2000 Nobel Prize in Physics with the inventors of the aforementioned heterojunction lasers(Noyce passed away by then). Figure 4 - 21 is a Pentium microprocessor integrated circuit manufactured by Intel Corporation, which integrates tens of millions of electronic components on a 1 cm^2 sized chip. As you can imagine, for such a small component, even the smallest dust particles can seriously impact it, so integrated circuits need to be wrapped in a coating, and their producting environment needs to be extremely pure.

Figure 4 - 21 Internal structure of the Pentium microprocessor integrated circuit

Integrated circuits enable electronic devices to be extremely miniaturized, modular, and therefore inexpensive and convenient to use. Today, the development of integrated circuits is extremely rapid, with more and more components being integrated, and the speed of processing and transmitting data getting faster and faster. From computers to communication equipments, from household appliances to mobile phones, almost all electronic devices are inseparable from integrated circuits. Integrated circuits have become the most important core components of modern electronic technology, greatly affecting human life and profoundly changing the social presence. This impact and change are far from over. It can be said that the use of integrated circuits is a leap in scientific development, opening a real electronic revolution.

6. Charge Coupled Device(CCD)

CCD is a kind of digital image sensor that is very sensitive to light, and has the advantages of high efficiency, high speed, and easy digital processing. A typical CCD can convert 40% – 80% of the incident light intensity to electronic signals, so it can be used in the situation of weak light and short exposure time. For a conventional film, only 2% to 3% of the light causes the chemical reaction of the film grain to image. The famous Hubble telescope has a CCD installed, which has captured amazing images of space nebulae and stars, as well as the spectacular red desert images of Mars. CCDs have been widely used in webcams, cameras and video camcorder (as shown in Figure 4 – 22). In the past 20 years, CCDs have completely replaced traditional films, and have ended the film era in the history of photography.

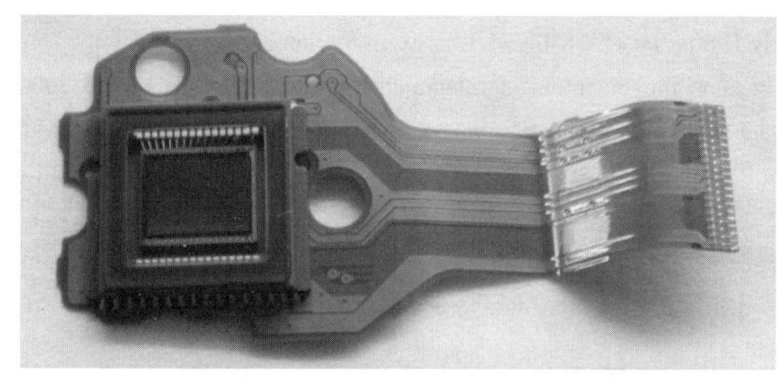

Figure 4 – 22 CCD and its application in digital cameras

CCDs use the photovoltaic effect to image. It consists of a three-layer structure: the upper layer is a series of metal electrodes (gates), the lower layer is a semiconductor silicon wafer, and they are separated by an insulating layer. When an appropriate voltage is applied to the gates of the two-dimensional array, the induced electric field produces a potential well, as a pixel, at the corresponding position of the silicon wafer. If light shines the silicon wafer, free electrons are generated and stored in potential wells. By alternately changing the gate voltage, the electrons in each potential well are transferred through the silicon surface channel to the processor for processing, so that the electrical signals on the two-dimensional array of pixels are obtained. Then the electrical signals are compiled into digital ones and zeros. By inputting the digital signals into an imaging device, a digital image whose brightness is relative to the intensity of the incident light can be displayed.

The 2009 Nobel Prize in Physics was awarded to the inventors of CCD, Willard S. Boyle of USA and Canada and George E. Smith of USA.

Summary

1. Free-electron-gas model of metals

The valence electrons in metals are regarded as free electrons in the three-dimensional infinitely-deep square potential well. The electrons follow the Pauli exclusion principle and the lowest energy principle when filling the quantum states. Quantum states are described by 4 quantum numbers(n_x, n_y, n_z, m_s), where n_x, n_y, n_z represent space states and m_s represents spin states.

Fermi energy at $T=0$ K $\quad E_F = (3\pi^2)^{2/3}\dfrac{\hbar^2}{2m}n^{2/3}$

Density of states of quantum states per unit volume in metals with energy

$$g(E) = \frac{dn_s}{dE} = \frac{(2m)^{3/2}}{2\pi^2\hbar^3}E^{1/2}$$

Fermi-Dirac distribution, the probability of an electron occupying a quantum state

$$f(E) = \frac{1}{1 + e^{(E-E_F)/kT}}$$

The number density of free electrons in metals by energy $\quad \dfrac{dn}{dE} = g(E)f(E)$

2. Energy bands of solids

When N atoms are aggregated into a solid, each energy level of the isolated atom is split into N energy levels that are extremely close to each other. The band-shaped energy range with quasi-continuous energy values is called energy band.

Electrons fill the energy levels in energy bands according to the Pauli exclusion principle and the lowest energy principle. The energy band into which an energy level with an orbital angular quantum number l is split can accommodate the electrons of the maximum number $2(2l+1)N$.

3. Valence, conduction and forbidden bands

At $T=0$ K the energy band where the electron filling is just completed is the highest energy band in which electrons exist, and generally occupied by valence electrons, thus called the valence band. The energy band with the empty quantum states at $T=0$ K is called the conduction band. The energy region without energy levels between different energy bands is called the forbidden band.

4. Conductor, Insulator and Semiconductor

Conductor: The valence band is partially filled with electrons, and the valence band is also the conduction band.

Insulator: The valence band is full, and the forbidden band between the valence band and the conduction band is wider.

Semiconductor: The valence band is full, and the forbidden band between the valence band and the conduction band is narrower.

5. Semiconductor conduction

Semiconductors have two kinds of current carriers: conduction band electrons and valence band holes. Semiconductors are divided into intrinsic semiconductors and impurity semiconductors, and impurity semiconductors are further divided into N – type (electron-type) and P – type (hole-type) semiconductors. A N – type semiconductor and a P – type semiconductor contacts to form a PN junction. PN junctions have a rectification effect.

The main applications of semiconductors are LEDs, transistors, photocells, junction lasers, integrated circuits, charge-coupled devices, etc.

Questions

4 – 1 Why can the free electrons in a metal be regarded to be in a three-dimensional square potential well?

4 – 2 What are Fermi energy, Fermi speed and Fermi temperature?

4 – 3 What is density of states? What factors are related to the probability of quantum states being filled by electrons?

4 – 4 What is energy band? For the energy band with orbital angular quantum number l, what is the maximum number of electrons that can be accommodated?

4 – 5 What are conduction band, valence band, and forbidden band? Try to explain the difference in the conductivity of conductors, insulators and semiconductors from their different energy band structures.

4 – 6 Some people said that in an intrinsic semiconductors, there are as the same many electrons with a negative basic charge as holes with a positive basic charge. They will soon neutralize and annihilate, no carriers exsiting, so the intrinsic semiconductors do not conduct electricity. What's the reason of the fault?

4 – 7 Are the numbers of two kinds of carriers in an impurity semiconductor the same? If not, which is more?

4 – 8 Why does PN junction have unidirectional conductivity?

4 – 9 The control of TV set can be realized by infrared ray. For it, semiconductor devices are used inside the remote control and TV set. What device is in the remote control? What device is in the TV set?

Problems

4 - 1 For copper, the molar mass is 63.54 g/mol and the density is 8,960 kg/m^3. If each copper atom contributes one valence electron, please calculate the number density of free electrons in copper. How many times of the molecular number density of ideal gas under the standard condition is it?

4 - 2 Zinc is a divalent metal with a molar mass of 65.37 g/mol and a density of 6,506 kg/m^3. Please calculate Fermi energy, Fermi velocity and Fermi temperature of zinc. What is the de Broglie wavelength of the electron with this Fermi energy?

4 - 3 Neutron star is composed of Fermi neutron gas. The density of a typical neutron star is about 5×10^{16} kg/m^3. Calculate the Fermi energy and Fermi velocity of the neutrons in the neutron star.

4 - 4 The electrons in a cube metal grain with side length of a can be regarded to be in a three-dimensional infinitely-deep square potential well.

(1) What conditions should be met for de Broglie wavelengths $\lambda_x, \lambda_y, \lambda_z$ in three directions?

(2) Derive the formula of the electronic energy of the system.

(3) If the grain contains 9 electrons, try to calculate Fermi energy (expressed by a formula).

4 - 5 In the free-electron-gas model, calculate the speed distribution function of free electron from the energy distribution function of free electron at $T = 0$ K (see example 4 - 1). Using this speed distribution function, calculate the average speed, root mean square speed, and average energy. Fermi energy E_F and Fermi speed v_F are known.

4 - 6 Using the data in Problem 4 - 1 and the results in Problem 4 - 5, calculate the energy released by the free electrons of 1 kg copper if the Pauli exclusion principle were suddenly canceled (of course, there is no way to cancel the Pauli exclusion principle).

4 - 7 Taking Fermi electron gas as an ideal gas, calculate the pressure generated by the free electrons of copper, using the pressure formula of ideal gas and the data in Table 4 - 1. How many times of the atmospheric pressure under the standard condition is it?

4 - 8 What is the probability that a quantum state at 50 meV above Fermi energy will be filled at $T = 0$ K and 300 K?

4 - 9 At a certain temperature, the quantum state at 10 meV above Fermi energy is filled with probability of 0.09. What probability is the quantum state at 10 meV below Fermi energy filled with?

4 - 10 The band gap width of diamond and silicon crystals are 5.5 eV and 1.2 eV respectively.

(1) The ratio of the electron numbers N_2 and N_1 on the energy level of E_2 at the upper edge of the band gap (i.e., the bottom of the conduction band) and on the energy level of E_1 at the lower edge (i.e., the top of the valence band) is approximately consistent with the Boltzmann distribution $N_2/N_1 = e^{-(E_2-E_1)/kT}$. Find the ratio at 300K. What does the result

mean?

(2) What are the maximum wavelengths of light required to make valence band electrons cross the forbidden bands to enter the conduction bands for two crystals? Which band of electromagnetic wave are they located?

4 – 11 Fermi-Dirac distribution function [Formula (4 – 11)] is applicable to metals, also to semiconductors and insulators. In an intrinsic semiconductor, the Fermi level is at the middle of the band gap (the Fermi level does not have to be an energy level that can be occupied).

(1) The band gap width of semiconductor germanium is known to be 0.67eV. Calculate respectively the probability of the conduction band bottom being occupied and the probability of valence band top not being occupied at 300K.

(2) The band gap width of the insulator diamond is known to be 5.5 eV. Calculate respectively the probability of the conduction band bottom being occupied and the probability of valence band top not being occupied at 300K.

Compared with the results of Problem 4 – 10, what do the results mean?

4 – 12 Potassium chloride crystal is transparent to visible light. Is it transparent or opaque to the ultraviolet light with a wavelength of 140nm? The band gap width of potassium chloride crystal is known to be 7.6 eV.

4 – 13 How many electron-hole pairs can be generated when gamma rays of 660keV pass through germanium (band gap width is 0.67eV)?

4 – 14 The band gap width of silicon crystal is 1.2 eV. After appropriate amount of phosphorus is added, the donor energy level is different by 0.045eV from the conduction band bottom of silicon. Calculate the maximum wavelength of light that can be absorbed by the doped semiconductor.

4 – 15 The number densities of free electrons and free holes in pure silicon at room temperature are about $n_0 = 10^{16}\ \text{m}^{-3}$. If the number density of free holes is increased by 10^6 times by doping aluminum, what proportion of silicon atoms should be replaced by aluminum atoms? How much aluminum should be added to 1g silicon? The density of silicon is known to be 2330 kg/m^3.

4 – 16 The semiconductor compound cadmium selenide (CdSe) is a material widely used for making light-emitting diodes. Its energy gap width is 1.8eV. What is the wavelength of the light emitted by this light-emitting diode? What color is the light?

Chapter 5

Nuclear Physics

In the early twentieth century, Rutherford used α - particle scattering experiments to demonstrate that there is a tiny nucleus at the center of each atom, which concentrates almost all the mass of the atom. In fact, the first human exposure to nuclear phenomena was due to the discovery of radioactivity. In 1896, the French physicist Becquerel discovered natural radioactivity, which is essentially a nuclear decay phenomenon. The radiation emitted by the atoms of radioactive elements through nuclear decay includes three types of rays: alpha(α), beta(β) and gamma(γ). These high-energy rays, especially α rays (or α particles), provide powerful tools for exploring the atomic structure and the composition of atomic nuclei. Nuclear energy is the energy contained in the nucleus of an atom, and fission and fusion are two effective ways to use nuclear energy. The use of nuclear energy can not only alleviate the shortage of conventional energy, but also an important way to reduce environmental pollution.

This chapter briefly introduces the basic properties of the atomic nucleus, including its composition, size, spin, etc., and then discusses the binding energy of the nucleus and the nuclear force that keeps the nucleus stable. After that, the law of radioactive decay and the characteristics of α, β and γ decay are introduced. Finally, the basic knowledge of nuclear reaction is introduced.

5.1 Properties of the Nucleus

5.1.1 Composition of the Nucleus

In the early twentieth century, an important question to physicists was whether the nucleus of an atom had a structure? If the nucleus has a structure, what is this structure? In 1919, Rutherford bombarded the nitrogen nucleus with α particles, knocking out a new particle from the nitrogen nucleus. According to the deflection of this particle in electric and magnetic fields, Rutherford measured its mass and charge, determined that it was the nucleus of a hydrogen atom, and named it a proton, denoted by the symbol p. Later, scientists studied more elements and used the same method to knock out protons from the nuclei of boron, fluorine, aluminum, phosphorus, etc., and concluded that protons are constituents of the nucleus.

The proton has a positive charge of $e(=1.602\times10^{-19}\,\mathrm{C})$. The mass of the proton is

$$m_p = 1.672,621,898 \times 10^{-27}\,\text{kg}$$

In 1920, Rutherford proposed the neutron hypothesis, that is, there is another particle in the nucleus, whose mass is very close to that of the proton and is electrically neutral. In 1932, Rutherford's student, the British physicist Chadwick, confirmed the existence of neutrons through experiments. The neutron is uncharged ($q = 0$) and is represented by the symbol n. The mass of the neutron is

$$m_n = 1.674,927,471 \times 10^{-27}\,\text{kg}$$

After the neutron was discovered, scientists put forward a model of the nucleus, which consists of protons and neutrons. Because protons and neutrons have very little mass difference, and they are both particles that make up the nucleus, they are collectively called nucleons. The nucleus of hydrogen consists of only one proton, while the nuclei of all other elements consist of both protons and neutrons. The different types of nuclei are often referred to as nuclides. The number of protons in a nucleus (or nuclide) is called the atomic number, which is represented by the symbol Z. Since neutrons are uncharged, the electric charge of the nucleus is Ze, and Z is also called the charge number of the nucleus. The number of nucleons, that is, the sum of the proton number Z and the neutron number N, is called the mass number of the nucleus, which is represented by the symbol A, i. e. ,

$$A = Z + N \tag{5-1}$$

Since the masses of protons and neutrons are almost equal, the mass of a nucleus is almost equal to A times the mass of a single nucleon, which is where the name mass number comes from.

A nuclide is commonly represented by the symbol ${}_Z^AX$. X is the chemical symbol for the element, the superscript A is the mass number of the nucleus, and the subscript Z is the atomic number. For example, ${}_7^{15}\text{N}$ means a nitrogen nucleus containing 7 protons and 8 neutrons for a total of 15 nucleons. Since the charge of a proton is equal to the absolute value of the charge of an electron, in a neutral atom, the number of electrons outside the nucleus is equal to the atomic number Z. The properties of an atom, and how it interacts with other atoms, depend largely on the number of electrons outside the nucleus. Hence Z determines which element the atom belongs to. Based on this, when the element symbol is given, the Z in the lower left corner can be omitted, and ${}_7^{15}\text{N}$ can be simplified as ${}^{15}\text{N}$.

For atoms of the same element, the nuclei have the same number of protons, but they may have different numbers of neutrons. For example, carbon nuclei always have 6 protons, but they may contain 5, 6, 7, 8, 9, or 10 neutrons. Nuclei with the same number of protons but different numbers of neutrons are called isotopes. Therefore, ${}^{11}\text{C}$, ${}^{12}\text{C}$, ${}^{13}\text{C}$, ${}^{14}\text{C}$, ${}^{15}\text{C}$ and ${}^{16}\text{C}$ are all isotopes of carbon. The lightest element, hydrogen, has three isotopes, namely ${}^{1}\text{H}$ (protium, also commonly known as hydrogen), ${}^{2}\text{H}$ (deuterium) and ${}^{3}\text{H}$ (tritium). The most abundant isotope of helium is ${}^{4}\text{He}$, which is also known as an α particle. Another isotope of helium is ${}^{3}\text{He}$.

Atomic and nuclear masses are usually expressed in unified atomic mass units (u). The unified atomic mass unit is defined as one-twelfth the mass of the neutral ${}^{12}\text{C}$ atom, i. e. ,

$$1\text{u} = 1.660,539,040 \times 10^{-27}\text{kg} = 931.494\text{MeV}/c^2$$

Thus, a proton has a mass of 1. 007,276u, a neutron 1. 008,665u, and a neutral hydrogen atom, ^{1}H (proton plus electron) 1. 007,825u. The atomic masses of several isotopes are listed in Table 5 – 1.

Table 5 – 1 Atomic Masses of Several Isotopes

Isotope	Z	Atomic mass/u	Isotope	Z	Atomic mass/u
^{1}H	1	1. 007,825	^{16}O	8	15. 994,915
^{2}H	1	2. 014,102	^{23}Na	11	22. 989,771
^{3}H	1	3. 016,050	^{39}K	19	38. 963,710
^{3}He	2	3. 016,030	^{56}Fe	26	55. 939,395
^{4}He	2	4. 002,603	^{63}Cu	29	62. 929,592
^{6}Li	3	6. 015,125	^{107}Ag	47	106. 905,094
^{7}Li	3	7. 016,004	^{197}Au	79	196. 966,541
^{10}B	5	10. 012,939	^{208}Pb	82	207. 976,650
^{12}C	6	12. 000,000	^{212}Po	84	211. 989,629
^{13}C	6	13. 003,354	^{222}Rn	86	222. 017,531
^{14}C	6	14. 003,242	^{226}Ra	88	226. 025,360
^{13}N	7	13. 005,738	^{238}U	92	238. 048,608
^{14}N	7	14. 003,074	^{242}Pu	94	242. 058,725

5. 1. 2 Shape and Size of the Nucleus

The size of nuclei was estimated originally by Rutherford through α – particle scattering experiments. Of course, due to wave-particle duality, we cannot determine the exact size of nuclei. However, a series of experiments show that the shape of most nuclei is approximately spherical, and the relationship that the radius increases with the increase of the mass number A can be approximately expressed as:

$$R = r_0 A^{1/3} \tag{5-2}$$

where r_0 is the proportionality constant, and it measures as $r_0 \approx 1.2 \times 10^{-15}\text{m} = 1.2\text{fm}$. Since the volume of a sphere is $V = \frac{4}{3}\pi R^3$, the volume of a nucleus is proportional to the number of nucleons, that is, $V \propto A$. Since the mass of the nucleus is also approximately proportional to A, all nuclei have approximately the same mass density. This property is very similar to that of a drop of liquid, whose density is independent of its size, indicating that atomic nuclei, like drops of liquid, are essentially incompressible. The liquid-drop model of the nucleus can explain some properties of nuclei, especially the fission of heavy nuclei.

According to Equation 5 – 2, the radius of the nuclei of ^{1}H, ^{40}Ca, ^{208}Pb and ^{235}U can be calculated as

$$^{1}\mathrm{H}: R \approx 1.2\mathrm{fm} \times 1^{1/3} = 1.2\mathrm{fm}$$
$$^{40}\mathrm{Ca}: R \approx 1.2\mathrm{fm} \times 40^{1/3} = 4.1\mathrm{fm}$$
$$^{208}\mathrm{Pb}: R \approx 1.2\mathrm{fm} \times 208^{1/3} = 7.1\mathrm{fm}$$
$$^{235}\mathrm{U}: R \approx 1.2\mathrm{fm} \times 235^{1/3} = 7.4\mathrm{fm}$$

Example 5 - 1: The most abundant iron nucleus in nature has a mass number of 56. Calculate its radius, mass, and density.

Solution: The radius of the iron nucleus is

$$R \approx 1.2\mathrm{fm} \times 56^{1/3} = 4.6\mathrm{fm}$$

Since $A = 56$, the mass of the iron nucleus is approximately 56u, or

$$m \approx 56 \times 1.66 \times 10^{-27}\mathrm{kg} = 9.3 \times 10^{-27}\mathrm{kg}$$

Its volume is

$$V = \frac{4}{3}\pi R^3 = \frac{4}{3} \times 3.14 \times (4.6 \times 10^{-15}\mathrm{m})^3 = 4.1 \times 10^{-43}\mathrm{m}^3$$

Its density is approximately

$$\rho = \frac{m}{V} = \frac{9.3 \times 10^{-26}\mathrm{kg}}{4.1 \times 10^{-43}\mathrm{m}^3} = 2.3 \times 10^{17}\ \mathrm{kg/m^3}$$

The density of solid iron is about $7000\mathrm{kg/m^3}$, so the density of iron nuclei is about 10^{13} times greater than that of solid iron. The density of nuclear matter is comparable to that of neutron stars. A cubic centimeter of matter with this density has a mass of 2.3×10^{11} kg, or 230 million tons.

5.1.3 Spin and Magnetic Moment of the Nucleus

Similar to the fact that electrons in atoms have orbital angular momentum and spin angular momentum, the nucleons that make up the nucleus also have orbital angular momentum and spin angular momentum. The total angular momentum of a nucleus is equal to the vector sum of the orbital angular momentum and spin angular momentum of all nucleons, which is also customarily called the nuclear spin angular momentum, or nuclear spin for short. The spin quantum numbers of the proton and the neutron are both 1/2, so the spin quantum number I of a nucleus composed of protons and neutrons has the following value rules when the nucleus is in the ground state: (1) The spin quantum numbers of even-even nuclei (both proton number Z and neutron number N are even) are equal to zero; (2) The spin quantum numbers of odd-odd nuclei (Z and N are both odd) are integers; (3) The spin quantum numbers of odd-even nuclei (one of Z and N is odd and the other is even) are half-integers. The magnitude of the spin angular momentum of the nucleus is $\sqrt{I(I+1)}\hbar$. Its projection in the given z direction is

$$I_z = m_I\hbar,\ m_I = -I,\ -I+1, \cdots, I-1, I \tag{5-3}$$

Here, m_I is called the magnetic quantum number of the nucleus, which can take $2I+1$ different values in total.

In relation to angular momentum, protons, neutrons and the nuclei that consist of them all have magnetic moments. When discussing the magnetic moment of the electron, we introduced the Bohr magneton $\mu_B = e\hbar/2m_e$ as the unit of the electron's magnetic moment, and the projection of the electron's spin magnetic moment in the z direction is approximately equal to one Bohr magneton. Similarly, when discussing the magnetic moment of an atomic nucleus, we introduce the nuclear magneton

$$\mu_{\mathrm{N}} = \frac{e\hbar}{2m_{\mathrm{p}}} = 5.050,78 \times 10^{-27}\,\mathrm{J/T} = 3.152,45 \times 10^{-8}\,\mathrm{eV/T} \tag{5-4}$$

as the unit of magnetic moment of the nucleus, where m_{p} is the mass of the proton. Since the mass of the proton is 1 836 times the mass of the electron, the nuclear magneton is 1 / 1 836 of the Bohr magneton. However, unlike the electron, the projection of the proton's spin magnetic moment in the z direction is not equal to a nuclear magneton, but

$$\mu_{\mathrm{p},z} = 2.792,847\mu_{\mathrm{N}} \tag{5-5}$$

Although the neutron is not charged, the projection of its spin magnetic moment in the z direction is

$$\mu_{\mathrm{n},z} = -1.913,044\mu_{\mathrm{N}} \tag{5-6}$$

The proton is positively charged, and its spin magnetic moment is in the same direction as the spin angular momentum. The spin magnetic moment of the neutron is in the opposite direction to the spin angular momentum, which is similar to the case of the electron. This indicates that although the neutron is not charged as a whole, there is a distribution of electric charge inside.

The magnetic moment of the entire nucleus is about a few nuclear magnetons, and its projection in the z direction is

$$\mu_z = g_I \mu_{\mathrm{N}} m_I \tag{5-7}$$

where g_I is called the g-factor of the nucleus, which is a pure number, and different nuclei have different g-factors. When a nucleus is placed in an external magnetic field $\boldsymbol{B}$, the interaction of its magnetic moment with the external magnetic field will generate an additional energy:

$$U = -\boldsymbol{\mu}_I \cdot \boldsymbol{B} = -\mu_z B \tag{5-8}$$

Since the magnetic quantum number m_I of the nucleus has $2I+1$ values, there are $2I+1$ different additional energies. Therefore, a nuclear energy level is split into $2I+1$ energy levels in the external magnetic field.

Example 5 – 2: Proton spin flip Places a proton in an external magnetic field oriented along the z-axis with a magnitude of 2. 30T. (1) What is the energy difference when the proton is in two states with its spin angular momentum parallel and antiparallel to the external magnetic field? (2) The proton can make a transition between these two energy states by emitting or absorbing a photon. What is the frequency and wavelength of this photon?

Solution: (1) When the proton's spin angular momentum is parallel to the external magnetic field, that is, its spin magnetic moment is parallel to the external magnetic field, the additional energy generated by the interaction between the magnetic moment and the external magnetic field is

$$U = -\mu_{\mathrm{p},z}B = -2.792,8 \times (3.152 \times 10^{-8}\,\mathrm{eV/T}) \times (2.30\mathrm{T}) = -2.025 \times 10^{-7}\,\mathrm{eV}$$

When the proton's spin is antiparallel to the external magnetic field, the additional energy is $+2.025 \times 10^{-7}$ eV. Therefore, the energy difference between these two energy states is

$$\Delta E = 2 \times (2.025 \times 10^{-7}\,\mathrm{eV}) = 4.05 \times 10^{-7}\,\mathrm{eV}$$

(2) When the proton makes a transition between these two energy states, the frequency and wavelength of the photon emitted or absorbed are respectively

$$\nu = \frac{\Delta E}{h} = \frac{4.05 \times 10^{-7}\,\text{eV}}{4.136 \times 10^{-15}\,\text{eV} \cdot \text{s}} = 9.79 \times 10^{7}\,\text{Hz} = 97.9\,\text{MHz}$$

$$\lambda = \frac{c}{\nu} = \frac{3.00 \times 10^{8}\,\text{m/s}}{9.79 \times 10^{7}\,\text{s}^{-1}} = 3.06\,\text{m}$$

This frequency is in the FM radio frequency band. Put a sample containing hydrogen into a magnetic field of 2.30T, and irradiate the sample with the electromagnetic wave of this frequency. When the sample absorbs the energy of the electromagnetic wave, the spin of protons is flipped.

The phenomenon in which atomic nuclei resonate and absorb electromagnetic waves of a certain frequency under the action of an external magnetic field is called nuclear magnetic resonance (NMR). Since the magnetic field and the frequency of electromagnetic waves can be measured accurately, the magnetic moment of the nucleus can be precisely determined using this technique. This technique is also applied to magnetic resonance imaging (MRI). Figure 5 – 1 shows the apparatus of MRI. Because human tissues contain a large amount of water and hydrocarbons, the densities of hydrogen nuclei are different in different tissue environments of human body, and the signal strength of NMR is also different. Using this difference, various tissues can be distinguished, and after computer processing, a very clear three-dimensional image of the interior of the human body can be drawn. This is a medical imaging technology that is harmless to the human body, and has been widely used in the diagnosis of diseases of various systems in the whole body, especially the diagnosis of early tumors. Figure 5 – 2 is an MRI image of the longitudinal section of a human head.

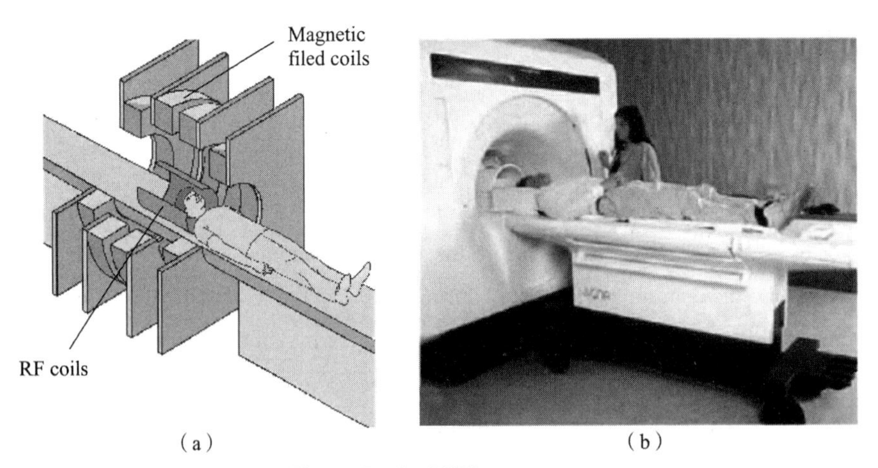

Figure 5 – 1 MRI apparatus

(a) Sketch; (b) Photo

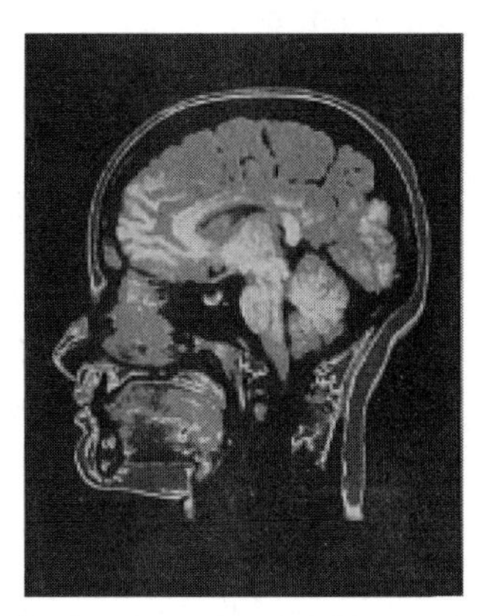

Figure 5 – 2 MRI image of the longitudinal section of a human head

5.2 Binding Energy and Nuclear Force

5.2.1 Binding Energy

The atomic nucleus is composed of protons and neutrons, and when the atomic nucleus is broken down into individual protons and neutrons, energy must be supplied to the nucleus. This energy is called the binding energy of the nucleus, denoted by E_B; it is also the energy released when individual nucleons combine to form a nucleus. Therefore, the total rest energy of Z protons and N neutrons that are far apart is greater than the rest energy of the nucleus composed of them, and this difference is the binding energy E_B.

The concept of binding energy is also applied to other systems. For example, the total rest energy of a proton and an electron far from each other is 13.6eV higher than the rest energy of the ground state hydrogen atom composed of the two. Therefore, an energy of 13.6eV is required to ionize the ground state hydrogen atom, that is, to separate the electron from the hydrogen atom. This is the binding energy of the hydrogen atom, often referred to as the ionization energy of the hydrogen atom.

According to Einstein's mass-energy relationship

$$E = mc^2$$

and from the conservation of energy, we know

$$(Zm_p + Nm_n)c^2 = m_A c^2 + E_B \tag{5-9}$$

Here, m_A is the rest mass of the nucleus with mass number A and charge number Z. Therefore,

$$E_B = (Zm_p + Nm_n - m_A)c^2 = \Delta mc^2 \tag{5-10}$$

where

$$\Delta m = Zm_p + Nm_n - m_A \tag{5-11}$$

which is called the mass defect of the atomic nucleus, so the rest mass of the atomic nucleus is less than the sum of the rest masses of Z protons and N neutrons that make up the nucleus. Since the mass m'_A of a neutral atom is usually given in the general data table instead of the mass m_A of the atomic nucleus, Equation 5 – 11 can be written as

$$\Delta m = Zm'_H + Nm_n - m'_A \qquad (5-12)$$

where m'_H is the rest mass of the neutral hydrogen atom ^1H, so the masses of the Z electrons in the first and third two terms on the right side of Equation 5 – 12 just cancel each other out. The binding energy of the nucleus can be written as

$$E_B = (Zm'_H + Nm_n - m'_A)c^2 \qquad (5-13)$$

Example 5 – 3: Calculate the binding energy of the last neutron in the ^4He nucleus.

Solution: The mass of the neutron is $m_n = 1.008,665u$. According to Table 5 – 1, the mass of the ^3He atom is $m'_{^3He} = 3.016,030u$ and the mass of the ^4He atom is $m'_{^4He} = 4.002,603u$. According to Equation 5 – 13, the binding energy of the last neutron in the ^4He nucleus is

$$E_B = \Delta mc^2 = (m'_{^3He} + m_n - m'_{^4He})c^2$$
$$= (3.016,030u + 1.008,665\ u - 4.002,603\ u) \times (931.5\ \text{MeV/uc}^2) \times c^2$$
$$= 20.58\text{MeV}$$

That is, it would require 20.58MeV input of energy to remove one neutron from the ^4He nucleus.

Different atomic nuclei have different binding energies, and the more nucleons that make up a nucleus, the higher its binding energy. It is therefore meaningful to have the average binding energy per nucleon, that is, the ratio of the total binding energy of a nucleus to the number of nucleons, also known as the specific binding energy. It reflects the tightness of atomic nucleus binding. The greater the average binding energy per nucleon, the stronger the nucleons in the nucleus are bound, and the more stable the nucleus is. Figure 5 – 3 shows a graph of the average binding energy per nucleon as a function of nucleon number (mass number) A. It can be seen from the figure that when the number

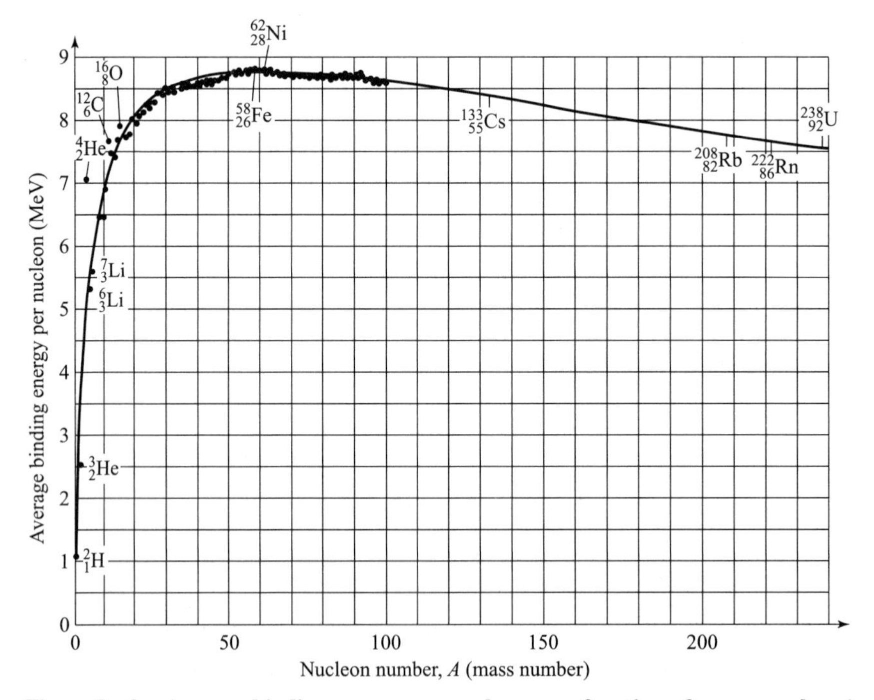

Figure 5 – 3 Average binding energy per nucleon as a function of mass number A

of nucleons A is small, the average binding energy per nucleon increases sharply with the increase of A. However, the data points (black dots) for ^{4}He, ^{12}C and ^{16}O lie significantly above the trend line (black curve), showing that these nuclei are more stable than their adjacent nuclei. When A reaches about 40, the curve tends to be flat, and the average binding energy per nucleon is basically independent of A, which is about 8.7MeV, indicating that the binding energy E_B of these nuclei is roughly proportional to the nucleon number A. When A is greater than 80, the curve decreases slowly, indicating that heavy nuclei are not as stable as those with intermediate mass. These characteristics of binding energy reflect the properties of nuclear force.

5.2.2 Nuclear Force

Before people knew the atomic nucleus, they only knew that there were two types of force in nature, one was the gravitational force and the other was the electromagnetic force. The protons are positively charged, and the electromagnetic force causes the protons to repel each other. The gravitational force is so weak that it can be completely ignored in the nucleus. Therefore, there is another attractive force between the nucleons that make up the nucleus, which is enough to counter the Coulomb repulsion, and bind the protons and neutrons tightly in the nucleus. This force is called the strong nuclear force. Experiments have proved that the strong nuclear force has the following important properties:

1. The nuclear force is independent of electric charge. Whether it is between protons and protons, or between neutrons and neutrons, or between protons and neutrons, the nuclear force of their interaction is the same, regardless of whether the nucleons are charged or not.

2. The nuclear force is a strong interaction force. Within the dimension of the atomic nucleus, the nuclear force is about two or three orders of magnitude greater than the electromagnetic force.

3. The nuclear force is a short-range force. It acts over a distance of about 2×10^{-15} m. When the distance between the two nucleons is 8×10^{-16} m ~ 2×10^{-15} m, the nuclear force acts as an attractive force, which decreases as the distance increases; when the distance is less than 8×10^{-16} m, it acts as a repulsive force, and increases rapidly as the distance decreases; when the distance exceeds 2×10^{-15} m, the nuclear force decreases sharply and almost disappears. However, the gravitational force and electromagnetic force are long-range forces, and their range of action is infinite.

4. There is saturation of nuclear forces in nuclei. A nucleon can only interact with its nearest neighbors by the nuclear force, but not with all other nucleons in the nucleus. This is different from the electromagnetic force, where every proton in the nucleus repels all other protons. These two competing interactions determine whether the nucleus is stable or not.

This property of the nuclear force can explain why the binding energy E_B of a nucleus with intermediate mass is approximately proportional to the nucleon number A. If there were no saturation of nuclear forces and each nucleon bonded to the other $(A-1)$ nucleons, there would be $A(A-1)/2$ pairs of interaction. To break up the atomic nucleus, it is necessary to provide enough energy to destroy the $A(A-1)/2$ pairs of interaction, that is, the binding energy E_B of the nucleus would then be proportional to $A(A-1)/2$, that is, to A^2, rather than A. In Figure 5 – 3, when A is small, the curve of the average binding energy per nucleon rises steeply with the increase of A. This is due to the increase in the number of nearest neighbors of each nucleon, which causes the average binding energy per nucleon to increase sharply. When A is larger, the Coulomb repulsion between protons is proportional to Z^2, so the average binding energy per nucleon decreases. The decrease of binding energy degrades the stability of the nucleus. But neutrons are not charged, not affected by Coulomb repulsion, and exert only the attractive nuclear force. Therefore, while increasing protons in the nucleus, adding more neutrons can maintain the stability of the nucleus. Figure 5 – 4 shows the relationship between the number of protons and the number of neutrons for stable nuclei. It can be seen from the figure that for lighter nuclei, the number of protons is nearly equal to the number of neutrons, but for heavier nuclei, the number of neutrons is greater than the number of protons. The heavier the nuclide, the greater the difference between the two. When the number of protons is very large ($Z > 82$), the distance between some nucleons is so large that there is no nuclear force at all, and stable nuclides do not exist.

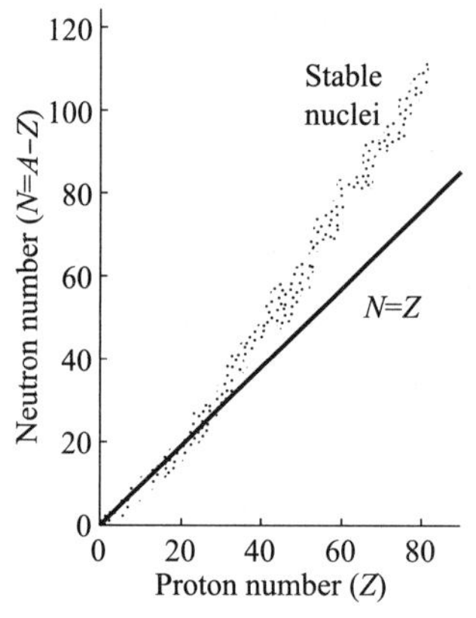

Figure 5 – 4 Number of neutrons versus number of protons for stable nuclei, which are represented by dots. The straight line represents $N = Z$

5. The nuclear force is related to spin. The nuclear force when two nucleons have parallel spins is greater than the nuclear force when their spins are opposite. The spin of deuteron (^2H) in nature is 1. The total spin can be 1 only if the proton and neutron spins are parallel. This shows that when the spins of the proton and the neutron are parallel, there is a strong nuclear force, which can combine the proton and the neutron to form a stable deuteron.

In addition to the strong nuclear force, scientists have discovered the fourth interaction in nature—the weak nuclear force—in the atomic nucleus. It is the cause of the β decay of the atomic nucleus, and it is also a short-range force. The range of action of the weak nuclear force is shorter than that of the strong nuclear force, and its strength of action is smaller than that of the electromagnetic force.

5.3 Radioactive Decay of the Nucleus

5.3.1 Radioactivity

In 1896, when the French physicist Becquerel was studying the phenomenon of fluorescence, he found that a uranium salt could emit radiation without any external source of energy. This kind of radiation could penetrate black paper and expose a photographic plate. Marie Curie believed that the phenomenon discovered by Becquerel was universal, and it was Marie Curie who proposed to call this phenomenon "radioactivity". In 1898, Marie Curie and her husband, the French physicist Pierre Curie, discovered two new elements that were far more radioactive than uranium, and named them polonium and radium respectively. Becquerel and the Curies shared the Nobel Prize in Physics in 1903, and Marie Curie also won the Nobel Prize in Chemistry in 1911, becoming the first scientist to win two Nobel Prizes.

Studies have found that the radioactivity is not affected by various physical and chemical treatments, such as heating, cooling, or the action of chemical reagents. This suggests that the source of radioactivity must be deep within the atom, emanating from the nucleus. Radioactivity is the result of the decay of an unstable nucleus. Some nuclides are not stable under the action of the nuclear force, and they spontaneously emit various types of radiation and decay to form another nuclide. Many unstable nuclides occur in nature, and such radioactivity is called natural radioactivity. Other unstable nuclides can be produced in the laboratory by nuclear reactions, and their radioactivity is called artificial radioactivity.

Shortly after the discovery of radioactivity, Rutherford and his collaborators classified the radiation emitted by the decay of various radioactive elements into three types according to the penetrating power of the radiation. One type of radiation has the weakest penetrating power and can barely pass through a piece of paper; the second type can penetrate a 3mm aluminum plate; the third is extremely penetrating and can pass through several centimeters of lead plate. Rutherford named these three types of radiation alpha (α) rays, beta (β) rays, and gamma (γ) rays. Further research found that the three types of rays split into three beams in a magnetic field (as shown in Figure 5 - 5), showing that α rays are positively charged particle streams, β rays are negatively charged particle streams, while γ rays are electrically neutral and not bent in a magnetic field. Finally, it is confirmed that α rays (or α particles) are the nuclei of helium atoms, ${}^{4}He$, each consisting of two protons and two neutrons; β rays are electrons; γ rays are high-energy photons whose energy is even higher than that of X - rays.

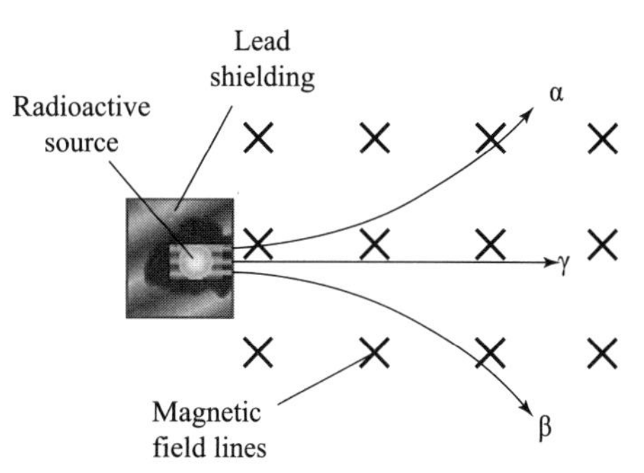

Figure 5 - 5 Schematic diagram of the trajectories of three rays in a magnetic field

5. 3. 2 Radioactive Decay Law

A macroscopic sample of any radioactive nuclide consists of a vast number of radioactive nuclei. These nuclei do not all decay at the same time. Rather, they decay one by one over a period of time. We cannot predict exactly when a particular nucleus will decay, but the decay of the entire sample is a statistical process.

Assume that N_0 is the number of radioactive nuclei at $t = 0$, and $N(t)$ is the number of radioactive nuclei remaining at some time t. The number of decays $-\mathrm{d}N(t)$ that occur in a time interval $\mathrm{d}t$ is proportional to $N(t)$ and $\mathrm{d}t$, that is

$$-\mathrm{d}N(t) = \lambda N(t)\mathrm{d}t \tag{5-14}$$

where λ is a constant of proportionality. Since $\mathrm{d}N(t)$ represents the decrease in $N(t)$, a negative sign is necessary to be added in front of it. After integrating Equation 5 – 14, we can get

$$N(t) = N_0 \mathrm{e}^{-\lambda t} \tag{5-15}$$

Equation 5 – 15 is called the radioactive decay law, indicating that the number of radioactive nuclei in a given sample decreases exponentially in time. Equation 5 – 14 can be rewritten as

$$\lambda = \frac{-\mathrm{d}N(t)/\mathrm{d}t}{N(t)} \tag{5-16}$$

In the above equation, the numerator represents the number of nuclei decaying per unit time, and the denominator represents the total number of nuclei at that time. Therefore, λ represents the probability of a nucleus decaying in unit time, called the decay constant (unit: s^{-1}). Different nuclides have different decay constants.

The time it takes for the number of nuclei to decrease to half of the original number N_0 due to decay is called the half-life and is represented by $T_{1/2}$. According to Equation 5 – 15, we can get

$$\frac{N_0}{2} = N_0 \mathrm{e}^{-\lambda T_{1/2}}$$

$$T_{1/2} = \frac{\ln 2}{\lambda} = \frac{0.693}{\lambda} \tag{5-17}$$

Both $T_{1/2}$ and λ are characteristic constants of radioactive nuclides. The larger λ is, the smaller $T_{1/2}$ is. Figure 5 – 6 shows the exponential decay of the number $N(t)$ of radioactive nuclides as a function of time. As a statistical process, there must be fluctuations.

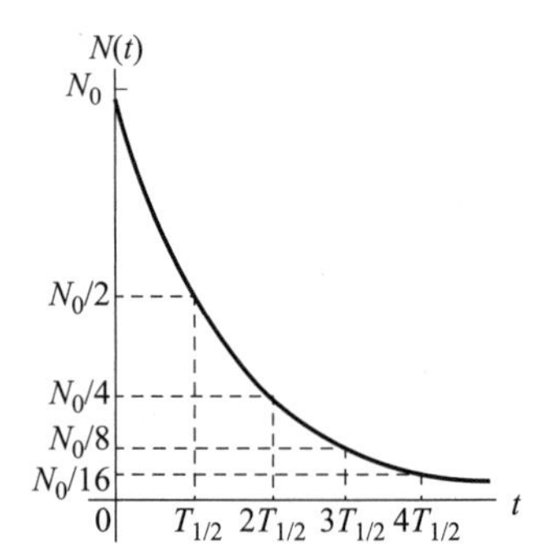

Figure 5 – 6 Exponential radioactive decay

For a given sample of radioactive nuclides, some nuclei decay early and some decay late, and the mean lifetime τ can be used to express the rate of decay. The number of nuclei decaying in the time interval $\mathrm{d}t$ is $-\mathrm{d}N(t)$, where the lifetime of each nucleus is t, then the mean lifetime of all radioactive nuclides is

$$\tau = \frac{1}{N_0}\int_0^{\infty} t[-\mathrm{d}N(t)]$$

Substituting Equation 5 – 14 and Equation 5 – 15 into it, we obtain

$$\tau = \frac{1}{N_0}\int_0^\infty t\lambda N(t)\,dt = \int_0^\infty t\lambda e^{-\lambda t}dt$$

The integral result is

$$\tau = \frac{1}{\lambda} = \frac{T_{1/2}}{\ln 2} = 1.44T_{1/2} \qquad (5-18)$$

Therefore, the mean lifetime is the reciprocal of the decay constant and 1.44 times the half-life. Substituting Equation 5 – 18 into Equation 5 – 15, we get

$$N(\tau) = N_0 e^{-1} \approx 37\%\ N_0$$

It can be seen that the number of remaining radioactive nuclides is about 37% of the original value after a time interval τ.

In the application of radioactive substances, the physical quantity of activity is often used. The number of atomic nuclei decaying per unit time is called the activity of a given sample, which is expressed as $A(t)$, that is

$$A(t) = -\frac{dN(t)}{dt} = \lambda N(t) = \lambda N_0 e^{-\lambda t} = A_0 e^{-\lambda t} \qquad (5-19)$$

where $A_0 = \lambda N_0$ is the activity at $t = 0$. From Equation 5 – 19, it can be seen that the activity is proportional to the decay constant λ and the number $N(t)$ of radioactive nuclei. Therefore, $A(t)$ also decreases exponentially in time, and its graph of change with time is the same as that in Figure 5 – 6. After a half-life, the activity decreases to half of the original value.

The SI unit of activity is the becquerel, denoted by Bq,

$$1\,\text{Bq} = 1\ \text{s}^{-1}$$

Another unit of activity in common use is the curie named for Marie Curie, and denoted by Ci,

$$1\,\text{Ci} = 3.7\times10^{10}\ \text{Bq} = 3.7\times10^{10}\ \text{s}^{-1}$$

One Curie is approximately the activity of 1g of radium.

Example 5 – 4: The activity of ^{57}Co. The radioactive nuclide ^{57}Co has a half-life of 272d. (1) Calculate its decay constant and mean lifetime; (2) If the activity of a sample of ^{57}Co is 2.00μCi, how many radioactive nuclei contained in the sample? (3) What is the activity of the sample after one year?

Solution: (1) The half-life of ^{57}Co is

$$T_{1/2} = (272\text{d})\times(86{,}400\text{s/d}) = 2.35\times10^7\text{s}$$

The mean lifetime is

$$\tau = \frac{T_{1/2}}{\ln 2} = \frac{2.35\times10^7\text{s}}{0.693} = 3.39\times10^7\text{s}$$

The decay constant is

$$\lambda = \frac{1}{\tau} = \frac{1}{3.39\times10^7\text{s}} = 2.95\times10^{-8}\text{s}^{-1}$$

(2) At $t = 0$, the activity of the sample of ^{57}Co is

$$A_0 = 2.00\mu\text{Ci} = (2.00\times10^{-6})\times(3.70\times10^{10}\text{s}^{-1}) = 7.40\times10^4\text{s}^{-1}$$

Then, at $t = 0$, the number of radioactive nuclei is

$$N_0 = \frac{A_0}{\lambda} = \frac{7.40 \times 10^4 \mathrm{s}^{-1}}{2.95 \times 10^{-8} \mathrm{s}^{-1}} = 2.51 \times 10^{12}$$

(3) After one year(3.156×10^7 s), the activity of the sample will be

$$A(t) = A_0 e^{-\lambda t} = (7.40 \times 10^4 \mathrm{s}^{-1}) \times e^{-(2.95\times10^{-8}\mathrm{s}^{-1})\times(3.156\times10^7 \mathrm{s})}$$
$$= 2.915 \times 10^4 \mathrm{s}^{-1} = 0.788 \mu\mathrm{Ci}$$

5.3.3 Alpha Decay

Heavy nuclei are unstable and can be stabilized by alpha decay. When an atomic nucleus spontaneously emits an α particle, it becomes a different nucleus due to the loss of two protons and two neutrons. The original nucleus is called the parent nucleus, and the new nucleus produced is called the daughter nucleus. For example, when Radium 226 ($^{226}_{88}$Ra) emits an α particle, it decays to a radon nucleus with $Z = 88 - 2 = 86$ and $A = 226 - 4 = 222$. This process is shown in Figure 5-7 and written as

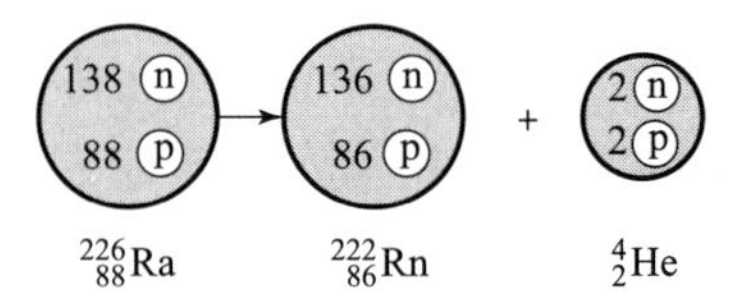

Figure 5-7 Alpha decay of radium 226

$$^{226}_{88}\mathrm{Ra} \rightarrow ^{222}_{86}\mathrm{Rn} + ^{4}_{2}\mathrm{He}$$

Alpha decay can be expressed generally as

$$^{A}_{Z}\mathrm{X} \rightarrow ^{A-4}_{Z-2}\mathrm{Y} + \alpha \tag{5-20}$$

where X is the parent, Y is the daughter, and A and Z are the mass number and atomic number, respectively, of the parent. The mass number of the daughter nucleus is 4 less than that of the parent nucleus, and the atomic number is 2 less than that of the parent nucleus.

The parent nucleus X can be regarded as at rest before decay. From conservation of energy,

$$m_X c^2 = m_Y c^2 + m_\alpha c^2 + E_\alpha + E_Y$$

where m_X, m_Y and m_α are the rest masses of the parent, daughter and α particle, respectively. E_α and E_Y are the kinetic energies of the α particle and the recoiling daughter nucleus, respectively. The total energy released in α decay is the kinetic energy carried away by the α particle and the recoiling daughter nucleus, which is called the disintegration energy, denoted by E_0, that is

$$E_0 = E_\alpha + E_Y = [m_X - (m_Y + m_\alpha)]c^2 \tag{5-21}$$

Since the mass m' of a neutral atom is usually given in the general data table instead of the mass m of the atomic nucleus, Equation 5-21 can be written as

$$E_0 = [m'_X - (m'_Y + m'_{He})]c^2 \tag{5-22}$$

where m'_X, m'_Y and m'_{He} are the atomic masses of X, Y and helium, respectively. The masses of the electrons in the equation just cancel each other out. For alpha decay to occur, the disintegration energy must be greater than zero, so the rest mass of the parent atom must be greater than the sum of the rest masses of the daughter atom and the helium atom.

Example 5-5: The ^{232}U nucleus decays with the emission of an α particle:

$$^{232}\mathrm{U} \rightarrow ^{228}\mathrm{Th} + ^{4}\mathrm{He}$$

The atomic masses of ^{232}U, ^{228}Th and ^{4}He are 232.037,146u, 228.028,731u and 4.002,603u, respectively. Calculate the disintegration energy in this process.

Solution: The total mass of the reaction products is

$$228.028,731\text{u} + 4.002,603\text{u} = 232.031,334\text{u}$$

The mass lost when the ^{232}U decays is

$$232.037,146\text{u} - 232.031,334\text{u} = 0.005,812\text{u}$$

Since 1u = 931.5MeV, the disintegration energy is

$$E_0 = 0.005,812\text{u} \times 931.5\text{MeV/u} = 5.414\text{MeV}$$

This energy is the sum of the kinetic energies of the α particle and the recoiling daughter ^{228}Th nucleus. Using conservation of momentum, it can be calculated that the α particle emitted by a ^{232}U nucleus at rest has a kinetic energy of about 5.3MeV. Therefore, the recoiling daughter ^{228}Th nucleus has about 0.1MeV of kinetic energy.

Since the parent nucleus can spontaneously decay when the sum of the masses of the daughter nucleus and the α particle is less than the mass of the parent nucleus, why do parent nuclei still exist in nature? That is, why haven't radioactive nuclides all decayed long ago, right after they were formed in the early stages of the universe? In fact, the half-lives of radioactive nuclides that decay by α decay range from 10^{-5}s to 10^{10} a. In 1928, George Gamow explained α decay according to quantum mechanics. Alpha decay is a process in which an α particle is first formed inside a nucleus. As shown in Figure 5 - 8, inside the nucleus ($r < R$, R is the nuclear radius), the α particle is attracted by the nuclear force, and the nucleus forms a potential well with a depth of tens of MeV for the α particle. Outside the nucleus ($r > R$), the α particle experiences the Coulomb repulsion by the daughter nucleus, and its potential energy is

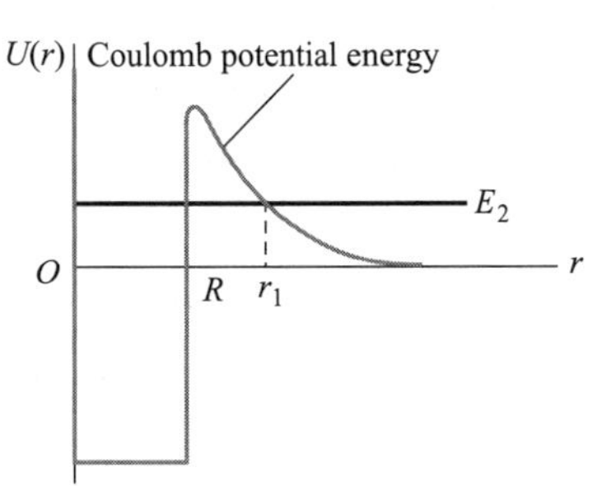

Figure 5 - 8 Potential energy for an α particle and a (daughter) nucleus

$$U(r) = \frac{2Ze^2}{4\pi\varepsilon_0 r} \tag{5-23}$$

where Z is the charge number of the daughter nucleus. The kinetic energy E_α of the α particle released by α decay is generally much lower than the height of the Coulomb barrier. The α particle could not escape the nucleus if it were governed by classical physics. But according to the barrier penetration theory of quantum mechanics, there is a certain probability that the α particle can tunnel through the Coulomb barrier. It can be known from Figure 5 - 8 that the greater the kinetic energy E_α of the escaping α particle, the smaller the thickness of the potential barrier it will penetrate, and thus the greater the probability of it tunnelling through the potential barrier, and the corresponding half-life of α decay is shorter. The expression for the half-life as a function of the kinetic energy of the α particle derived by Gamow is in excellent agreement with experimental results.

5.3.4 Beta Decay

Nuclei are stabilized by β decay when they contain too many or too few neutrons compared to the number of protons in the nucleus. At the very beginning, β decay just referred to the emission of an electron (β^- particle) from the nucleus. In this process, the number of nucleons in the nucleus does not decrease, but the charge of the nucleus increases due to the release of the electron. Therefore, the daughter nucleus has the same mass number A as the parent nucleus, while the atomic number Z is increased by 1. If the decay energy were shared by only the daughter nucleus and the emitted electron, the kinetic energy of the electron would be uniquely determined according to the conservation of energy and momentum. However, experiments have found that the electrons emitted in the β decay have a continuous range of kinetic energies from zero up to a certain maximum value, as shown in Figure 5 – 9.

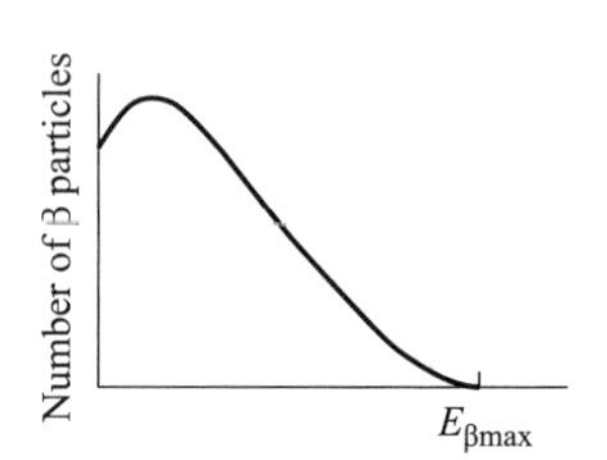

Figure 5 – 9 Kinetic energy of β particles

In order to explain this experimental phenomenon, Pauli proposed in 1930: an undetected particle was emitted during β decay in addition to the electron. This hypothesized particle could be carrying off the energy, momentum, and angular momentum required to maintain the conservation laws in β decay. This new particle was named the neutrino by Fermi, which is denoted by ν_e. The neutrino has no charge and almost zero mass. It is far more difficult to observe in experiments and was not directly detected until 1956.

There are no electrons in the nucleus, where do the electrons emitted in β decay come from? Fermi believed that in the process of decay, a neutron in the nucleus changes into a proton, and an electron and a neutrino are released at the same time. Further analysis confirmed that what is emitted in this process is an antineutrino $\bar{\nu}_e$, which is the antiparticle of the neutrino. This process can be expressed as

$$\mathrm{n} \longrightarrow \mathrm{p} + \mathrm{e}^- + \bar{\nu}_e \tag{5-24}$$

where n, p, and e^- represent the neutron, proton, and electron, respectively. In Figure 5 – 4, the nuclides that lie above the stable nuclei are nuclides that have too many neutrons compared to the number of protons. They tend to be stable by electron emission. The nuclides that fall below the stable nuclei of Figure 5 – 4 are nuclides that have too few neutrons compared to their number of protons. These tend to be stable by emitting a positron e^+ (β^+ particle). In this decay process, a proton in the nucleus changes into a neutron, releasing a positron and a neutrino at the same time, which can be expressed as

$$\mathrm{p} \longrightarrow \mathrm{n} + \mathrm{e}^+ + \nu_e \tag{5-25}$$

The β decay corresponding to Equation 5 – 24 is called β^- decay, which can be generally expressed as

$$^{A}_{Z}\mathrm{X} \longrightarrow {}^{A}_{Z+1}\mathrm{Y} + \mathrm{e}^- + \bar{\nu}_e \tag{5-26}$$

where X and Y represent the parent and daughter nucleus, respectively. The β decay corresponding to Equation 5 – 25 is called β^+ decay, which can be generally expressed as

$$^{A}_{Z}\mathrm{X} \longrightarrow {}^{A}_{Z-1}\mathrm{Y} + \mathrm{e}^+ + \nu_e \tag{5-27}$$

In β^+ decay, the daughter nucleus has the same mass number A as the parent nucleus, while the atomic number Z is decreased by 1.

Any atomic nucleus that can decay via β^+ can also absorb one of its orbital electrons, converting a proton in the nucleus into a neutron, and emitting a neutrino. This process is called electron capture (EC), and its basic reaction is

$$p + e^- \rightarrow n + \nu_e \tag{5-28}$$

Electron capture can be generally expressed as

$${}_Z^A X + e^- \rightarrow {}_{Z-1}^{A} Y + \nu_e \tag{5-29}$$

Since the K shell electrons are closest to the nucleus, usually it is an electron in the K shell that is captured. After the nucleus captures an electron, a hole appears in the shell, and when the outer electron jumps down to fill the hole, X-rays are emitted.

In 1934, Fermi proposed the theory of β decay. In this theory, Fermi pointed out that it is the weak nuclear force that causes β decay to produce an electron and a neutrino. The neutrino interacts with matter only through the weak nuclear force, which is why it is so hard to detect.

5.3.5 Gamma Decay

Gamma rays are high-energy photons. Like an atom, an atomic nucleus in an excited state emits a photon when it makes a transition to a lower excited state, or to the ground state. The spacing between the energy levels of a nucleus is on the order of keV to MeV, which is much larger than those of an atom of only a few eV. Therefore, the emitted γ-ray photons have energies that can range from a few keV to several MeV. Since a γ ray is electrically neutral, emission of a γ ray does not change the nucleus into a different nuclide.

Alpha and beta decay do not always leave the daughter nucleus in its ground state. Sometimes the daughter nucleus is left in an excited state. Figure 5-10 shows the energy level diagram of a decay process. ^{12}B can decay directly to the ground state of ^{12}C through β decay, or it can decay by β decay to an excited state $^{12}C^*$ of ^{12}C, and then emits a 4.4MeV photon to decay to the ground state through γ decay.

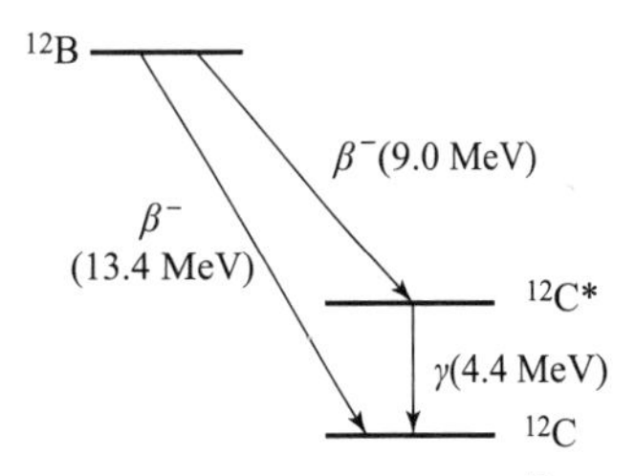

Figure 5-10　Decay of ^{12}B

We can write γ decay as

$${}_Z^A X^* \rightarrow {}_Z^A X + \gamma \tag{5-30}$$

where "∗" means the excited state of that nucleus.

In some cases, a nucleus may remain in an excited state for some time before it emits a γ ray. Nuclear energy states that have long lifetimes are called metastable states, and a nuclide in a metastable state is called an isomer.

An excited nucleus can sometimes make a transition to the ground state by another process known as internal conversion with no γ ray emitted. In this process, the excited nucleus interacts with one of the orbital electrons and ejects this electron from the atom. The kinetic energy of the ejected electron is the energy that an emitted γ-ray photon would have had minus the binding energy of the electron.

5.3.6 Applications of Radioactivity

Radioactivity has a wide range of applications in industry, agriculture, medicine, and scientific research. For example, isotope tracing can be used for disease diagnosis and research on the absorption of chemical fertilizers by crops. Gamma rays can be used for metal non-destructive testing and treatment of malignant tumors. Here is a brief introduction to the application of radioactive dating in archaeological and geological research.

All living plants absorb CO_2 from the air to synthesize organic molecules. The vast majority of these carbon atoms are ^{12}C, but a small fraction (about 1.3×10^{-12}) is the radioactive isotope ^{14}C. Although ^{14}C decays with a half-life of about 5730 a, the ratio of ^{14}C to ^{12}C in the atmosphere remains roughly constant. This is because a large number of neutrons are produced when high-energy particles in cosmic rays bombard atoms in the atmosphere. These neutrons react with nitrogen nuclei in the atmosphere to produce ^{14}C, and release protons, which can be expressed as

$$n + {}^{14}N \rightarrow {}^{14}C + p$$

This continued production of ^{14}C in the atmosphere is roughly in balance with the radioactive decay of ^{14}C, so that its number remains constant. As long as a plant is alive, it constantly absorbs carbon dioxide from the air for metabolism. Animals eat plants, so they are also constantly exchanging carbon with the environment. Organisms cannot distinguish ^{14}C from ^{12}C, so the ratio of ^{14}C to ^{12}C in a living organism remains nearly constant, the same as that in the atmosphere. But when an organism dies, the exchange of carbon with the environment stops, and ^{14}C decreases due to radioactive decay, so the ratio of ^{14}C to ^{12}C in a dead organism decreases in time. Since the half-life of ^{14}C is about 5730 a, the ratio of ^{14}C to ^{12}C decreases by half every 5730 years. For example, if an ancient wooden tool has half the ratio of ^{14}C to ^{12}C in living trees, then the tool must have been made from a tree that was felled about 5,700 years ago. In fact, the ratio of ^{14}C to ^{12}C in the atmosphere is not precisely constant, so the results of the radioactive dating method need to be calibrated.

Example 5-6: Age of an ancient animal. An animal bone fragment found in an archeological site has a carbon mass of 200g. Its activity is 16Bq. What is the age of the bone fragment?

Solution: When the animal was alive, the ratio of ^{14}C to ^{12}C was 1.3×10^{-12}. The number of ^{14}C nuclei of the 200-g piece of bone at that time was

$$N_0 = \frac{6.02 \times 10^{23}\,\mathrm{mol}^{-1}}{12\,\mathrm{g/mol}} \times (200\mathrm{g}) \times 1.3 \times 10^{-12} = 1.3 \times 10^{13}$$

The half-life of ^{14}C is

$$T_{1/2} = (5{,}730\ \mathrm{a}) \times (3.156 \times 10^{7}\,\mathrm{s/a}) = 1.81 \times 10^{8}\,\mathrm{s}$$

The decay constant of ^{14}C is

$$\lambda = \frac{0.693}{T_{1/2}} = \frac{0.693}{1.81 \times 10^{8}\,\mathrm{s}} = 3.83 \times 10^{-12}\,\mathrm{s}^{-1}$$

So the initial activity of ^{14}C in the animal bone is

$$A_0 = \lambda N_0 = (3.83 \times 10^{-12}\,\mathrm{s}^{-1}) \times 1.3 \times 10^{13} = 50\mathrm{s}^{-1}$$

From $A(t) = A_0 e^{-\lambda t}$, we can get

$$t = \frac{1}{\lambda}\ln\frac{A_0}{A} = \frac{1}{3.83\times10^{-12}\text{s}^{-1}}\ln\frac{50\text{s}^{-1}}{16\text{s}^{-1}}$$
$$=2.98\times10^{11}\text{s}=9400\text{a}$$

So the age of the bone fragment is 9400 years old.

Carbon dating is valid only for determining the age of objects less than about 60,000 years old. The older the object, the smaller the amount of ^{14}C it contains, and the less precise the measurement. In some cases, radioactive isotopes with longer half-lives can be used to determine the age of these older objects. For example, the half-life of ^{238}U is up to 4.5×10^9 a, it can be used to determine the geological age of rocks. The age of the oldest rocks on Earth determined by using ^{238}U and other radioactive isotopes is about 4×10^9 a. The age of rocks in which the oldest fossilized organisms are embedded indicates that life appeared at least 3.5 billion years ago. The earliest fossilized remains of mammals are found in rocks about 200 million years old, while the first humanlike creatures appeared 2 million years ago. Radioactive dating plays an indispensable role in the reconstruction of Earth's history.

5.4 Nuclear Reactions

5.4.1 Artificial Nuclear Reactions

Radioactive decay is the spontaneous change of an atomic nucleus during which energy is released. A nuclear reaction usually refers to the process in which particles with certain energy (including nuclei, protons, neutrons, α particles or γ-ray photons, etc.) bombard the nucleus to cause changes in the nucleus.

In 1919, Rutherford observed that protons were produced when α particles emitted by radioactive decay of ^{212}Po struck nitrogen nuclei, and nitrogen nuclei were transformed into oxygen nuclei at the same time, namely

$$\alpha + {}^{14}\text{N} \rightarrow {}^{17}\text{O} + \text{p}$$

This is the first nuclear reaction to be achieved artificially. A shortened form is often used for nuclear reactions, for example, the above reaction can be written as $^{14}N(\alpha, p)^{17}O$. The symbol on the left side of the parentheses represents the nucleus (target nucleus) before the reaction, and the symbol on the right side of the parentheses represents the nucleus after the reaction. The symbols inside the parentheses represent the incident particle before the reaction (first) and the lighter particle emitted after the reaction (second), respectively.

In 1932, Chadwick discovered that the nuclear reaction of neutrons is

$$\alpha + {}^{9}\text{Be} \rightarrow {}^{12}\text{C} + \text{n}$$

In 1932, the British physicist Cockcroft and the Irish physicist Walton used artificially accelerated particles to produce a nuclear reaction for the first time,

$$\text{p} + {}^{7}\text{Li} \rightarrow \alpha + \alpha$$

In 1934, the French Joliot-Curie couple discovered that the aluminum foil bombarded by α

particles contained radioactive ^{30}P, namely

$$\alpha + {}^{27}Al \rightarrow {}^{30}P + n$$

^{30}P is unstable and undergoes β^+ decay after generation

$$^{30}P \rightarrow {}^{30}Si + e^+ + \nu_e$$

^{30}P does not exist naturally and it was the first artificial radioactive nuclide produced by a nuclear reaction.

In nuclear reactions, the transformation and production of particles must obey some conservation laws, such as the conservation of charge, the conservation of nucleon number, the conservation of angular momentum, the conservation of momentum and the conservation of energy.

A nuclear reaction can generally be expressed as

$$a + X \rightarrow Y + b \qquad (5-31)$$

where a is an incident particle that strikes nucleus X, producing nucleus Y and particle b (typically, p, n, α, γ). We define the reaction energy, or Q – value as

$$Q = (m'_a + m'_X - m'_b - m'_Y)c^2 \qquad (5-32)$$

Since energy is conserved, Q is equal to the increase in kinetic energy:

$$Q = E_{kb} + E_{kY} - E_{ka} - E_{kX} \qquad (5-33)$$

where E_{ka}, E_{kX}, E_{kY}, and E_{kb} are the kinetic energies of particles a, X, Y, and b, respectively. For different nuclear reactions, Q can be positive or negative. A nuclear reaction with $Q>0$ is called an exoergic reaction, in which energy is released, so the total kinetic energy of the particles after the reaction is greater than that before the reaction. A nuclear reaction with $Q<0$ is called an endoergic reaction, and the total kinetic energy of the particles after the reaction is less than that before the reaction. To produce a nuclear reaction in this case, energy must be input, which comes from the kinetic energy of the initial colliding particles (a and X).

Example 5 – 7: Will the nuclear reaction "go"? Can the nuclear reaction $^{13}C(p,n)^{13}N$ occur when ^{13}C is bombarded by 2.0 – MeV protons?

Solution: The mass of the neutron is $m_n = 1.008,665u$. According to Table 5 – 1, the mass of the ^{13}C atom is $m'_{^{13}C} = 13.003,354u$, the mass of the ^{13}N atom is $m'_{^{13}N} = 13.005,738u$, and the mass of the hydrogen atom is $m'_{^1H} = 1.007,825u$. Here, we must use the mass of the hydrogen atom rather than that of the proton because the masses of ^{13}C and ^{13}N include the electrons. Since electrons cannot be produced or destroyed during the reaction, the number of electrons before and after the reaction must be equal.

From equation 5 – 32, the Q – value of this nuclear reaction is

$$Q = (13.003,354u + 1.007,825u - 13.005,738u - 1.008,665u)c^2$$
$$= -(0.003,224u) \times (931.5MeV/uc^2) \times c^2 = -3.00MeV$$

Q is negative, and this nuclear reaction is endoergic and requires energy. The kinetic energy of the protons is 2.0 MeV, which is not enough to make the reaction go.

For the proton in the above example to make this nuclear reaction go, its kinetic energy must be slightly greater than 3.0 MeV. This is because 3.0MeV would be enough to conserve energy, but a proton of this energy would produce the ^{13}N and neutron with no kinetic energy and hence no momentum. But an incident 3.0 – MeV proton has momentum, so the reaction would not satisfy the law of conservation of momentum. A more complicated calculation shows that the minimum kinetic energy of a proton that satisfies the conservation of energy and momentum before and after the reaction $^{13}C(p,n)^{13}N$ is 3.23 MeV. This minimum kinetic energy of the incident particle required to initiate an endoergic reaction is called the threshold energy of the reaction (see Problem 5.14).

5.4.2 Nuclear Fission

In 1938, the German scientists Hahn and Strassmann discovered that when uranium nuclei were bombarded with neutrons, they could produce lighter nuclei with a mass about half that of uranium nuclei. Subsequently, the Austrian scientists Meitner and Frisch made a correct explanation for this: after the uranium nucleus captures neutrons, it splits into two pieces of almost equal mass. This new phenomenon is called nuclear fission because of its similarity to cell division.

Nuclear fission is more likely to occur for ^{235}U than for the more abundant ^{238}U. As shown in Figure 5 – 11, according to the liquid-drop model of the nucleus, the capture of a neutron by the ^{235}U nucleus increases its internal energy (like heating a drop of water). This intermediate state, or compound nucleus, is the ^{236}U nucleus in an excited state. The shape of the excited nucleus is elongated. When the nucleus elongates into the shape shown in Figure 5 – 11 (c), the increased separation distance between the two ends greatly weakens the attraction produced by the short-range nuclear force, and the Coulomb repulsion becomes the dominant interaction, so the nucleus splits into two parts. The two resulting nuclei, X_1 and X_2, are called fission fragments. In this process, typically two or three neutrons are also released. The mass of the two fission fragments is usually 40% ~ 60% of the mass of the uranium nucleus, rather than 50% each. Figure 5 – 12 shows the distribution of fission fragments with the mass number A. Most of the fission fragments are distributed in the two intervals of $85 < A < 105$ and $130 < A < 150$, while the equal mass fragments ($A = 118$) are very few. Two typical fission processes can be expressed as

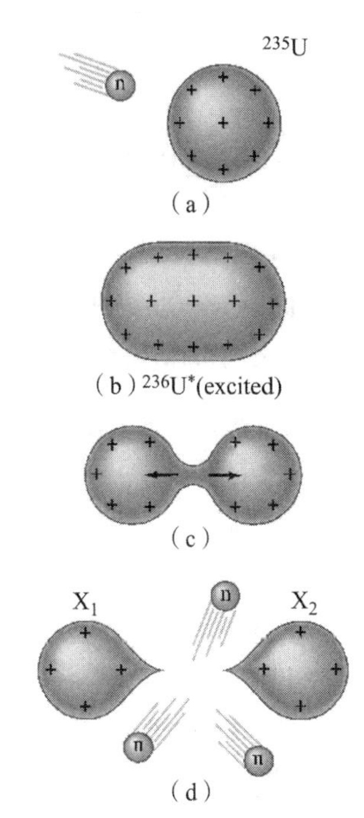

Figure 5 – 11 Fission of a ^{235}U nucleus according to the liquid-drop model

$$n + ^{235}U \rightarrow ^{236}U^* \rightarrow ^{141}Ba + ^{92}Kr + 3n \tag{5-34}$$

$$n + ^{235}U \rightarrow ^{236}U^* \rightarrow ^{139}Xe + ^{95}Sr + 2n \tag{5-35}$$

The compound nucleus, $^{236}U^*$, exists for less than 10^{-12} s, so this process occurs extremely quickly. The fission fragments are often neutron-rich nuclides, which are also unstable, and undergo β decay one or more times to form a stable nuclide.

Because the mass of the ^{235}U nucleus is considerably greater than the total mass of the fission

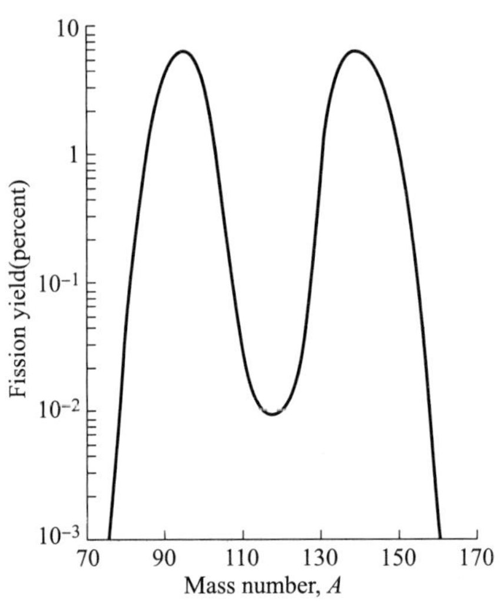

Figure 5 – 12 Distribution of the fission fragments of ^{235}U. Note that the vertical scale is logarithmic

fragments and the neutrons released in the fission, the fission reaction can release a tremendous amount of energy. It can also be seen from Figure 5 – 3 that the average binding energy per nucleon for uranium nuclei is about 7.6MeV/nucleon, while the average binding energy per nucleon for fission fragments with intermediate mass($A \approx 100$) is about 8.5MeV/nucleon. The mass difference, or energy difference, between the original uranium nucleus and the fission fragments is

$$(8.5\text{MeV/nucleon} - 7.6\text{MeV/nucleon}) \times (236 \text{ nucleons}) \approx 200\text{MeV}$$

This is an enormous amount of energy for the split of one single nucleus. At a practical level, the energy from one fission is insignificant. But if a large number of fissions could occur in a very short period of time, a macroscopically significant amount of energy would be released. For example, the energy released by the complete fission of 1kg of ^{235}U is equivalent to the energy released by the complete combustion of 2800 tons of standard coal. A number of scientists, including Fermi, recognized that the neutrons released in each fission could be used to create a chain reaction. That is, a neutron initially induces one fission of a uranium nucleus, and the two or three neutrons released in the reaction can go on to induce additional fissions, thus producing a continuous generation-by-generation fission reaction as shown in Figure 5 – 13. An uncontrolled chain reaction is the basis of nuclear bombs. To make peaceful use of the energy released in the fission reaction, the chain reaction must be controlled. In 1942, Fermi presided over the establishment of the world's first nuclear reactor at the University of Chicago, realizing the release of nuclear energy through a controlled chain reaction for the first time.

Several problems have to be overcome to achieve a self-sustaining chain reaction. First, what are emitted during a nuclear fission are fast neutrons with a very high speed (the kinetic energy is mostly on the order of MeV), and what are suitable for inducing a nuclear fission of ^{235}U are thermal neutrons whose speed is equivalent to the speed of thermal motion, or called slow neutrons (the

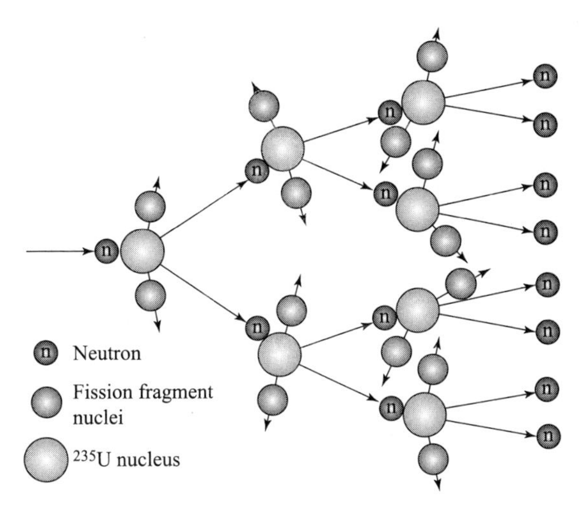

Figure 5 - 13 Chain reaction

kinetic energy is less than 1eV). Therefore, a moderator must be used to slow down the neutrons. Hydrogen nuclei, ^{1}H, in water have the same mass as neutrons and are therefore ideal for slowing down the neutrons. Unfortunately, ^{1}H tends to absorb neutrons, while its isotope called deuterium, ^{2}H, absorb less neutrons, so heavy water containing deuterium nuclei is the moderator commonly used at present. Another common moderator is graphite, which contains ^{12}C nuclei.

Second, if light water (ordinary water) is used as a moderator, not only the ^{1}H nuclei absorb neutrons, but also the ^{238}U nuclei absorb neutrons without fission through the reaction

$$n + {}^{238}U \rightarrow {}^{239}U + \gamma$$

Naturally occurring uranium contains 99.27% ^{238}U and only 0.72% fissionable ^{235}U. In order to increase the probability of fission of ^{235}U nuclei, it is necessary to enrich natural uranium to increase the percentage of ^{235}U. If heavy water is used as moderator, since heavy water is not easy to absorb neutrons, and the probability of ^{238}U capturing thermal neutrons is small, natural uranium can be dircctly used as nuclear fuel in this case.

Third, some neutrons will escape through the surface of the reactor core before the chain reaction can occur, so the mass of nuclear fuel must be sufficiently large to maintain a self-sustaining chain reaction. The minimum mass of fissile material required to sustain a chain reaction is called the critical mass. The value of the critical mass depends on the moderator, the type of nuclear fuel (^{239}Pu may be used instead of ^{235}U), and how much the fuel is enriched.

Whether the chain reaction can continue can be described by the multiplication factor, f. The multiplication factor is defined as the average number of neutrons produced by each fission that can cause a subsequent fission. For a self-sustaining chain reaction, $f \geqslant 1$. If $f < 1$, the reactor is subcritical. If $f > 1$, the reactor is supercritical. A key component of a reactor is the movable control rod, usually made of cadmium or boron. Its function is to absorb neutrons and maintain the reactor at just critical, $f = 1$. Since the chain reaction proceeds very fast, about 1000 generations of neutrons can be produced in about 1 second, it is very difficult to maintain $f = 1$ by manipulating the control rods. This problem is solved by a very small percentage ($\approx 1\%$) of delayed neutrons. Delayed neutrons come from the decay of neutron-rich fission fragments, and it takes several seconds or minutes before they are released from the fragments, which ensures the time for operating the control rods and maintaining $f = 1$.

There are many types of nuclear reactors and they have a wide variety of applications. A power reactor uses the nuclear energy released by nuclear fission in the reactor to heat water and produce steam to drive a turbine connected to an electric generator for power generation. Figure 5 – 14 is the schematic diagram of a power reactor. The core of a nuclear reactor consists of the fuel and a moderator. Thc fuel is usually enriched uranium containing 2% to 4% ^{235}U. When light water at high pressure flows through the core, the thermal energy is absorbed, and steam is produced in the heat exchanger, driving the turbine to generate electricity.

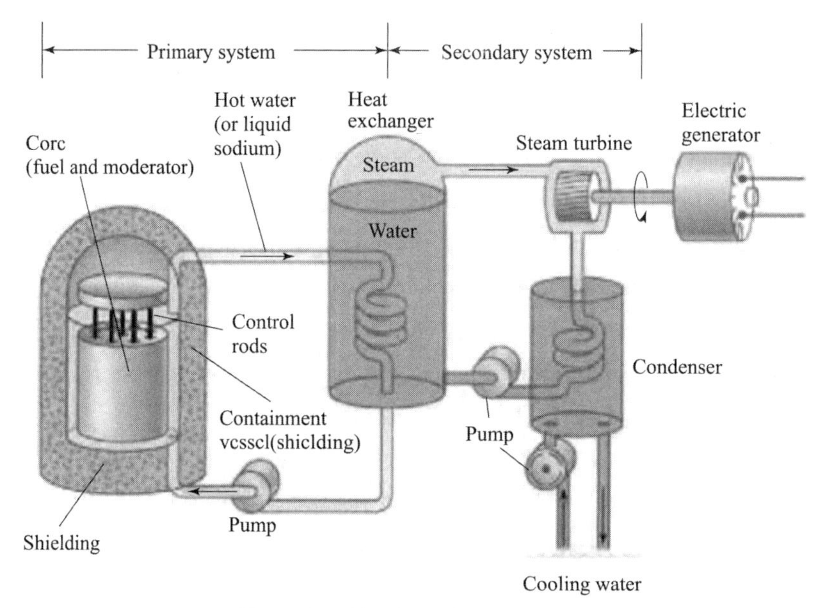

Figure 5 – 14 A nuclear reactor

Nuclear reactors have other uses besides generating electricity. For example, fission produces a large number of neutrons, which can be used to produce various radioactive isotopes, to produce high-quality semiconductor materials, and for cancer treatment. Nuclear energy can also provide heat, and can be used as special power for rockets, spacecraft, artificial satellites, submarines, aircraft carriers, etc.

5.4.3 Nuclear Fusion

The combination of two light nuclei into a larger nucleus is called nuclear fusion. It can be seen from Figure 5 – 3 that the average binding energy per nucleon for light nuclei increases rapidly as the mass number A increases, and reaches the peak of the average binding energy per nucleon curve around $A = 60$. Therefore, energy is released during the fusion of small nuclei into larger ones.

The fusion of light nuclei is the source of energy for the Sun and other stars. In 1938, the

American physicist Hans Bethe proposed two fusion cycles to explain energy production in stars. The energy of the Sun mainly comes from the following fusion reactions:

$$p + p \rightarrow {}^{2}H + e^{+} + \nu(0.44\ MeV) \tag{5-36a}$$

$$p + {}^{2}H \rightarrow {}^{3}He + \gamma(5.48\ MeV) \tag{5-36b}$$

$${}^{3}He + {}^{3}He \rightarrow {}^{4}He + 2p(12.86\ MeV) \tag{5-36c}$$

where the energy released (Q – value) for each nuclear reaction is given in parentheses. This sequence of fusion reactions is called the proton-proton cycle. Its net effect is that four protons fuse into a ${}^{4}He$ nucleus and produce two positrons, two neutrinos and two γ – ray photons. The three steps can be summarized as

$$4p \rightarrow {}^{4}He + 2e^{+} + 2\nu + 2\gamma \tag{5-37}$$

Note that the first two reactions (Equation 5 – 36a and b) must each take place twice to form the two ${}^{3}He$ nuclei needed for the third reaction (Equation 5 – 36c), so the total energy released in the net reaction, Equation 5 – 37, is

$$2 \times 0.44MeV + 2 \times 5.48MeV + 12.86MeV = 24.7MeV$$

However, each positron formed in (Equation 5 – 36a) quickly annihilates with an electron, and release an energy of $2m_e c^2 = 1.02MeV$. Therefore, the total energy released in the proton – proton cycle is

$$24.7MeV + 2 \times 1.02MeV = 26.7MeV$$

The first reaction, the formation of the ${}^{2}H$ nucleus from two protons (Equation 5 – 36a), has a very low probability, so it determines the consumption rate of hydrogen in the Sun.

In stars hotter than the Sun, the more likely fusion reaction is called the carbon cycle, which comprises the following sequence of reactions

$$p + {}^{12}C \rightarrow {}^{13}N + \gamma \tag{5-38a}$$

$${}^{13}N \rightarrow {}^{13}C + e^{+} + \nu \tag{5-38b}$$

$$p + {}^{13}C \rightarrow {}^{14}N + \gamma \tag{5-38c}$$

$$p + {}^{14}N \rightarrow {}^{15}O + \gamma \tag{5-38d}$$

$${}^{15}O \rightarrow {}^{15}N + e^{+} + \nu \tag{5-38c}$$

$${}^{15}N + p \rightarrow {}^{12}C + {}^{4}He \tag{5-38f}$$

It can be seen that no carbon is consumed in the carbon cycle, and its net effect is the same as that of the proton-proton cycle, with the same total energy released.

It is very attractive to utilize the energy released in fusion to build a power reactor. The fusion reactions most likely to be achieved in a reactor are those involving the isotopes of hydrogen, deuterium (${}^{2}H$) and tritium (${}^{3}H$), namely

$${}^{2}H + {}^{2}H \rightarrow {}^{3}H + p(4.00MeV) \tag{5-39a}$$

$${}^{2}H + {}^{2}H \rightarrow {}^{3}He + n(3.23MeV) \tag{5-39b}$$

$${}^{2}H + {}^{3}H \rightarrow {}^{4}He + n(17.57MeV) \tag{5-39c}$$

Comparing the energy released in these reactions (the value given in parentheses) with the energy released in fission of ${}^{235}U$, we can see that, for the same mass of nuclear fuel, the energy released in fusion reactions is much greater than that in fission reactions. Furthermore, the fuel required in fusion reactions is deuterium, which is abundant in the water of the oceans (0.0156% of water molecules in seawater contain deuterium atoms).

To achieve a controlled nuclear fusion reaction, several problems must first be solved. All atomic nuclei are positively charged, and Coulomb repulsion exists between two positive charges. However, if the two nuclei can be brought close enough so that the short-range attractive nuclear force can counteract the Coulomb repulsion and pull the two nuclei together, fusion will occur. In order for the nuclei to get close enough together, they must have rather large kinetic energy to overcome the Coulomb repulsion. High temperatures can make atomic nuclei have great kinetic energy, so the fusion reaction needs to be carried out at high temperatures, which is also called thermonuclear reaction. The temperature at the core of the Sun is as high as 1.5×10^{7} K, which can produce fusion reactions. The released energy can keep the temperature high so that further thermonuclear reactions can occur. The temperature produced within a fission (or atomic) bomb is close to 10^{8} K, which can be used to ignite a thermonuclear weapon, that is, a hydrogen bomb, made according to the principle of fusion reactions for light nuclei such as deuterium and tritium, releasing the huge energy of fusion. At the high temperatures required for fusion reactions, all atoms are completely ionized, forming the fourth state of matter: plasma.

In addition to high temperatures, in order to achieve self-sustained fusion reactions and obtain sufficient usable energy, the following requirements must be met: the density of the plasma must be large enough; the required temperature and density must be maintained for a long enough time. In 1957, the British physicist J. D. Lawson derived the relation between the ion density n and confinement time τ of the plasma:

$$n\tau > 10^{20}\ \mathrm{s/m^3} \tag{5-40}$$

which is known as Lawson's criterion. It is a necessary condition for realizing self-sustaining fusion reactions and obtaining energy gain.

A plasma at high temperatures is difficult to be stably confined. Ordinary materials have vaporized at a maximum of a few thousand degrees, and hence cannot be used to confine a high-temperature plasma of hundreds of millions of degrees. Thermonuclear fusion in the Sun is confined by the gravitational force. The gravitational force produced by the huge mass of the star confines the high-temperature plasma together to undergo thermonuclear fusion reactions. This cannot be achieved on Earth. Two principal confinement schemes are being investigated at present: magnetic confinement and inertial confinement.

In Magnetic confinement, magnetic fields are used to confine the high-temperature plasma. The magnetic bottle described in the section "Motion of a charged particle in a magnetic field" can confine the charged particles between the strong magnetic fields (magnetic mirrors) at both ends, but some of the charged particles will still escape before sufficient fusion reactions take place. In the 1950s, scientists in the USSR invented the tokamak, also known as the circulator, which is one of the most promising magnetic confinement devices at present. Figure 5 – 15 is a schematic diagram of a tokamak device. It consists of the toroidal vacuum chamber, toroidal field coils, poloidal field coils and other components. The toroidal magnetic field produced by the toroidal field coils confines the plasma, and the poloidal magnetic field produced by the poloidal field coils controls the position and shape of the plasma. In this way, the fusion fuel in the extremely high-temperature plasma state is confined in the toroidal vacuum chamber to realize the fusion reaction. In order to maintain a strong confinement magnetic field, the intensity of the current is very large, and the coil will heat up after a long time. In order to solve this problem, people have introduced superconducting technology into the tokamak device. Since the process of solar radiation and heating is a thermonuclear fusion reaction, the controlled thermonuclear fusion device is also vividly called "artificial sun". Figure 5 - 16 shows the "artificial sun" EAST, the Experimental Advanced Superconducting Tokamak, which took 8 years and cost 200 million yuan to independently design and build in China. EAST successfully completed the discharge experiment for the first time on September 28, 2006, becoming the first fully superconducting non-circular cross-section nuclear fusion experimental device in the world that has been built and actually operated. In February 2016, EAST successfully realized a long-pulse high-temperature plasma discharge with an electron temperature exceeding 50 million degrees Celsius and a duration of 102 seconds. This is the plasma discharge with the longest lasting time in the international tokamak experimental device when the electron temperature reaches 50 million degrees Celsius. It is an important stage of research progress, making China's research on controlled thermonuclear fusion reaction continue to lead the world.

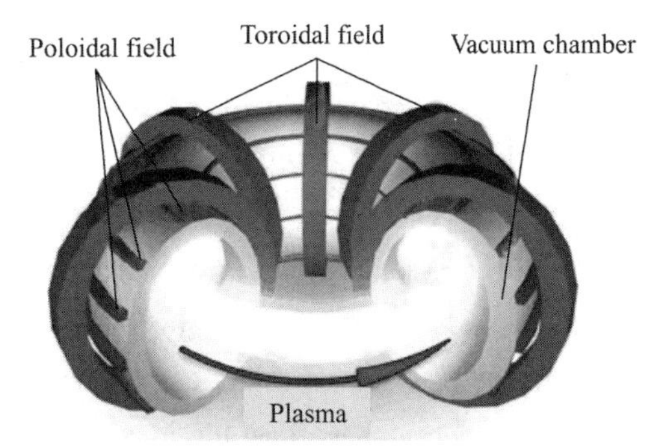

Figure 5 - 15 Schematic diagram of a tokamak device

Figure 5 - 16 "Artificial sun" EAST

In inertial confinement, the inertia of the fusion fuel itself is used to confine the plasma. Inertial confinement is realized in hydrogen bombs, but it is not controlled. Controlled inertial-confinement nuclear fusion is to put deuterium and tritium into a pellet with a diameter on the order of mm, and irradiate the pellet uniformly from all sides by high-power pulsed laser beams or particle beams. The fusion fuel is heated and ionized into plasma under high-energy irradiation, and the implosion occurs so that the pellet is compressed to an ultra-high density and reaches the reaction temperatures required for nuclear fusion. The time of inertia confinement is extremely short, on the order of 10^{-11} to 10^{-9} s. In such a short time, the ions do not fly around because of their own inertia, and very soon thereafter fusion takes place.

Controlled nuclear fusion has the advantages of sufficient fuel, safety and reliability, excellent economic performance, and no environmental pollution, etc. It is an inexhaustible ideal energy source for human beings. However, due to its enormous technical difficulty, there is still a long way to go to achieve controlled nuclear fusion.

Summary

1. Properties of the Nucleus

The atomic nucleus is composed of protons and neutrons. Proton number Z, neutron number N and mass number A satisfy $A = Z + N$.

Radius of the nucleus: $R = r_0 A^{1/3}$, $r_0 = 1.2\text{fm}$

Spin of the nucleus: The spin quantum number of the nucleus is I. The projection of the spin angular momentum of the nucleus in the z direction is

$$I_z = m_I \hbar,\ m_I = -I,\ -I+1, \cdots, I-1, I$$

Nuclear magneton: $\mu_N = \dfrac{e\hbar}{2m_p}$

The projection of the magnetic moment of the nucleus in the z direction is

$$\mu_z = g_I \mu_N m_I$$

2. Binding energy

The energy that must be supplied to separate an atomic nucleus into individual protons and neutrons is called the binding energy of the nucleus, which is also the energy released when individual nucleons combine to form a nucleus. The ratio of the total binding energy of a nucleus to the number of nucleons is called the average binding energy per nucleon, also known as the specific binding energy.

3. Nuclear force

The nuclear force is independent of electric charge. It is a strong interaction and short-range force. There is saturation of nuclear forces in nuclei. The nuclear force is related to spin.

4. Radioactive Decay Law

Radioactive nuclides decay exponentially:

$$N(t) = N_0 e^{-\lambda t}$$

where λ is the decay constant, which represents the probability of a nucleus decaying in unit time.

The relationship between half-life $T_{1/2}$, mean lifetime τ and decay constant λ satisfies

$$\tau = \frac{1}{\lambda} = 1.44T_{1/2}$$

The activity A is the number of atomic nuclei decaying per unit time in a given sample:

$$A(t) = -\frac{dN(t)}{dt} = \lambda N(t) = \lambda N_0 e^{-\lambda t} = A_0 e^{-\lambda t}$$

The commonly used unit of activity: $1\,\mathrm{Ci} = 3.7 \times 10^{10}\,\mathrm{Bq}$

5. Alpha decay

Alpha decay can be generally expressed as

$${}_Z^AX \rightarrow {}_{Z-2}^{A-4}Y + \alpha$$

The parent nucleus X emits an α particle, that is, a ${}^4He$ nucleus, and produces the daughter nucleus Y. This is the phenomenon that an α particle in the atomic nucleus tunnels through the Coulomb barrier. The greater the kinetic energy E_α of the escaping α particle, the shorter the half-life of α decay.

6. Beta decay

Beta decay includes β^- decay, which can be generally expressed as

$${}_Z^AX \rightarrow {}_{Z+1}^{A}Y + e^- + \bar{\nu}_e$$

β^+ decay, which can be generally expressed as

$${}_Z^AX \rightarrow {}_{Z-1}^{A}Y + e^+ + \nu_e$$

and electron capture, which can be generally expressed as

$${}_Z^AX + e^- \rightarrow {}_{Z-1}^{A}Y + \nu_e$$

Since there is no single electron or positron in the atomic nucleus, β decay is the result of the mutual conversion of protons and neutrons in the nucleus.

7. Gamma decay

An atomic nucleus in an excited state emits a γ-ray photon when it makes a transition to a lower excited state, or to the ground state. It is generally expressed as

$${}_Z^AX^* \rightarrow {}_Z^AX + \gamma$$

where " * " means the excited state of that nucleus.

8. Nuclear reaction

A nuclear reaction usually refers to the process in which particles with certain energy (including nuclei, protons, neutrons, α particles or γ-ray photons, etc.) bombard the nucleus to cause changes in the nucleus. It is generally expressed as

$$a + X \rightarrow Y + b$$

Q - value is the energy released in a nuclear reaction. A nuclear reaction with $Q > 0$ is called an exoergic reaction, and a nuclear reaction with $Q < 0$ is called an endoergic reaction.

9. Nuclear fission

Fission occurs when a heavy nucleus is bombarded by neutrons and split apart into two more tightly bound medium-mass nuclei, releasing a large amount of energy. A chain reaction is possible because several neutrons are emitted by a nucleus when it undergoes fission. The minimum mass of fissile material required to sustain a chain reaction is called the critical mass. In a nuclear reactor, a moderator must be used to slow down the neutrons.

10. Nuclear Fusion

Nuclear fusion is the process of combining two light nuclei into a larger nucleus and releasing energy. Fusion reactions need to be carried out at high temperatures, so they are also called thermonuclear reactions. The fusion of light nuclei is the source of energy for the Sun and other stars. A high-temperature plasma can realize controlled nuclear fusion by magnetic confinement and inertial confinement.

Questions

5 – 1 Why are the densities of various nuclei roughly equal?

5 – 2 What is the basic principle of nuclear magnetic resonance?

5 – 3 What are the similarities and the differences between the strong nuclear force and the electric force?

5 – 4 Describe the difference between α, β, and γ rays.

5 – 5 The isotope ^{64}Cu can decay by γ, β^-, and β^+ emission. What is the resulting nuclide for each case?

5 – 6 The α particles from a given α – emitting nuclide all have the same kinetic energy. But the β particles from a β – emitting nuclide have a spectrum of energies. Explain the difference between these two cases.

5 – 7 An isotope has a half-life of one month. After two months, will a given sample of this isotope have completely decayed? If not, how much remains?

5 – 8 Can ^{14}C dating be used to measure the age of stone walls and tablets of ancient civilizations?

5 – 9 What assumptions are made in the carbon dating? What do you think could affect these assumptions?

5 – 10 A proton strikes a ^{20}Ne nucleus, and an α particle is released. What is the residual nucleus? Write down the reaction equation.

5 – 11 A reactor that uses highly enriched uranium can use ordinary water (instead of heavy water) as a moderator and still have a self-sustaining chain reaction. Explain.

5 – 12 Discuss the advantages and disadvantages, including pollution and safety, of power generation by fossil fuels, nuclear fission, and nuclear fusion.

5 – 13 Light energy emitted by the Sun and stars comes from the fusion process. What conditions in the interior of stars makes this possible?

5 – 14 What is the basic difference between fission and fusion?

Problems

5 – 1 The mass of the Earth is 5.98×10^{24} kg. (1) Determine the density of nuclear matter in kg/m^3. (2) What would be the radius of the Earth if it had the density of nuclear matter?

5 – 2 How much energy must an α particle have to just "touch" the surface of a ^{238}U nucleus?

5 – 3 The mass of a ^{14}N atom is 14.003,074u. Calculate the average binding energy per nucleon for a ^{14}N nucleus.

5 – 4 The atomic masses of ^{4}He and ^{8}Be are 4.002,603u and 8.005,305u, respectively. (1) Show that the nucleus ^{8}Be is unstable to decay into two α particles. (2) Is the nucleus ^{12}C stable against decay into three α particles? Show why or why not.

5 – 5 The atomic masses of ^{4}He and ^{232}U are 4.002,603u and 232.037,146u, respectively. When a ^{232}U nucleus emits an α particle with a kinetic energy of 5.32MeV, what is the daughter nucleus and what is the approximate atomic mass (in u) of the daughter atom? Ignore recoil of the daughter nucleus.

5 – 6 The atomic masses of ^{23}Ne and ^{23}Na are 22.994,5u and 22.989,8u, respectively. When ^{23}Ne decays to ^{23}Na, what is the maximum kinetic energy of the emitted electron? What is the minimum energy? What is the energy of the neutrino in each case?

5 – 7 A radioactive material produces 1280 decays per minute at one time, and 6h later produces 320 decays per minute. What is its half-life?

5 – 8 The iodine isotope ^{131}I is used in hospitals for diagnosis of thyroid function. If 632μg of ^{131}I are ingested by a patient, determine the activity (1) immediately, (2) 1.0h later when the thyroid is being tested, and (3) 180d later. (The half-life of ^{131}I is 8.0207d, and its molar mass is 131g/mol.)

5 – 9 The rubidium isotope ^{87}Rb decays to stable ^{87}Sr via β decay with a half-life of 4.75×10^{10}a, which can be used to determine the age of rocks and fossils. Rocks containing fossils of early animals contain a ratio of ^{87}Sr to ^{87}Rb of 0.0160. Assuming that there was no ^{87}Sr present when the rocks were formed, calculate the age of these fossils.

5 – 10 ^{7}Be is produced in the upper atmosphere, and filters down onto the Earth's surface with a half-life of about 53d. If the activity of ^{7}Be on a plant leaf is detected to be 250Bq, how long do we have to wait for the activity to drop to 10Bq? Estimate the initial mass of ^{7}Be on the leaf.

5 – 11 An ancient club is found that contains 190g of carbon and has an activity of 5.0Bq. Determine its age assuming that the ratio of ^{14}C to ^{12}C is 1.3×10^{-12} in the atmosphere.

5 – 12 In the reaction $^{14}N(\alpha, p)^{17}O$, the incident α particles have 7.68MeV of kinetic energy. (1) Can this reaction occur? (2) If so, what is the total kinetic energy of the products? The mass of ^{17}O is 16.999,131u.

5 – 13 In the reaction $^{6}Li(d, p)X$, (1) what is X, the resulting nucleus? (2) What is the Q – value

of this reaction? Is the reaction endothermic or exothermic?

5 – 14 Use conservation of energy and momentum to show that a bombarding proton must have an energy of 3.23MeV in order to make the reaction $^{13}C(p,n)^{13}N$ occur. (See Example 5 – 7.)

5 – 15 Suppose that the average electric power consumption in a typical house is 300W. What initial mass of ^{235}U would have to undergo fission to supply the electrical needs of such a house for a year? (Assume 200MeV is released per fission, as well as 100% efficiency.)

5 – 16 If a 1.0 – MeV neutron emitted in a fission reaction loses one-half of its kinetic energy in each collision with moderator nuclei, how many collisions must it make to reach thermal energy $\left(\frac{3}{2}kT = 0.040\text{eV}\right)$?

5 – 17 If a typical house requires 300W of electric power on average, what minimum amount of deuterium fuel would have to be used in a year to supply these electrical needs? Assume the reaction of Equation 5 – 38b.

5 – 18 How much energy is contained in 1.00kg of water if its natural deuterium is used in the fusion reaction of Equation 5 – 38a? Compare to the energy obtained from the burning of 1.0kg of gasoline, about 5×10^7J. (Assume 0.015% of water molecules contain deuterium atoms.)

References

[1] TIPLER P A. Physics for Scientists and Engineers[M]. 4th ed. New York: W. H. Freeman and Company, 1999.

[2] GIANCOLI D C. Physics for Scientists and Engineers with Modern Physics[M]. 4th ed. New Jersey: Pearson Education, Inc., 2009.

[3] GIAMBATTISTA A, RICHARDSON B M, RICHARDSON R C. College Physics (Volume 2): With an Integrated Approach to Forces and Kinematics[M]. 4th ed. New York: McGraw-Hill Companies, Inc., 2013.

[4] SEARS F W, ZEMANSKY M W, YOUNG H D. University Physics[M]. Boston: Addison-Wesley Publishing Company, 1987.

[5] SERWAY R A, JEWETT J W. Principle of Physics: A Calculus-Based Text[M]. 3rd ed. Fort Worth, TX: Harcourt College Publishers, 2002.

[6] EINSTEIN A, translated by Yang Runyin. The Special and the General Theory (A Popular Exposition) (in Chinese)[M]. Beijing: Beijing University Press, 2006.

[7] ALAN G. RICHARDON B M, RICHARDON R C. College Physics (Volume 5): Quantum and Particle Physics and Relativity[M]. Asia: McGraw-Hill Education and Beijing: China Machine Press, 2013.

[8] KITTEL Charles. Introduction to Solid State Physics[M]. 8th ed. New York: John Wiley & Sons Inc., 2005.

[9] HALLIDAY David, RESNICK Robert, WALKER Jearl. Fundamentals of Physics[M]. 6th ed. New York: John Wiley & Sons Inc., 2000.

[10] GUO Yiling, SHEN Huijun. History of Physics (in Chinese)[M]. Beijing: Tsinghua University Press, 1999.

[11] ZHANG Sanhui. College Physics (in Chinese)[M]. 3rd ed. Beijing: Tsinghua University Press, 2014.

[12] Zhang Zongsui. Electrodynamics and special relativity (in Chinese)[M]. Beijing: Beijing University Press, 2004.

[13] CAI Bolian. Special Theory of Relativity (in Chinese)[M]. Beijing: Higher Education Press, 1991.

[14] GUO Shikun. Introduction to General Relativity (in Chinese)[M]. Chengdu: UEST Press, 2005.

[15] ZHAO Kaihua. Physics Enlightens the World (in Chinese)[M]. Beijing: Beijing University

Press,2005.

[16] GUO Yiling, SHEN Huijun. The Nobel Prize in Physics 1901 – 2010 (in Chinese) [M]. Beijing: Tsinghua University Press, 2012.

[17] GOU Bingcong, HU Haiyun. College Physics (Volume Ⅱ) (in Chinese) [M]. 2nd ed. Beijing: National Defense Industry Press, 2011.

[18] HU Haiyun, MIAO Jinsong, FENG Yanquan, WU Xiaoli. College Physics (Volume IV) Modern Physics (in Chinese) [M]. Beijing: Higher Education Press, 2017.